D0151640

ALGORITHMS +

DATA STRUCTURES =

PROGRAMS

Prentice-Hall
Series in Automatic Computation

AHO, ed., *Currents in the Theory of Computing*
AHO AND ULLMAN, *The Theory of Parsing, Translation, and Compiling,* Volume I: *Parsing*; Volume II: *Compiling*
ANDREE, *Computer Programming: Techniques, Analysis, and Mathematics*
ANSELONE, *Collectively Compact Operator Approximation Theory and Applications to Integral Equations*
BATES AND DOUGLAS, *Programming Language/One*, 2nd ed.
BLUMENTHAL, *Management Information Systems*
BRENT, *Algorithms for Minimization without Derivatives*
BRINCH HANSEN, *Operating System Principles*
COFFMAN AND DENNING, *Operating Systems Theory*
CRESS, et al., *FORTRAN IV with WATFOR and WATFIV*
DAHLQUIST, BJORCK, AND ANDERSON, *Numerical Methods*
DANIEL, *The Approximate Minimization of Functionals*
DEO, *Graph Theory with Applications to Engineering and Computer Science*
DESMONDE, *Computers and Their Uses*, 2nd ed.
DRUMMOND, *Evaluation and Measurement Techniques for Digital Computer Systems*
ECKHOUSE, *Minicomputer Systems: Organization and Programming* (*PDP-11*)
FIKE, *Computer Evaluation of Mathematical Functions*
FIKE, *PL/1 for Scientific Programmers*
FORSYTHE AND MOLER, *Computer Solution of Linear Algebraic Systems*
GEAR, *Numerical Initial Value Problems in Ordinary Differential Equations*
GORDON, *System Simulation*
GRISWOLD, *String and List Processing in SNOBOL4: Techniques and Applications*
HANSEN, *A Table of Series and Products*
HARTMANIS AND STEARNS, *Algebraic Structure Theory of Sequential Machines*
JACOBY, et al., *Iterative Methods for Nonlinear Optimization Problems*
JOHNSON, *System Structure in Data, Programs, and Computers*
KIVIAT, et al., *The SIMSCRIPT II Programming Language*
LAWSON AND HANSON, *Solving Least Squares Problems*
LORIN, *Parallelism in Hardware and Software: Real and Apparent Concurrency*
LOUDEN AND LEDIN, *Programming the IBM 1130*, 2nd ed.
MARTIN, *Computer Data-Base Organization*
MARTIN, *Design of Man-Computer Dialogues*
MARTIN, *Design of Real-Time Computer Systems*
MARTIN, *Future Developments in Telecommunications*
MARTIN, *Programming Real-Time Computing Systems*
MARTIN, *Security, Accuracy, and Privacy in Computer Systems*
MARTIN, *Systems Analysis for Data Transmission*
MARTIN, *Telecommunications and the Computer*

MARTIN, *Teleprocessing Network Organization*
MARTIN AND NORMAN, *The Computerized Society*
MCKEEMAN, et al., *A Compiler Generator*
MYERS, *Time-Sharing Computation in the Social Sciences*
MINSKY, *Computation: Finite and Infinite Machines*
NIEVERGELT, et al., *Computer Approaches to Mathematical Problems*
PLANE AND MCMILLAN, *Discrete Optimization: Integer Programming and Network Analysis for Management Decisions*
POLIVKA AND PAKIN, *APL: The Language and Its Usage*
PRITSKER AND KIVIAT, *Simulation with GASP II: A FORTRAN-based Simulation Language*
PYLYSHYN, ed., *Perspectives on the Computer Revolution*
RICH, *Internal Sorting Methods Illustrated with PL/1 Programs*
SACKMAN AND CITRENBAUM, eds., *On-Line Planning: Towards Creative Problem-Solving*
SALTON, ed., *The SMART Retrieval System: Experiments in Automatic Document Processing*
SAMMET, *Programming Languages: History and Fundamentals*
SCHAEFER, *A Mathematical Theory of Global Program Optimization*
SCHULTZ, *Spline Analysis*
SCHWARZ, et al., *Numerical Analysis of Symmetric Matrices*
SHAH, *Engineering Simulation Using Small Scientific Computers*
SHAW, *The Logical Design of Operating Systems*
SHERMAN, *Techniques in Computer Programming*
SIMON AND SIKLOSSY, eds., *Representation and Meaning: Experiments with Information Processing Systems*
STERBENZ, *Floating-Point Computation*
STOUTEMYER, *PL/1 Programming for Engineering and Science*
STRANG AND FIX, *An Analysis of the Finite Element Method*
STROUD, *Approximate Calculation of Multiple Integrals*
TANENBAUM, *Structured Computer Organization*
TAVISS, ed., *The Computer Impact*
UHR, *Pattern Recognition, Learning, and Thought: Computer-Programmed Models of Higher Mental Processes*
VAN TASSEL, *Computer Security Management*
VARGA, *Matrix Iterative Analysis*
WAITE, *Implementing Software for Non-Numeric Application*
WILKINSON, *Rounding Errors in Algebraic Processes*
WIRTH, *Algorithms + Data Structures = Programs*
WIRTH, *Systematic Programming: An Introduction*
YEH, ed., *Applied Computation Theory: Analysis, Design, Modeling*

ALGORITHMS + DATA STRUCTURES = PROGRAMS

NIKLAUS WIRTH

Eidgenossische Technische Hochschule
Zurich, Switzerland

PRENTICE-HALL, INC.

ENGLEWOOD CLIFFS, N.J.

Library of Congress Cataloging in Publication Data

WIRTH, NIKLAUS.
Algorithms + data structures = programs.

Bibliography: p.
Includes index.
1. Electronic digital computers—Programming.
2. Data structures (Computer science) 3. Algorithms.
I. Title.
QA76.6.W56 001.6'42 75-11599
ISBN 0-13-022418-9

10 9 8

Printed in the United States of America

PRENTICE-HALL INTERNATIONAL, INC., *London*
PRENTICE-HALL OF AUSTRALIA, PTY., LTD., *Sydney*
PRENTICE-HALL OF CANADA, LTD., *Toronto*
PRENTICE-HALL OF INDIA PRIVATE LIMITED, *New Delhi*
PRENTICE-HALL OF JAPAN, INC., *Tokyo*
PRENTICE-HALL OF SOUTHEAST ASIA (PTE.) LTD., *Singapore*

To Nani

CONTENTS

2 SORTING 56

3 RECURSIVE ALGORITHMS 125

4 DYNAMIC INFORMATION STRUCTURES 162

PREFACE

In recent years the subject of *computer programming* has been recognized as a discipline whose mastery is fundamental and crucial to the success of many engineering projects and which is amenable to scientific treatment and presentation. It has advanced from a craft to an academic discipline. The initial outstanding contributions toward this development were made by E. W. Dijkstra and C. A. R. Hoare. Dijkstra's "Notes on Structured Programming"* opened a new view of programming as a scientific subject and an intellectual challenge, and it coined the title for a "revolution" in programming. Hoare's "Axiomatic Basis of Computer Programming"† showed in a lucid manner that programs are amenable to an exacting analysis based on mathematical reasoning. Both these papers argue convincingly that many programming errors can be prevented by making programmers aware of the methods and techniques which they hitherto applied intuitively and often unconsciously. These papers focused their attention on the aspects of composition and analysis of programs, or, more explicitly, on the structure of algorithms represented by program texts. Yet, it is abundantly clear that a systematic and scientific approach to program construction primarily has a bearing in the case of large, complex programs which involve complicated sets of data. Hence, a methodology of programming is also bound to include all aspects of data structuring. *Programs*, after all, are concrete formulations of abstract *algorithms* based on particular representations and structures of *data*. An outstanding contribution to bring order into the bewildering variety of terminology and concepts on data structures was made by Hoare through his "Notes on Data Structuring."‡ It made clear that decisions

*In *Structured Programming* by Dahl, Djkstra, and Hoare (New York: Academic Press, 1972), pp. 1–82.

†In *Comm. ACM*, **12**, No. 10 (1969), 576–83.

‡In *Structured Programming*, pp. 83–174

about structuring data cannot be made without knowledge of the algorithms applied to the data and that, vice versa, the structure and choice of algorithms often strongly depend on the structure of the underlying data. In short, the subjects of program composition and data structures are inseparably intertwined.

Yet, this book starts with a chapter on data structure for two reasons. First, one has an intuitive feeling that data precede algorithms: you must have some objects before you can perform operations on them. Second, and this is the more immediate reason, this book assumes that the reader is familiar with the basic notions of computer programming. Traditionally and sensibly, however, introductory programming courses concentrate on algorithms operating on relatively simple structures of data. Hence, an introductory chapter on data structures seems appropriate.

Throughout the book, and particularly in Chap. 1, we follow the theory and terminology expounded by Hoare* and realized in the programming language PASCAL.† The essence of this theory is that data in the first instance represent abstractions of real phenomena and are preferably formulated as abstract structures not necessarily realized in common programming languages. In the process of program construction the data representation is gradually refined—in step with the refinement of the algorithm—to comply more and more with the constraints imposed by an available programming system.‡ We therefore postulate a number of basic building principles of data structures, called the *fundamental structures*. It is most important that they are constructs that are known to be quite easily implementable on actual computers, for only in this case can they be considered the true elements of an actual data representation, as the *molecules* emerging from the final step of refinements of the data description. They are the *record*, the *array* (with fixed size), and the *set*. Not surprisingly, these basic building principles correspond to mathematical notions which are fundamental as well.

A cornerstone of this theory of data structures is the distinction between fundamental and "advanced" structures. The former are the molecules—themselves built out of atoms—which are the components of the latter. Variables of a fundamental structure change only their value, but never their structure and never the set of values they can assume. As a consequence, the size of the store they occupy remains constant. "Advanced" structures, however, are characterized by their change of value *and* structure during

*"Notes of Data Structuring."

†N. Wirth, "The Programming Language Pascal," *Acta Informatica,* **1**, No. 1 (1971), 35–63.

‡N. Wirth, "Program Development by Stepwise Refinement," *Comm. ACM,* **14**, No. 4 (1971), 221–27.

the execution of a program. More sophisticated techniques are therefore needed for their implementation.

The sequential file—or simply the sequence—appears as a hybrid in this classification. It certainly varies its length; but that change in structure is of a trivial nature. Since the sequential file plays a truly fundamental role in practically all computer systems, it is included among the fundamental structures in Chap. 1.

The second chapter treats *sorting algorithms.* It displays a variety of different methods, all serving the same purpose. Mathematical analysis of some of these algorithms shows the advantages and disadvantages of the methods, and it makes the programmer aware of the importance of analysis in the choice of good solutions for a given problem. The partitioning into methods for sorting arrays and methods for sorting files (often called internal and external sorting) exhibits the crucial influence of data representation on the choice of applicable algorithms and on their complexity. The space allocated to sorting would not be so large were it not for the fact that sorting constitutes an ideal vehicle for illustrating so many principles of programming and situations occurring in most other applications. It often seems that one could compose an entire programming course by selecting examples from sorting only.

Another topic that is usually omitted in introductory programming courses but one that plays an important role in the conception of many algorithmic solutions is recursion. Therefore, the third chapter is devoted to *recursive algorithms.* Recursion is shown to be a generalization of repetition (iteration), and as such it is an important and powerful concept in programming. In many programming tutorials it is unfortunately exemplified by cases in which simple iteration would suffice. Instead, Chap. 3 concentrates on several examples of problems in which recursion allows for a most natural formulation of a solution, whereas use of iteration would lead to obscure and cumbersome programs. The class of *backtracking* algorithms emerges as an ideal application of recursion, but the most obvious candidates for the use of recursion are algorithms operating on data whose structure is defined recursively. These cases are treated in the last two chapters, for which the third chapter provides a welcome background.

Chapter 4 deals with *dynamic data structures*, i.e., with data that change their structure during the execution of the program. It is shown that the recursive data structures are an important subclass of the dynamic structures commonly used. Although a recursive definition is both natural and possible in these cases, it is usually not used in practice. Instead, the mechanism used in its implementation is made evident to the programmer by forcing him to use explicit reference or *pointer* variables. This book follows this technique and reflects the present state of the art: Chapter 4 is devoted to programming with pointers, to lists, trees, and to examples involving even more complicated meshes of data. It presents what is often (and somewhat inappropriately) called "list processing." A fair amount of space is devoted to tree organizations, and in particular to search trees. The chapter ends with a presentation of scatter tables, also called "hash" codes, which are often preferred to search trees. This provides the possibility of comparing two fundamentally different techniques for a frequently encountered application.

The last chapter consists of a concise introduction to the definition of *formal languages* and the problem of *parsing*, and of the construction of a *compiler* for a small and simple language for a simple computer. The motivation to include this chapter is threefold. First, the successful programmer should have at least some insight into the basic problems and techniques of the compilation process of programming languages. Second, the number of applications which require the definition of a simple input or control language for their convenient operation is steadily growing. Third, formal languages define a recursive structure upon sequences of symbols; their processors are therefore excellent examples of the beneficial application of recursive techniques, which are crucial to obtaining a transparent structure in an area where programs tend to become large or even enormous. The choice of the sample language, called PL/0, was a balancing act between a language that is too trivial to be considered a valid example at all and a language whose compiler would clearly exceed the size of programs that can usefully be included in a book that is not directed only to the compiler specialist.

Programming is a constructive art. How can a constructive, inventive activity be taught? One method is to crystallize elementary composition principles out of many cases and exhibit them in a systematic manner. But programming is a field of vast variety often involving complex intellectual activities. The belief that it could ever be condensed into a sort of pure "recipe teaching" is mistaken. What remains in our arsenal of teaching methods is the careful selection and presentation of master examples. Naturally, we should not believe that every person is capable of gaining equally much from the study of examples. It is the characteristic of this approach that much is left to the student, to his diligence and intuition. This is particularly true of the relatively involved and long examples of programs. Their inclusion in this book is not accidental. Longer programs are the "normal" case in practice, and they are much more suitable for exhibiting that elusive but essential ingredient called style and orderly structure. They are also meant to serve as exercises in the art of program *reading*, which too often is neglected in favor of program writing. This is a primary motivation behind the inclusion of larger programs as examples in their entirety. The reader is led through a gradual development of the program; he is given various "snapshots" in the evolution of a program, whereby this development becomes manifest as a *stepwise refinement* of the details. I consider it essential that programs are shown in final form with sufficient attention to details, for in programming, the devil hides in the details. Although the mere presentation of an algorithm's principle and its mathematical analysis may be stimulating and challenging to the academic mind, it seems dishonest to the engineering practitioner. I have therefore strictly adhered to the rule of presenting the final programs in a language in which they can actually be run on a computer.

Of course, this raises the problem of finding a form which at the same time is both machine executable and sufficiently machine independent to be included in such a text. In this respect, neither widely used languages nor abstract notations proved to be adequate. The language PASCAL provides an appropriate compromise; it had been developed with exactly this aim in mind, and it is therefore used throughout this book. The programs can easily be understood by programmers who are familiar with some other high-level language, such as ALGOL 60 or PL/1, because it is easy to understand the PASCAL notation while proceeding through the text. However, this not to say that some preparation would not be beneficial. The book *Systematic Programming** provides an ideal background because it is also based on the PASCAL notation. The present book was, however, not intended as a manual on the language PASCAL; there exist more appropriate texts for this purpose.†

This book is a condensation—and at the same time an elaboration—of several courses on programming taught at the Federal Institute of Technology (ETH) at Zürich. I owe many ideas and views expressed in this book to discussions with my collaborators at ETH. In particular, I wish to thank Mr. H. Sandmayr for his careful reading of the manuscript, and Miss Heidi Theiler for her care and patience in typing the text. I should also like to mention the stimulating influence provided by meetings of the Working Groups 2.1 and 2.3 of IFIP, and particularly the many memorable arguments I had on these occasions with E. W. Dijkstra and C. A. R. Hoare. Last but not least, ETH generously provided the environment and the computing facilities without which the preparation of this text would have been impossible.

N. WIRTH

*N. Wirth (Englewood Cliffs, N.J.: Prentice-Hall, Inc., 1973.)

†K. Jensen and N. Wirth, "PASCAL—User Manual and Report" *Lecture Notes in Computer Science*, Vol. 18 (Berlin, New York; Springer-Verlag, 1974).

Our lieutenant general, L. Euler, issues, through our good offices, the following Declaration. He openly confesses:

. . .

III. that, although being the king of mathematicians, he will always blush over his offense against common sense and most ordinary knowledge, committed by deducing from his formulae that a body attracted by gravitational forces located at the center of a sphere, will suddenly reverse its direction at the center;

IV. that he will do his utmost not to let reason be betrayed by a wrong formula once more. He apologizes on his knees for once having postulated in view of a paradoxical result: "although it seems to contradict reality, we must trust our computations more than our good senses."

V. that in the future he will no more compute sixty pages (of output) for a result, that one can deduce in ten lines after some careful deliberations; and if he once again pushes up his sleeves in order to compute for three days and three nights in a row, that he will spend a quarter of an hour before to think which principles (of computation) shall be most appropriate.

Excerpt from Voltaire's "Diatribe du docteur Akakia,"
(November 1752)

1 FUNDAMENTAL DATA STRUCTURES

1.1. INTRODUCTION

The modern digital computer was invented and intended as a device that should facilitate and speed up complicated and time-consuming computations. In the majority of applications its capability to store and access large amounts of information plays the dominant part and is considered to be its primary characteristic, and its ability to compute, i.e., to calculate, to perform arithmetic, has in many cases become almost irrelevant.

In all these cases, the large amount of information that is to be processed in some sense represents an *abstraction* of a part of the real world. The information that is available to the computer consists of a selected set of *data* about the real world, namely, that set which is considered relevant to the problem at hand, that set from which it is believed that the desired results can be derived. The data represent an abstraction of reality in the sense that certain properties and characteristics of the real objects are ignored because they are peripheral and irrelevant to the particular problem. An abstraction is thereby also a simplification of facts.

We may regard a personnel file of an employer as an example. Every employee is represented (abstracted) on this file by a set of data relevant either to the employer or to his accounting procedures. This set may include some identification of the employee, for example, his name and his salary. But it will most probably not include irrelevant data such as the color of hair, weight, and height.

In solving a problem with or without a computer it is necessary to choose an abstraction of reality, i.e., to define a set of data that is to represent the real situation. This choice must be guided by the problem to be solved. Then follows a choice of representation of this information. This choice is guided by the tool that is to solve the problem, i.e., by the facilities offered by the computer. In most cases these two steps are not entirely independent.

The *choice of representation* of data is often a fairly difficult one, and it is not uniquely determined by the facilities available. It must always be taken in the light of the operations that are to be performed on the data. A good example is the representation of numbers, which are themselves abstractions of properties of objects to be characterized. If addition is the only (or at least the dominant) operation to be performed, then a good way to represent the number n is to write n strokes. The addition rule on this representation is indeed very obvious and simple. The Roman numerals are based on the same principle of simplicity, and the adding rules are similarly straightforward for small numbers. On the other hand, the representation by Arabic numerals requires rules that are far from obvious (for small numbers) and they must be memorized. However, the situation is inverse when we consider either addition of large numbers or multiplication and division. The decomposition of these operations into simpler ones is much easier in the case of representation by Arabic numerals because of its systematic structuring principle that is based on positional weight of the digits.

It is well-known that computers use an internal representation based on binary digits (bits). This representation is unsuitable for human beings because of the usually large number of digits involved, but it is most suitable for electronic circuits because the two values 0 and 1 can be represented conveniently and reliably by the presence or absence of electric currents, electric charge, and magnetic fields.

From this example we can also see that the question of representation often transcends several levels of detail. Given the problem of representing, say, the position of an object, the first decision may lead to the choice of a pair of real numbers in, say, either Cartesian or polar coordinates. The second decision may lead to a floating-point representation, where every real number x consists of a pair of integers denoting a fraction f and an exponent e to a certain base (say, $x = f \cdot 2^e$). The third decision, based on the knowledge that the data are to be stored in a computer, may lead to a binary, positional representation of integers, and the final decision could be to represent binary digits by the direction of the magnetic flux in a magnetic storage device. Evidently, the first decision in this chain is mainly influenced by the problem situation, and the later ones are progressively dependent on the tool and its technology. Thus, it can hardly be required that a programmer decide on the number representation to be employed or even on the storage device characteristics. These "lower-level decisions" can be left to the designers of computer equipment, who have the most information available on current technology with which to make a sensible choice that will be acceptable for all (or almost all) applications where numbers play a role.

In this context, the significance of *programming languages* becomes apparent. A programming language represents an abstract computer capable of understanding the terms used in this language, which may embody a certain level of abstraction from the objects used by the real machine. Thus, the programmer who uses such a "higher-level" language will be freed (and barred) from questions of number representation, if the number is an elementary object in the realm of this language.

The importance of using a language that offers a convenient set of basic abstractions common to most problems of data processing lies mainly in the area of reliability of the resulting programs. It is easier to design a program based on reasoning with familiar notions of numbers, sets, sequences, and repetitions than on bits, "words," and jumps. Of course, an actual computer will represent all data, whether numbers, sets, or sequences, as a large mass of bits. But this is irrelevant to the programmer as long as he does not have to worry about the details of representation of his chosen abstractions and as long as he can rest assured that the corresponding representation chosen by the computer (or compiler) is reasonable for his purposes.

The closer the abstractions are to a given computer, the easier it is to make a representation choice for the engineer or implementor of the language, and the higher is the probability that a single choice will be suitable for all (or almost all) conceivable applications. This fact sets definite limits on the degree of abstractions from a given real computer. For example, it would not make sense to include geometric objects as basic data items in a general-purpose language, since their proper representation will, because of its inherent complexity, be largely dependent on the operations to be applied to these objects. The nature and frequency of these operations will, however, not be known to the designer of a general-purpose language and its compiler, and any choice he makes may be inappropriate for some potential applications.

In this book these deliberations determine the choice of notation for the description of algorithms and their data. Clearly, we wish to use familiar notions of mathematics, such as numbers, sets, sequences, and so on, rather than computer-dependent entities such as bitstrings. But equally clearly we wish to use a notation for which efficient compilers are known to *exist*. It is equally unwise to use a closely machine-oriented and machine-dependent language, as it is unhelpful to describe computer programs in an abstract notation which leaves problems of representation widely open.

The programming language PASCAL has been designed in an attempt to find a compromise between these extremes, and it is used throughout this book [1.3 and 1.5]. This language has been successfully implemented on several computers, and it has been shown that the notation is sufficiently close to real machines that the chosen features and their representations can be clearly explained. The language is also sufficiently close to other languages, particularly ALGOL 60, that the lessons taught here may equally well be applied in their use.

1.2. THE CONCEPT OF DATA TYPE

In mathematics it is customary to classify variables according to certain important characteristics. Clear distinctions are made between real, complex, and logical variables or between variables representing individual values, or sets of values, or sets of sets, or between functions, functionals, sets of functions, and so on. This notion of classification is equally important, if not more important, in data processing. We will adhere to the principle that *every constant, variable, expression, or function is of a certain type*. This type essentially characterizes the set of values to which a constant belongs, or which can be assumed by a variable or expression, or which can be generated by a function.

In mathematical texts the type of a variable is usually deducible from the typeface without consideration of context; this is not feasible in computer programs. For there is usually one typeface commonly available on computer equipment (i.e., Latin letters). The rule is therefore widely accepted that the associated type is made explicit in a *declaration* of the constant, variable, or function, and that this declaration textually precedes the application of that constant, variable, or function. This rule is particularly sensible if one considers the fact that a compiler has to make a choice of representation of the object within the store of a computer. Evidently, the capacity of storage allocated to a variable will have to be chosen according to the size of the range of values that the variable may assume. If this information is known to a compiler, so-called dynamic storage allocation can be avoided. This is very often the key to an efficient realization of an algorithm.

The primary characteristics of the concept of type that is used throughout this text, and that is embodied in the programming language PASCAL, thus are the following [1.2]:

1. A data type determines the set of values to which a constant belongs, or which may be assumed by a variable or an expression, or which may be generated by an operator or a function.
2. The type of a value denoted by a constant, variable, or expression may be derived from its form or its declaration without the necessity of executing the computational process.
3. Each operator or function expects arguments of a fixed type and yields a result of a fixed type. If an operator admits arguments of several types (e.g., + is used for addition of both integers and real numbers), then the type of the result can be determined from specific language rules.

As a consequence, a compiler may use this information on types to check the compatibility and legality of various constructs. For example, the assignment of a Boolean (logical) value to an arithmetic (real) variable may be detected without executing the program. This kind of redundancy in the program text is extremely useful as an aid in the development of programs, and it must be considered as the primary advantage of good high-level languages over machine code (or symbolic assembly code). Evidently, the data will ultimately be represented by a large number of binary digits, irrespective of whether or not the program had initially been conceived in a high-level language using the concept of type or in a typeless assembly code. To the computer, the store is a homogeneous mass of bits without apparent structure. But it is exactly this abstract structure which alone is enabling human programmers to recognize meaning in the monotonous landscape of a computer store.

The theory presented in this book and the programming language PASCAL specify certain methods of defining data types. In most cases new data types are defined in terms of previously defined data types. Values of such a type are usually conglomerates of *component values* of the previously defined *constituent types*, and they are said to be *structured*. If there is only one constituent type, that is, if all components are of the same constituent type, then it is known as the *base type*.

The number of distinct values belonging to a type T is called the *cardinality* of T. The cardinality provides a measure for the amount of storage needed to represent a variable x of the type T, denoted by $x : T$.

Since constituent types may again be structured, entire hierarchies of structures may be built up, but, obviously, the ultimate components of a structure must be atomic. Therefore, it is necessary that a notation is provided to introduce such primitive, unstructured types as well. A straightforward method is that of *enumeration* of the values that are to constitute the type. For example, in a program concerned with plane geometric figures, there may be introduced a primitive type called *shape*, whose values may be denoted by the identifiers *rectangle*, *square*, *ellipse*, *circle*. But apart from such programmer defined types, there will have to be some *standard types* that are said to be predefined. They will usually include *numbers* and *logical values*. If an ordering exists among the individual values, then the type is said to be ordered or *scalar*. In PASCAL, all unstructured types are assumed to be ordered; in the case of explicit enumeration, the values are assumed to be ordered by their enumeration sequence.

With this tool in hand, it is possible to define primitive types and to build conglomerates, structured types up to an arbitrary degree of nesting. In practice, it is not sufficient to have only one general method of combining constituent types into a structure. With due regard to practical problems of representation and use, a general-purpose programming language must offer several *methods of structuring*. In a mathematical sense, they may all be equivalent; they differ in the operators available to construct their values and to select components of these values. The basic structuring methods presented here are the *array*, the *record*, the *set*, and the *sequence* (*file*). More complicated structures are not usually defined as "static" types, but are instead "dynamically" generated during the execution of the program during which they may vary in size and shape. Such structures are the subject of Chap. 4 and include lists, rings, trees, and general finite graphs.

Variables and data types are introduced in a program in order to be used for computation. To this end, a set of *operators* must be available. As with data types, programming languages offer a certain number of primitive, standard (atomic) operators, and a number of structuring methods by which composite operations can be defined in terms of the primitive operators. The task of composition of operations is often considered the heart of the art of programming. However, it will become evident that the appropriate composition of data is equally fundamental and essential.

The most important basic operators are *comparison* and *assignment*, i.e., the test for equality (and order in the case of ordered types) and the command to enforce equality. The fundamental difference between these two operations is emphasized by the clear distinction in their denotation throughout this text (although it is unfortunately obscured in such widely used programming languages as Fortran and PL/I, which use the equal sign as assignment operator).

$$\text{Test for equality:} \quad x = y$$

$$\text{Assignment to } x\text{:} \quad x := y$$

These fundamental operators are defined for most data types, but it should be noted that their execution may involve a substantial amount of computational effort if the data are large and highly structured.

Apart from test of equality (or order) and assignment, a class of fundamental and implicitly defined operators are the so-called *transfer operators*. They are mapping data types onto other data types. They are particularly important in connection with structured types. Structured values are generated from their component values by so-called *constructors*, and the component values are extracted by so-called *selectors*. Constructors and selectors are thus transfer operators mapping constituent types into structured types and vice versa. Every structuring method owns its particular pair of constructors and selectors that clearly differ in their denotation.

Standard primitive data types also require a set of standard primitive operators. Thus, along with the standard data types of numbers and logical values, we also introduce the conventional operations of arithmetic and propositional logic.

1.3. PRIMITIVE DATA TYPES

In many programs integers are used when numerical properties are not involved and when the integer represents a choice from a small number of alternatives. In these cases we introduce a new primitive, unstructured data type T by enumerating the set of all possible values $c_1, c_2 \ldots c_n$.

$$\boxed{\textbf{type } T = (c_1, c_2, \ldots, c_n)} \tag{1.1}$$

The cardinality of T is $card(T) = n$.

EXAMPLES

type *shape* = (*rectangle*, *square*, *ellipse*, *circle*)
type *color* = (*red*, *yellow*, *green*)
type *sex* = (*male*, *female*)
type *Boolean* = (*false*, *true*)
type *weekday* = (*Monday*, *Tuesday*, *Wednesday*, *Thursday*, *Friday*, *Saturday*, *Sunday*)
type *currency* = (*franc*, *mark*, *pound*, *dollar*, *shilling*, *lira*, *guilder*, *krone*, *ruble*, *cruzeiro*, *yen*)
type *destination* = (*hell*, *purgatory*, *heaven*)
type *vehicle* = (*train*, *bus*, *automobile*, *boat*, *airplane*)
type *rank* = (*private*, *corporal*, *sergeant*, *lieutenant*, *captain*, *major*, *colonel*, *general*)
type *object* = (*constant*, *type*, *variable*, *procedure*, *function*)
type *structure* = (*file*, *array*, *record*, *set*)
type *condition* = (*manual*, *unloaded*, *parity*, *skew*)

The definition of such types introduces not only a new type identifier, but at the same time the set of identifiers denoting the values of the new type. These identifiers may then be used as constants throughout the program, and they enhance its understandability considerably. If, as an example, we introduce variables s, d, r, and b

var *s*: *sex*
var *d*: *weekday*
var *r*: *rank*
var *b*: *Boolean*

then the following assignment statements are possible:

s := *male*
d := *Sunday*
r := *major*
b := *true*

Evidently, they are considerably more informative than their counterparts

$$s := 1 \qquad d := 7 \qquad r := 6 \qquad b := 2$$

which are based on the assumption that c, d, r, and b are defined as type *integer* and that the constants are mapped onto the natural numbers in the order of their enumeration. Furthermore, a compiler can check against the inconsiderate use of arithmetic operators on such non-numeric types as, for example,

$$s := s+1$$

If, however, we consider a type as ordered, then it is sensible to introduce functions that generate the successor and predecessor of their argument. These functions are denoted by *succ*(x) and *pred*(x). The ordering among values of T is defined by the rule

$$(c_i < c_j) \equiv (i < j) \tag{1.2}$$

1.4. STANDARD PRIMITIVE TYPES

Standard primitive types are those types that are available on most computers as built-in features. They include the whole numbers, the logical truth values, and a set of printable characters. On larger computers fractional numbers are also incorporated, together with an adequate set of primitive operators. We denote these types by the identifiers

integer, *Boolean*, *char*, *real*

The type *integer* comprises a subset of the whole numbers whose size may vary between individual computer systems. It is assumed, however, that all operations on data of this type are exact and correspond to the ordinary laws of arithmetic and that the computation will be interrupted in the case of a result lying outside the representable subset. The standard operators are the four basic arithmetic operations of addition (+), subtraction (−), multiplication (*), and division (**div**). The latter is understood to yield an integer result, ignoring a possible remainder, such that, for positive m and n

$$m-n < (m \textbf{ div } n)*n \leq m \tag{1.3}$$

The modulus operator is defined in terms of division by the equation

$$(m \textbf{ div } n)*n+(m \textbf{ mod } n) = m \tag{1.4}$$

Thus, m **div** n is the integer quotient of m and n, and m **mod** n is the associated remainder.

The type *real* denotes a subset of the real numbers. Whereas arithmetic involving integers only is assumed to yield exact results, arithmetic on values of type real is permitted to be inaccurate within the limits of round-off errors caused by computation on a finite number of digits. This is the principal reason for the explicit distinction between the types *integer* and *real* made in most programming languages.

We denote division of real numbers yielding a real valued quotient by a slash (/) in contrast to the **div** of integer division.

The two values of the standard type *Boolean* are denoted by the identifiers *true* and *false*. The Boolean operators are the logical conjunction, union, and negation whose values are defined in Table 1.1. The logical conjunction

p	q	$p \vee q$	$p \wedge q$	$\neg p$
true	*true*	*true*	*true*	*false*
true	*false*	*true*	*false*	*false*
false	*true*	*true*	*false*	*true*
false	*false*	*false*	*false*	*true*

Table 1.1 Boolean Operators.

is denoted by the symbol $\wedge$ (or **and**), the logical union by $\vee$ (or **or**), and negation by $\neg$ (or **not**). Note that comparisons are operators yielding a result of type *Boolean*. Thus, the result of a comparison may be assigned to a variable, or it may be used as an operand of a logical operator in a Boolean expression. For instance, given Boolean variables p and q and integer variables $x = 5$, $y = 8$, $z = 10$, the two assignments

$$p := x = y$$
$$q := (x<y) \wedge (y \leq z)$$

yield $p = false$ and $q = true$.

The standard type *char* comprises a set of printable characters. Unfortunately, there is no generally accepted standard character set used on all computer systems. Therefore, the use of the predicate "standard" may in this case be almost misleading; it is to be understood in the sense of "standard on the computer system on which a certain program is to be executed."

The character set defined by the International Standards Organization (ISO), and particularly its American version ASCII (American Standard Code for Information Interchange) is probably the most widely accepted set. The ASCII set is therefore tabulated in Appendix A. It consists of 95 printable (*graphic*) characters and 33 control characters, the latter mainly being used in data transmission and for the control of printing equipment. A subset with 64 printable characters (capital letters only) is widely used and called the *restricted* ASCII set.

In order to be able to design algorithms involving characters (i.e., values of type *char*) that are computer system independent, we should like to be able to assume certain properties of character sets as binding, namely,

1. The type *char* contains the 26 Latin *letters*, the 10 Arabic *digits*, and a number of other graphic characters, such as punctuation marks.
2. The subsets of letters and digits are *ordered* and *coherent*, i.e.,

$$\begin{aligned} (\text{'A'} \leq x) \wedge (x \leq \text{'Z'}) &\equiv x \text{ is a letter} \\ (\text{'0'} \leq x) \wedge (x \leq \text{'9'}) &\equiv x \text{ is a digit} \end{aligned} \qquad (1.5)$$

3. The type *char* contains a non-printing, blank character which may be used as a separator. (Blanks are denoted by ␣ in Fig. 1.1).

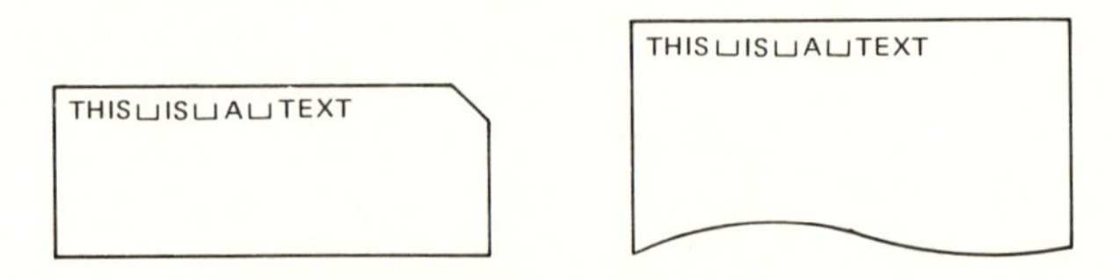

Fig. 1.1 Representations of a text.

The availability of two standard type transfer functions between the types *char* and *integer* is particularly important in the quest to write programs in a machine independent form. We will call them $ord(c)$, denoting the ordinal number of character c in the set *char*, and $chr(i)$, denoting the ith character in the set *char*. Thus, *chr* is the inverse function of *ord*, and vice versa, that is,

$$\begin{aligned} ord(chr(i)) &= i \qquad \text{(if } chr(i) \text{ is defined)} \\ chr(ord(c)) &= c \end{aligned} \tag{1.6}$$

Particularly noteworthy are the functions

$$\begin{aligned} f(c) &= ord(c) - ord(\text{'0'}) = \text{the position of } c \text{ among the digits} \\ g(i) &= chr(i+ord(\text{'0'})) \quad = \text{the } i\text{'th digit} \end{aligned} \tag{1.7}$$

For example, $f(\text{'3'}) = 3$, $g(5) = \text{'5'}$. f is the inverse function of g, and vice versa, i.e.,

$$\begin{aligned} f(g(i)) &= i \qquad (0 \le i \le 9) \\ g(f(c)) &= c \qquad (\text{'0'} \le c \le \text{'9'}) \end{aligned} \tag{1.8}$$

These transfer functions are used in the conversion of internal representations of numbers into sequences of digits, and vice versa. In fact, they represent these conversions on the most elementary level, namely that of a single digit.

1.5. SUBRANGE TYPES

It is often the case that a variable will assume values of a certain type within a specific interval only. This can be expressed by defining the variable to be of a *subrange* type according to the format

$$\boxed{\textbf{type } T = min \,..\, max} \tag{1.9}$$

where *min* and *max* are the limits of the interval.

EXAMPLES

type *year* = 1900 . . 1999
type *letter* = 'A' . . 'Z'
type *digit* = '0' . . '9'
type *officer* = *lieutenant* . . *general*

Given the variables

var y: *year*
var L: *letter*

the assignments $y := 1973$ and $L :=$ 'W' are permissible, but $y := 1291$ and $L :=$ '9' are not. The legality of such assignments cannot be verified by a compiler unless the value to be assigned is denoted by a constant or a variable of the same type. However, the admissibility of assignments of the kind

$$y := i \qquad \text{and} \qquad L := c$$

where i is of type *integer* and c is of type *char*, can be checked only upon program execution. In practice, systems performing these checks have proved enormously valuable in program development. Their utilization of redundant information to detect possible errors is again a prime motivation for using high-level language.

1.6. THE ARRAY STRUCTURE

The array is probably the most widely known structure of data because in many languages, including FORTRAN and ALGOL 60, it is the only explicitly available structure. An array is a *homogeneous* structure; it consists of components which are all of the same type, called the *base type*. The array is also a so-called *random-access* structure; all components can be selected at random and are equally accessible. In order to denote an individual component, the name of the entire structure is augmented by the so-called *index* selecting the component. This index is to be a value of the type defined as the *index type* of the array. The definition of an array type T therefore specifies both a base type T_0 and an index type I.

type T = **array**[I] **of** T_0 (1.10)

EXAMPLES

type *Row* = **array**[1 . . 5] **of** *real*
type *Card* = **array**[1 . . 80] **of** *char*
type *alfa* = **array**[1 . . 10] **of** *char*

A particular value of a variable

var x: *Row*

with every component satisfying the equation $x_i = 2^{-i}$ may be visualized as shown in Fig. 1.2.

x_1	0.5
x_2	0.25
x_3	0.125
x_4	0.0625
x_5	0.03125

Fig. 1.2 Array of type *row*.

A structured value x of type T with component values $c_1, \ldots, c_n$ may be denoted by an array *constructor* and an assignment statement:

$$x := T(c_1, \ldots, c_n) \tag{1.11}$$

The inverse operator of the constructor is the *selector*. It selects an individual component from the array. Given an array variable x, we denote an array selector by the array name augmented with the respective component index i:

$$\boxed{x[i]} \tag{1.12}$$

The common way of operating with arrays, particularly with large arrays, is to selectively update single components rather than to construct entirely new structured values. This is expressed by considering an array variable as representing an array of component variables and by permitting assignments to selected components.

EXAMPLE

$$x[i] := 0.125$$

Although selective updating causes only a single component value to change, from a conceptual point of view we must regard the entire composite value as having changed too.

The fact that array indices, i.e., "names" of array components, must be of a defined (scalar) data type has a most important consequence. Indices may be computed; an index *expression* may be substituted in place of an index constant. This expression is to be evaluated, and the result determines the selected component. This generality not only provides a most significant and powerful programming facility, but at the same time it also gives rise to one of the most frequently encountered programming mistakes: The resulting value may be outside the interval specified as the range of indices of the array. We will assume that adequate computing systems provide a warning in the case of such a mistaken access to a non-existent array component.

Normally, an index type will have to be a scalar type, i.e., an unstructured type upon which an ordering relation is defined. If the base type of an array is also an ordered type, then a natural ordering relation is given upon that array type. The natural ordering of two arrays is determined by the two corresponding unequal components with least indices. This is expressed formally as follows:

Given two arrays x and y, the relation $x < y$ holds if and only if there exists an index k such that $x[k] < y[k]$ and $x[i] = y[i]$ for all $i < k$. (1.13)

For example,

$$(2, 3, 5, 7, 9) < (2, 3, 5, 7, 11)$$

'LABEL' < 'LIBEL'

In most applications, however, no ordering is presumed to exist on array types.

The cardinality of a structured type is the product of the cardinality of its components. Since all components of an array type A are of the same base type B, we obtain

$$\text{cardinality}(A) = (\text{cardinality}(B))^n \tag{1.14}$$

where $n = \text{cardinality}(I)$, and I is the array's index type.

The following short piece of program shows the use of the array selector. The purpose of the program is to find the least index i of a component with value x. The search is performed by a sequential scan of the array a, declared as

```
var a: array [1 .. N] of T;   {N>0}

i := 0;
repeat i := i+1 until (a[i]=x) ∨ (i=N);                    (1.15)
if a[i] ≠ x then "there is no such element in a"
```

A variant of this program uses the common technique of a *sentinel* posted at the end of the array. The purpose of the sentinel is to allow a simplification of the termination condition of the repetition.

```
var a: array[1 .. N+1] of T;

i := 0; a[N+1] := x;
repeat i := i+1 until a[i] = x;                             (1.16)
if i > N then "there is no such element in a"
```

The assignment $a[N+1] := x$ is an example of *selective updating*, i.e., of the alteration of a selected component of a structured variable. The essential condition that holds no matter how often the statement $i := i+1$ was repeated, is

$$a[j] \neq x, \qquad \text{for} \quad j = 1 \ldots i - 1$$

in both versions (1.15) and (1.16). This condition is therefore called a *loop invariant*.

The search can, of course, be speeded up considerably if the elements are already ordered (sorted). In this case, the principle of repeated halving of the interval in which the desired element must be searched is most common. It is called *bisection* or *binary search* and is shown in (1.17). In each repetition, the inspected interval between indices i and j is bisected. The number of required comparisons is therefore at most $[\log_2 (N)]$.

$$
\begin{array}{l}
i := 1;\ j := N; \\
\textbf{repeat}\ k := (i+j)\ \textbf{div}\ 2; \\
\quad \textbf{if}\ x > a[k]\ \textbf{then}\ i := k+1\ \textbf{else}\ j := k-1 \\
\textbf{until}\ (a[k]=x) \lor (i>j)
\end{array}
\qquad (1.17)
$$

(The relevant invariant condition at the entrance of the repeat statement is

$$a[h] < x \qquad \text{for} \quad h = 1 \ldots i-1$$

$$a[h] \geq x \qquad \text{for} \quad h = j+1 \ldots N$$

Hence, if the program terminates with $a[k] \neq x$, it implies that there exists no $a[h] = x$ with $1 \leq h \leq N$.)

Constituents of array types may themselves be structured. An array variable whose components are again arrays is called a *matrix*. For example,

$$M\text{: }\textbf{array}[1 \,..\, 10]\ \textbf{of}\ Row$$

is an array consisting of ten components (rows), each consisting of five components of type real, and is called a 10×5 matrix with real components. Selectors may be concatenated accordingly, such that

$$M[i][j]$$

denotes the jth component of row $M[i]$, which is the ith component of M. This is usually abbreviated as

$$M[i, j]$$

and in the same spirit the declaration

$$M\text{: }\textbf{array}[1 \,..\, 10]\ \textbf{of}\ \textbf{array}[1 \,..\, 5]\ \textbf{of}\ real$$

can be written more concisely as

$$M\text{: }\textbf{array}[1 \,..\, 10, 1 \,..\, 5]\ \textbf{of}\ real.$$

If a certain operation has to be performed on *all* components of an array or on adjacent components of a part of the array, then this fact may conveniently be emphasized by using the **for** statement, as shown in the following example.

Assume that a fraction f is represented by the array d such that

$$f = \sum_{i=1}^{k-1} d_i * 10^{-i}$$

i.e., by its decimal form with $k - 1$ digits. Now f is to be divided by 2. This is done by repeating the familiar division operation for *all* $k - 1$ digits d_i, starting with $i = 1$. It consists of dividing the digit by 2, taking into account a possible carry from the previous position, and of retaining a possible remainder r for the next step [see (1.18)].

```
r := 10*r+d[i];
d[i] := r div 2;                    (1.18)
r := r-2*d[i]
```

This process is applied in Program 1.1 to compute a table of negative powers of 2. The repetition of halving to compute $2^{-1}, 2^{-2}, \ldots, 2^{-n}$ is again appropriately expressed by a **for** statement, thus leading to a nesting of two **for** statements.

```
program power (output);
{decimal representation of negative powers of 2}
const n = 10;
type digit = 0 .. 9;
var i,k,r: integer;
   d: array [1 .. n] of digit;
begin  for k := 1 to n do
   begin write('.'); r := 0;
      for i := 1 to k-1 do
      begin r := 10*r+d[i]; d[i] := r div 2;
            r := r-2*d[i]; write(chr(d[i]+ord('0')))
      end ;
      d[k] := 5; writeln('5')
   end
end .
```

Program 1.1 Compute powers of 2.

The resulting output for $n = 10$ is

```
.5
.25
.125
.0625
.03125
.015625
.0078125
.00390625
.001953125
.0009765625
```

1.7. THE RECORD STRUCTURE

The most general method to obtain structured types is to join elements of arbitrary, possibly themselves structured, types into a compound. Examples from mathematics are complex numbers, composed of two real numbers, and coordinates of points, composed of two or more real numbers according to the dimensionality of the space spanned by the coordinate system. An example from data processing is describing people by a few relevant characteristics, such as their first and last names, their dates of birth, sex, and marital status.

In mathematics such a compound type is called the *Cartesian product* of its constituents types. This stems from the fact that the set of values defined by this compound type consists of all possible combinations of values, taken one from each set defined by each constituent type. Thus, the number of such combinations, also called *n-tuples*, is the product of the number of elements in each constituent set, that is, the cardinality of the compound type is the product of the cardinalities of the constituent types.

In data processing, composite types, such as descriptions of persons or objects, usually occur in files or "data banks" and record the relevant characteristics of a person or object. The word **record** has therefore become widely accepted to describe a compound of data of this nature, and we adopt this nomenclature in preference to the term Cartesian product.

In general, a record type T is defined as follows:

$$\textbf{type } T = \textbf{record } s_1: T_1; \; s_2: T_2; \; \ldots \; s_n: T_n \; \textbf{end} \tag{1.19}$$

Cardinality (T) = cardinality $(T_1)* \cdots *$cardinality (T_n)

EXAMPLES

```
type Complex = record re: real;
                      im: real
               end

type Date = record day: 1 .. 31;
                 month: 1 .. 12;
                  year: 1 .. 2000
            end
```

type *Person* = **record** *name*: *alfa*;
firstname: *alfa*;
birthdate: *Date*;
sex: (*male*, *female*);
marstatus: (*single*, *married*,
widowed, *divorced*)
end

A value of type T may be constructed by a *record constructor* and subsequently assigned to a variable of that type:

$$x := T(x_1, x_2, \ldots, x_n) \tag{1.20}$$

where the x_i's are values of the constituent types T_i.

Given record variables

z: *Complex*
d: *Date*
p: *Person*

particular values may be assigned, for example, as follows (see Fig. 1.3):

z := *Complex* (1.0,−1.0)
d := *Date* (1,4,1973)
p := *Person* ('WIRTH', 'CHRIS', *Date* (18,1,1966), *male*, *single*)

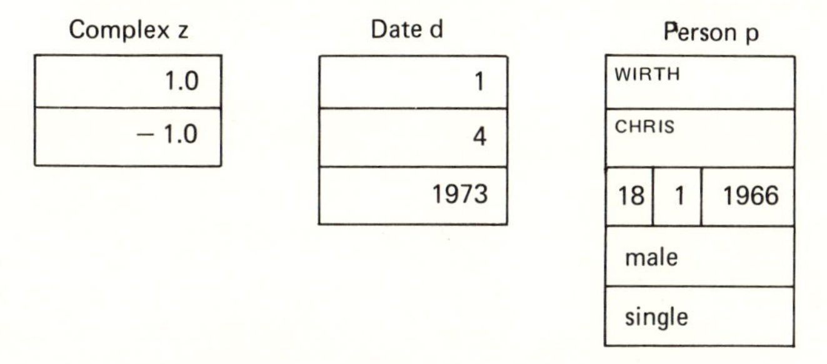

Fig. 1.3 Records of the types *complex*, *date*, and *person*.

The identifiers $s_1, \ldots, s_n$ introduced by a record type definition are the names given to the individual components of variables of that type, and they are used in *record selectors* applied to record structured variables. Given a variable x: T, its ith component is denoted by

$$\boxed{x.s_i} \tag{1.21}$$

Selective updating of x is achieved by using the same selector denotation on the left side in an assignment statement:

$$x.s_i := x_i$$

where x_i is a value (expression) of type T_i.

Given the record variables

z: *Complex*
d: *Date*
p: *Person*

the following are selectors of components of *z*, *d*, and *p*:

z.im	(of type *real*)
d.month	(of type 1 . . 12)
p.name	(of type *alfa*)
p.birthdate	(of type *Date*)
p.birthdate.day	(of type 1 . . 31)

The example of the type *Person* shows that a constituent of a record type may itself be structured. Thus, selectors may be concatenated. Naturally, different structuring types may also be used in a nested fashion. For example, the *i*th component of an array *a* being a component of a record variable *r* is denoted by

$$r.a[i]\ ,$$

and the component with the selector name *s* of the *i*th record structured component of the array *a* is denoted by

$$a[i].s$$

It is a characteristic of the Cartesian product that it contains *all* combinations of elements of the constituent types. But it must be noted that in practical application not all of them may be "legal," i.e., meaningful. For instance, the type *Date* as defined above contains the values

(31,4,1973) and (29,2,1815)

which are both dates of days that never occurred. Thus, the definition of this type does not mirror the real-world situation; but it is close enough for practical purposes, and it is the responsibility of the programmer to ensure that meaningless values never occur during the execution of his program.

The following short excerpt from a program shows the use of record variables. Its purpose is to count the number of "Persons" represented by the array variable *a* which are both female and single:

```
var a: array[1 .. N] of Person;
    count: integer;
count := 0;
for i := 1 to N do                                              (1.22)
    if (a[i].sex = female) ∧ (a[i].marstatus = single) then
        count := count+1
```

The relevant loop-invariant is

$$count = C(i)$$

where $C(i)$ is the number of single, female members of the subset $a_1 \ldots a_i$.

A notational variant of the above statement uses a construct that is called a *with-statement*:

```
for i := 1 to N do
    with a[i] do
        if (sex = female) ∧ (marstatus = single) then           (1.23)
            count := count+1
```

The meaning of **with** r **do** s is that selector names of the type of the variable r may be used without prefix within the statement s and are taken to refer to the variable r. The with-statement thus serves to abbreviate the program text as well as to prevent re-evaluation of the storage address of the indexed component $a[i]$.

In a further example, we assume that (possibly in order to find them more quickly) certain groups of persons in the array a are linked together. The linking information is represented by an additional component of the record structure *Person*, named *link*. The links connect records into a linear chain so that each person's successor and predecessor may be found easily. The interesting property of this linking technique is that the chain may be traversed in both directions on the basis of a single number stored in each record. The technique works as follows.

Assume that the indices of three consecutive members of the chain are i_{k-1}, i_k, i_{k+1}. The link value of the kth member is chosen to be $i_{k+1} - i_{k-1}$. Traversing the chain in the forward direction, i_{k+1} is determined from the two current index variables $x = i_{k-1}$ and $y = i_k$ as

$$i_{k+1} = x + a[y].link\ ,$$

whereas traversing the chain in the backward direction, i_{k-1} is determined from $x = i_{k+1}$ and $y = i_k$ as

$$i_{k-1} = x - a[y].link$$

An example is linking all persons of equal sex in a table (see Table 1.2).

	First Name	Sex	Link
1	Carolyn	F	2
2	Chris	M	2
3	Tina	F	5
4	Robert	M	3
5	Jonathan	M	3
6	Jennifer	F	5
7	Raytheon	M	5
8	Mary	F	3
9	Anne	F	1
10	Mathias	M	3

Table 1.2. Array with elements of type *Person*.

The record structure and the array stucture have the common property that both are "random-access" structures. The record is more general in the sense that there is no requirement that all constituent types must be identical. In turn, the array offers greater flexibility by allowing its component selectors to be computable values (represented by expressions), whereas the selectors of record components are fixed identifiers declared in the record type definition.

1.8. VARIANTS OF RECORD STRUCTURES

In practice, it is often convenient and natural to consider two types simply as *variants* of the same type. For example, the type *Coordinate* of the preceding section may be regarded as the union of its two variants of Cartesian and polar coordinates whose constituents are (a) two lengths and (b) a length and an angle, respectively. In order to identify the variant actually assumed by a variable, a third component will be introduced. It is called the *type discriminator* or *tag field.*

```
type Coordinate =
    record case kind: (Cartesian, polar) of
                Cartesian: (x, y: real);
                polar: (r: real; φ: real)
    end
```

Here, the name of the tag field is *kind*, and the names of the coordinates are either x and y in the case of a Cartesian value or they are r and φ in the case of a polar value.

The set of values denoted by this type Coordinate is the *union* of the two types

$$T_1 = (x, y\colon real)$$
$$T_2 = (r\colon real;\ \varphi\colon\ real)$$

and its cardinality is the sum of the cardinalities of T_1 and T_2.

Very often, however, there are not two entirely distinct types to be united, but rather two types with partly identical components. It is this predominant situation which gave rise to the term *variant record* structure. An example is that of the type *Person*, defined in the preceding section in which the relevant characteristics to be recorded in a file depend on the sex of the person. For example, for a male, his weight and whether or not he is bearded may be regarded as relevant in a particular situation, but for a female, three characteristic measurements may be taken as significant (whereas her weight may be confidential). Following is a type definition resulting from such considerations:

```
type Person =
     record name, firstname: alfa;
              birthdate: Date;
              marstatus: (single, married, widowed, divorced);
         case sex: (male, female) of
         male: (weight: real;
                   bearded: Boolean);
         female: (size: array[1 .. 3] of integer)
     end
```

The general form of a variant record type definition is

$$\begin{array}{l}\textbf{type } T = \\ \quad \textbf{record } s_1 : T_1; \ldots ; s_{n-1} : T_{n-1}; \\ \qquad \textbf{case } s_n : T_n \textbf{ of} \\ \qquad c_1\colon\ (s_{1,1} : T_{1,1}; \ldots ; s_{1,n_1} : T_{1,n_1}); \\ \qquad \qquad \cdots\cdots\cdots\cdots \\ \qquad c_m\colon\ (s_{m,1} : T_{m,1}; \ldots ; s_{m,n_m} : T_{m,n_m}) \\ \quad \textbf{end}\end{array} \tag{1.24}$$

The s_i and s_{ij} are the selector names of the components with constituent types T_i and T_{ij}, and s_n is the name of the discriminating tag field with type T_n. The constants $c_1 \ldots c_m$ denote the values of the (scalar) type T_n. A variable x of type T consists of the components

$$x.s_1, x.s_2, \ldots, x.s_n, x.s_{k,1}, \ldots, x.s_{k,n_k}$$

if and only if the (current) value of $x.s_n = c_k$. The components $x.s_1, \ldots, x.s_n$ constitute the *common part* of the m variants.

Consequently, using a component selector $x.s_{k,h}$ $(1 \leq h \leq n_k)$ when $x.s_n \neq c_k$ must be regarded as a serious programming mistake and (in reference to the type *Person* defined above) amounts to asking whether or not a lady is bearded or (in the case of selective updating) ordering her to be so!

In using variant records utmost care is therefore required, and corresponding operations on the individual variants are best grouped in a selective statement, the so-called *case statement*, whose structure mirrors that of the variant record type definition.

$$\begin{array}{l} \textbf{case } x.s_n \textbf{ of} \\ \quad c_1 : S_1; \\ \quad c_2 : S_2; \\ \quad \ldots \\ \quad c_m : S_m \\ \textbf{end} \end{array} \tag{1.25}$$

S_k stands for the statement catering for the case that x assumes the form of variant k, i.e., it is selected for execution if and only if the tag field $x.s_n$ has the value c_k. As a consequence, it is fairly easy to safeguard against the misuse of selector names by verifying that each S_k contains only selectors

$$x.s_1 \ldots x.s_{n-1}$$

and

$$x.s_{k,1} \ldots x.s_{k,n_k}$$

The purpose of the following short piece of program is to compute the distance between two points A and B given by the variables a and b of the variant record type *Coordinate*. The computational procedure differs according to the four possible combinations of Cartesian and polar coordinates (see Fig. 1.4).

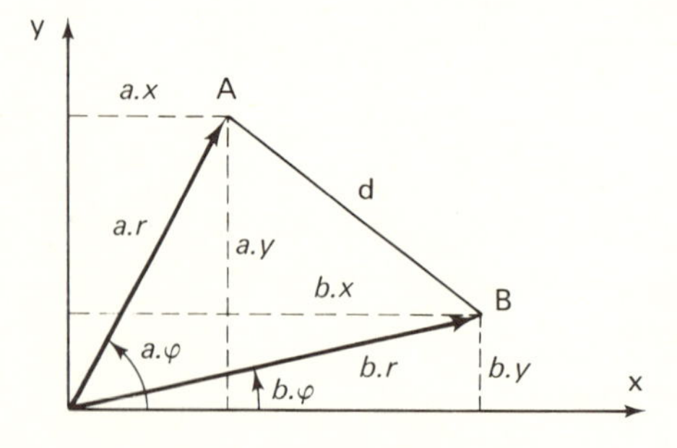

Fig. 1.4 Cartesian and polar coordinates.

```
case a.kind of
   Cartesian: case b.kind of
              Cartesian: d := sqrt(sqr(a.x—b.x)+sqr(a.y—b.y));
              Polar:     d := sqrt(sqr(a.x—b.r*cos(b.φ)
                                   +sqr(a.y—b.r*sin(b.φ))
              end;
   Polar:     case b.kind of
              Cartesian: d := sqrt(sqr(a.r*sin(a.φ)—b.x)
                                   +sqr(a.r*cos(a.φ)—b.y));
              Polar:     d := sqrt(sqr(a.r)+sqr(b.r)
                                   —2*a.r*b.r*cos(a.φ—b.φ))
              end
end
```

1.9. THE SET STRUCTURE

The third fundamental data structure—in addition to the array and the record—is the *set structure*. It is defined by the following declaration pattern:

$$\boxed{\textbf{type } T = \textbf{set of } T_0} \qquad (1.26)$$

The possible values of a variable x of type T are sets of elements of T_0. The set of all subsets of elements of a set T_0 is called the *powerset* of T_0. The type T thus comprises the powerset of its *base type* T_0.

EXAMPLES

type *intset* = **set of** 0 . . 30
type *charset* = **set of** *char*
type *tapestatus* = **set of** *exception*

The second example is based on the standard set of characters denoted by the type *char*; the third example is based on a set of exception conditions which might be defined as a scalar type

type *exception* = (*unloaded, manual, parity, skew*)

describing the various exceptional states that a magnetic tape unit may assume. Given the variables

is : *intset*
cs : *charset*
t : **array**[1 . . 6] **of** *tapestatus*

particular values of set types may be constructed and assigned, for example, as follows:†

$$
\begin{aligned}
is &:= [1, 4, 9, 16, 25] \\
cs &:= [\text{'+'}, \text{'−'}, \text{'*'}, \text{'/'}] \\
t[3] &:= [manual] \\
t[5] &:= [\ ] \\
t[6] &:= [unloaded\,..\,skew]
\end{aligned}
$$

Here, the value assigned to $t[3]$ is the singleton set consisting of the single element *manual*; to $t[5]$ is assigned the empty set, meaning that the fifth tape unit is returned to operational (non-exceptional) status, whereas $t[6]$ is assigned the set of all four exceptions.

The cardinality of a set type T is

$$\text{cardinality}(T) = 2^{\text{cardinality}(T_0)} \tag{1.27}$$

This can easily be derived from the fact that each of the cardinality (T_0) elements of T_0 must be represented by one of the two values "present" or "absent" and that all elements are independent of each other. It is evidently essential for an efficient and economical implementation that the base type be not only finite, but that its cardinality is reasonably small.

The following elementary operators are defined on all set types:

*	set intersection
+	set union
−	set difference
in	set membership

Constructing the intersection or union of two sets is often called *set multiplication* or *set addition*, respectively; the priorities of the set operators are defined accordingly, with the intersection operator having priority over the union and difference operators, which in turn have priority over the membership operator, which is classified as a relational operator. Following are examples of set expressions and their fully parenthesized equivalents:

$$
\begin{aligned}
r * s + t &= (r*s)+t \\
r - s * t &= r-(s*t) \\
r - s + t &= (r-s)+t \\
x \ \textbf{in} \ s + t &= x \ \textbf{in} \ (s+t)
\end{aligned}
$$

Our first example of an application of the set structure is the program of a simple scanner of a compiler. We assume that the purpose of the scanner is

†Contrary to conventional notation, we use brackets instead of braces for sets. We reserve braces to delimit comments within programs.

to translate a sequence of characters into a sequence of textual units of the language to be compiled, of so-called tokens or *symbols*. The scanner is to be represented as a procedure which each time it is called reads a sufficient number of input characters in order to generate the single next output symbol. The particular rules of translation will be the following:

1. The set of output symbols consists of the elements *identifier*, *number*, *lessequal*, *greaterequal*, *becomes*, and others which correspond to various single characters such as +, −, *, etc.
2. The symbol *identifier* is generated upon reading a sequence of letters and digits starting with a letter.
3. The symbol *number* is generated upon reading a sequence of digits.
3. The symbols *lessequal*, *greaterequal*, and *becomes* are generated upon reading the respective character pairs <= , >= , := .
4. Blanks and ends of lines are skipped.

We have at our disposal a primitive procedure *read*(x) which picks the next character off the input sequence and assigns it to the variable x. The resulting output symbol is to be assigned to a global variable called *sym*. Moreover, there are the global variables *id* and *num*, whose purpose will be evident from Program 1.2, and *ch* representing the currently scanned character in the input sequence. S denotes a mapping of characters to symbols, i.e., an array of symbols with an index domain over those characters which are neither digits nor letters. The use of sets of characters demonstrates how a scanner can be programmed independently of the order of the characters in the underlying character set.

A second example is drawn from constructing a school timetable. Suppose that M students have made their choices among N subjects. Now a timetable is to be constructed such that certain subjects are scheduled to be given at the same time in such a way that no conflicts arise [1.1].

In general, the construction of a timetable is a most difficult combinatorial problem, and a choice must be made under many constraints and with many factors to be considered. In this example we will simplify the problem drastically, without making the claim of a solution for a realistic timetable situation.

First of all, we realize that in order to find suitable choices for "parallel" sessions, we may base the decisions upon a set of data derived from the individual student's course registrations, namely, from the enumeration of courses that cannot be given simultaneously. Therefore, we first program a data reduction process, based on the following declarations and on the convention that students are numbered from 1 to M and courses from 1 to N.

```
var ch: char;
    sym: symbol;
    num: integer;
    id: record
            k: 0 .. maxk;
            a: array [1 .. maxk] of char
        end ;
procedure scanner;
   var ch1: char;
begin {skip blanks}
   while ch = '␣' do read(ch);
   if ch in ['A' .. 'Z'] then
      with id do
      begin sym := identifier; k := 0;
         repeat if k < maxk then
                  begin k := k+1; a[k] := ch
                  end ;
               read(ch)
         until ¬(ch in ['A' .. 'Z' , '0' .. '9'])
      end else
   if ch in ['0' .. '9'] then
      begin sym := number; num := 0;
         repeat num := 10*num+ord(ch)-ord('0');
            read(ch)
         until ¬(ch in ['0' .. '9'])
      end else
   if ch in ['<', ':', '>'] then
      begin ch1 := ch; read(ch);
         if ch = '=' then
         begin
            if ch1 = '<' then sym := leq else
            if ch1 = '>' then sym := geq else sym := becomes;
            read(ch)
         end
         else sym := S[ch1]
      end else
   begin {other symbols}
      sym := S[ch]; read(ch)
   end
end {scanner}
```

Program 1.2 A scanner.

```
type course    = 1 .. N;
     student   = 1 .. M;
     selection = set of course;
var s: course;
    i: student;
    registration: array[student] of selection;
    conflict: array[course] of selection;                          (1.28)
{Determine the sets of conflicting courses from the individual
 student's course registrations}
for s := 1 to N do  conflict[s] := [ ];
for i := 1 to M do
    for s := 1 to N do
    if s in registration[i] then
       conflict[s] := conflict[s] + registration[i]
```

(Note that *s* **in** *conflict*[*s*] is a consequence of this algorithm.)

The main task now consists of constructing a timetable, i.e., a list of sessions, each session being a selection of courses that do not conflict. From the whole set of courses we pick suitable, non-conflicting subsets of courses, subtracting them from a variable called *remaining*, until that set of remaining courses is empty.

```
var k: integer;
    remaining, session: selection;
    timetable: array[1 .. N] of selection;
k := 0; remaining := [1 .. N];
while remaining ≠ [ ] do                                          (1.29)
   begin session := next suitable selection;
         remaining := remaining − session;
         k := k+1; timetable[k] := session
   end
```

How do we make a "next suitable selection"? At the outset, we may select any single course from the set of remaining courses. Subsequently, the choice of further candidates may be restricted to the set of courses from the set of remaining courses which do not conflict with the ones initially selected. We call this set the *trialset*. When investigating a candidate from the trialset, we see that its choice depends upon whether or not the intersection of the already selected courses with the conflict-set of the candidate is empty. This leads to the following elaboration of the statement "*session* := *next suitable selection*":

Structure	Declaration	Selector	Access to Components by	Component Types	Cardinality
Array	a: **array**$[I]$ *of* T_0	$a[i] \quad (i \in I)$	Selector with computable index i	All identical (T_0)	$card(T_0)^{\mathrm{card}(I)}$
Record	r: **record** $s_1: T_1;$ $s_2: T_2;$ ... $s_n: T_n$ **end**	$r.s \quad (s \in \{s_1 \dots s_n\})$	Selector with declared component name s	May individually differ	$\prod_{i=1}^{n} card(T_i)$
Set	s: **set of** T_0	None	Membership test with relational operator **in**	All identical (and of scalar type T_0)	$2^{card(T_0)}$

Table 1.3 Fundamental data structures.

```
var s,t: course;
    trialset: selection;
begin s := 1;
    while ¬(s in remaining) do s := s+1;
    session := [s]; trialset := remaining − conflict[s];        (1.30)
    for t := 1 to N do
        if t in trialset then
        begin if conflict[t] * session = [ ] then
                session := session + [t]
        end
end
```

Evidently, this solution for selecting "suitable" sessions will not generate a timetable which is necessarily optimal in any specific sense. In unfortunate cases the number of sessions may be as large as that of courses, even if simultaneous scheduling were feasible.

1.10. REPRESENTATION OF ARRAY, RECORD, AND SET STRUCTURES

The essence of the use of abstractions in programming is that a program may be conceived, understood, and verified on the basis of the laws governing the abstractions and that it is not necessary to have further insight and knowledge about the ways in which the abstractions are implemented and represented in a particular computer. Nevertheless, it is helpful for a successful programmer to have an understanding of widely used techniques for representing the basic concepts of programming abstractions, such as the fundamental data structures. It is helpful in the sense that it might enable the programmer to make sensible decisions about program and data design in the light not only of the abstract properties of structures, but also of their realizations on actual computers, taking into account a computer's particular capabilities and limitations.

The problem of data representation is that of mapping the abstract structure into a computer store. Computer stores are—in a first approximation—arrays of individual storage cells called words. The indiceso f the words are called addresses.

```
var store: array[address] of word        (1.31)
```

The cardinalities of the types *address* and *word* vary from one computer to another. A particular problem is the great variability of the cardinality of the word. Its logarithm is called the *wordsize*, because it is the number of bits that a storage cell consists of.

1.10.1. Representation of Arrays

A representation of an array structure is a mapping of the (abstract) array with components of type T onto the store which is an array with components of type *word.*

The array should be mapped in such a way that the computation of addresses of array components is as simple (and therefore efficient) as possible. The address or store index i of the jth array component is computed by the linear mapping function

$$i = i_0 + j * s \tag{1.32}$$

where i_0 is the address of the first component, and s is the number of words that a component "occupies." Since the word is by definition the smallest individually accessible unit of store, it is evidently highly desirable that s be a whole number, the simplest case being $s = 1$. If s is not a whole number (and this is the normal case), then s is usually rounded up to the next larger integer $\lceil s \rceil$. Each array component then occupies $\lceil s \rceil$ words, whereby $\lceil s \rceil - s$ words are left unused (see Figs. 1.5 and 1.6). Rounding up of the number of

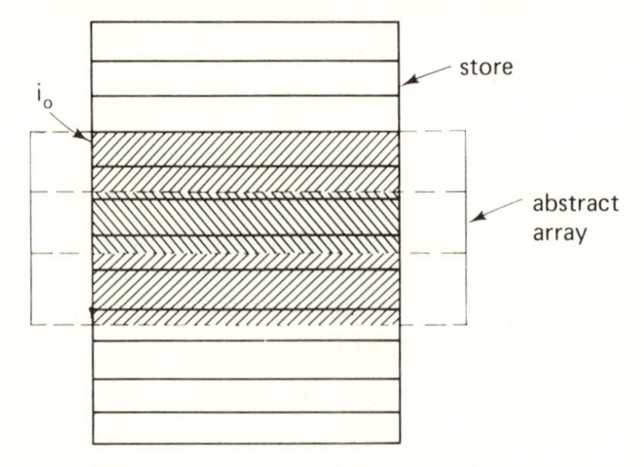

Fig. 1.5 Mapping an array onto a store.

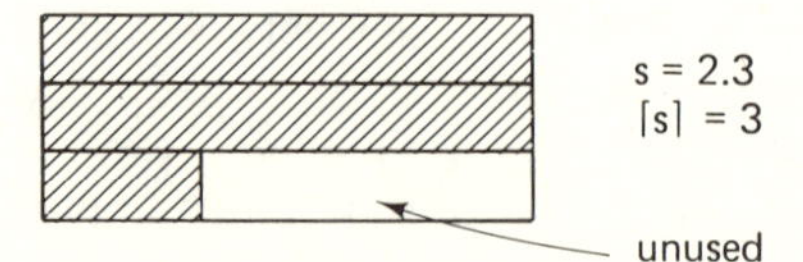

Fig. 1.6 Padded representation of a record.

words needed to the next whole number is called *padding*. The storage utilization factor u is the quotient of the minimal amounts of storage needed to represent a structure and of the amounts actually used:

$$u = \frac{s}{s'} = \frac{s}{\lceil s \rceil} \tag{1.33}$$

Since an implementor will have to aim for a storage utilization as close to

1 as possible, and since accessing parts of words is a cumbersome and relatively inefficient process, he will have to compromise. Following are the considerations to be made:

1. Padding will decrease storage utilization.
2. Omission of padding may necessitate inefficient partial word access.
3. Partial word access may cause the code (compiled program) to expand and therefore to counteract the gain obtained by omission of padding.

In fact, considerations 2 and 3 are usually so dominant that compilers will always use padding automatically. We notice that the utilization factor will always be $u > 0.5$, if $s > 0.5$. However, if $s \leq 0.5$, the utilization factor may be significantly increased by putting more than one array component into each word. This technique is called *packing*. If n components are packed into a word, the utilization factor is (see Fig. 1.7)

$$u = \frac{n \cdot s}{\lceil n \cdot s \rceil} \tag{1.34}$$

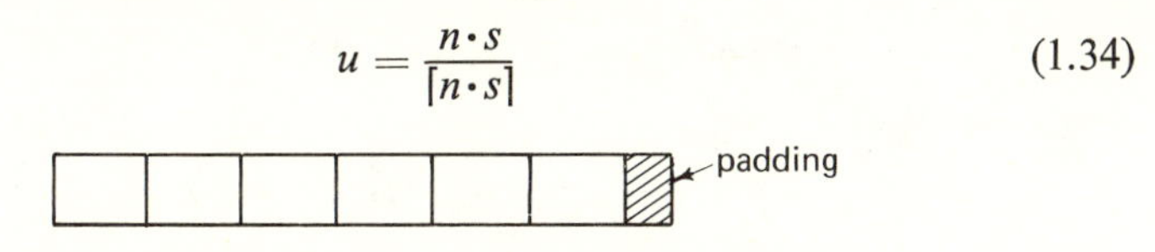

Fig. 1.7 Packing six components into one word.

Access to the ith component of a packed array involves the computation of the word address j in which the desired component is located and involves the computation of the respective component position k within the word.

$$\begin{aligned} j &= i \textbf{ div } n \\ k &= i \textbf{ mod } n = i - j * n \end{aligned} \tag{1.35}$$

In most programming languages the programmer is given no control over the representation of the abstract data structures. However, it should be possible to indicate the desirability of packing at least in those cases in which more than one component would fit into a single word, i.e., when a gain of storage economy by a factor of 2 and more could be achieved. We introduce the convention to indicate the desirability of packing by prefixing the symbol **array** (or **record**) in the declaration by the symbol **packed**.

EXAMPLE

type *alfa* = **packed array** [1 . . *n*] **of** *char*

This feature is particularly valuable on computers with large words and relatively convenient accessibility of partial fields of words. The essential property of this prefix is that it does in no way change the meaning (or correctness) of a program. This means that the choice of an alternative representation can be easily indicated with the implied guarantee that the meaning of the program remains unaffected.

The cost of accessing components of a packed array can in many cases be drastically reduced if the entire array is unpacked (or packed) at once. The reason is that an efficient sequential scan over the entire array is possible, making it unnecessary to evaluate a complicated mapping function for each individual component. We therefore postulate the existence of two standard procedures *pack* and *unpack* as defined below. Assume that there are variables

$$u : \textbf{array } [a\,..\,d] \textbf{ of } T$$
$$p : \textbf{packed array } [b\,..\,c] \textbf{ of } T$$

where $a \leq b \leq c \leq d$ are all of the same scalar type. Then

$$pack(u, i, p), \qquad (a \leq i \leq b-c+d) \tag{1.36}$$

is equivalent to

$$p[j] := u[j+i-b], \qquad j = b \ldots c$$

and

$$unpack(p, u, i), \qquad (a \leq i \leq b - c + d) \tag{1.37}$$

is equivalent to

$$u[j + i - b] := p[j], \qquad j = b \ldots c$$

1.10.2. Representation of Record Structures

Records are mapped onto a computer store (allocated) by simply juxtaposing their components. The address of a component (field) r_i relative to the origin address of the record r is called the component's *offset* k_i. It is computed as

$$k_i = s_1 + s_2 + \cdots + s_{i-1} \tag{1.38}$$

where s_j is the size (in words) of the jth component. The fact that all components of an array are of equal type has the consequence that

$$s_1 = s_2 = \cdots = s_n$$

and therefore

$$k_i = s_1 + \cdots + s_{i-1} = (i - 1)\cdot s$$

The generality of the record structure does not allow such a simple, linear function for offset address computation in general, and it is therefore the very reason for the requirement that record components be selectable only by fixed identifiers. This restriction has the desirable consequence that the respective offsets are known at compile time. The resulting greater efficiency of record field access is well-known.

The problem of packing may arise if several record components can be fitted into a single storage word (see Fig. 1.8). We will again assume that the desirability of packing may be indicated in a declaration by prefixing the symbol **record** by the symbol **packed**. Since offsets are computable by a compiler,

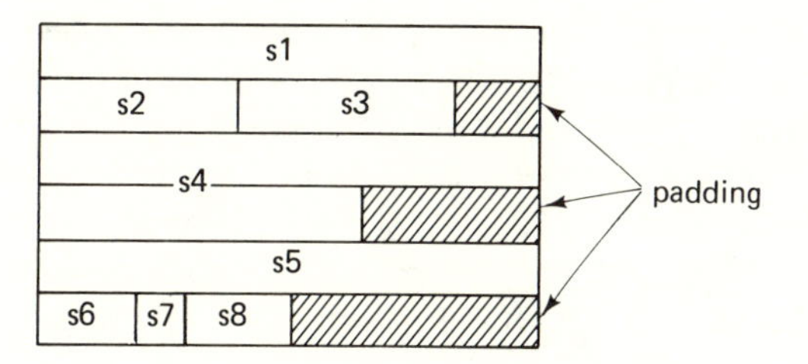

Fig. 1.8 Representation of a packed record.

the offset of a component within a word may also be determined by a compiler. This means that on many computers the packing of records will cause a decrease in access efficiency considerably smaller than that caused by the packing of arrays.

1.10.3. Representation of Sets

A set s is conveniently represented in a computer store by its *characteristic function* $C(s)$. This is an array of logical values whose ith component specifies the presence or absence of the value i in the set. The size of the array is determined by the set type's cardinality.

$$C(s_i) = (i \textbf{ in } s) \tag{1.39}$$

As an example, the set of small integers

$$s = [1, 4, 8, 9]$$

is representable by the sequence of logical values F (false) and T (true),

$$C(s) = (FTFFTFFFTT)$$

if the base type of s is the integer subrange 0 . . 9. In a computer store the sequence of logical values is represented as a so-called *bitstring* (see Fig. 1.9).

S	0	1	0	0	1	0	0	0	1	1
	0	1	2 ...							9

Fig. 1.9 Representation of a set as a bitstring.

The representation of sets by their characteristic function has the advantage that the operations of computing the union, intersection, and difference of two sets may be implemented as elementary logical operations. The following equivalences, which hold for all elements i of the base type of the sets x and y, relate logical operations with operations on sets:

$$\begin{aligned} i \textbf{ in } (x+y) &\equiv (i \textbf{ in } x) \lor (i \textbf{ in } y) \\ i \textbf{ in } (x*y) &\equiv (i \textbf{ in } x) \land (i \textbf{ in } y) \\ i \textbf{ in } (x-y) &\equiv (i \textbf{ in } x) \land \neg(i \textbf{ in } y) \end{aligned} \tag{1.40}$$

These logical operations are available on all digital computers, and moreover they operate *concurrently* on all corresponding elements (bits) of a word. It therefore appears that in order to be able to implement the basic set operations in an efficient manner, sets must be represented in a small, fixed number of words upon which not only the basic logical operations, but also those of shifting are available. Testing for membership is then implemented by a single shift and a subsequent (sign) bit test operation. As a consequence, a test of the form

$$x \textbf{ in } [c_1, c_2, \ldots, c_n]$$

can be implemented in a much more efficient manner than the equivalent conventional Boolean expression

$$(x = c_1) \lor (x = c_2) \lor \cdots \lor (x = c_n)$$

A corollary is that the set structure should be used only in the case of *small base types*. The limit of the cardinality of base types for which a reasonably efficient implementation can be guaranteed is determined by the wordlength of the underlying computer, and it is plain that computers with large wordlengths are preferred in this respect. If the wordsize is relatively small, a representation using multiple words for a set may be chosen.

1.11. THE SEQUENTIAL FILE STRUCTURE

The common characteristic of the data structures presented so far, namely, the array, the record, and the set structure, is that their *cardinality is finite* (provided that the cardinality of the types of their components is finite). Therefore, they present little difficulty for the implementor; suitable representations are readily found on any digital computer.

Most so-called advanced structures—sequences, trees, graphs, etc.—are characterized by their cardinality being infinite. This difference to the fundamental structures with finite cardinality is of profound importance, and it has significant practical consequences. As an example, we define the *sequence* structure as follows:

A sequence with base type T_0 is either the empty sequence or the concatenation of a sequence (with base type T_0) with a value of type T_0.

The sequence type T thus defined comprises an infinity of values. Each value itself contains a finite number of components of type T_0, but this number is unbounded, i.e., for every such sequence it is possible to construct a longer one.

Analogous considerations apply to all other "advanced" data structures. The prime consequence is that the necessary amount of store to represent a value of an advanced structural type is not known at compile time; in fact, it may vary during the execution of the program. This requires some scheme of *dynamic storage allocation* in which storage is occupied as the respective values "grow" and is possibly released for other uses when values "shrink." It is therefore plain that the question of representation of advanced structures is a subtle and difficult one and that its solution will crucially influence the efficiency of a process and the economy exercised on the usage of storage. A suitable choice can be made only on the basis of knowledge of the primitive operations to be performed on the structure and of the frequencies of their execution. Since none of this information is known to the designer of a language and its compiler, he is well advised to exclude advanced structures from a (general-purpose) language. It also follows that programmers should avoid their use whenever their problem can be treated by using fundamental structures only.

The dilemma of having to provide advanced data structuring facilities without information about their potential usage is circumvented in most languages and compilers by recognizing and using the fact that all advanced structures are composed either of unstructured elements or of fundamental structures. Arbitrary structures may then be generated by explicit, programmer specified operations, if facilities for the dynamic allocation of the components and for the dynamic linking and referencing of components are provided. Techniques for generating and manipulating these advanced structures are treated in Chapter 4.

There exists, however, one structure that is advanced in the sense that its cardinality is infinite, but which is used so widely and so frequently that its inclusion in the set of basic structures is almost mandatory: the *sequence*. In order to describe the abstract notion of sequence, the following terminology and notation are introduced:

1. $\langle\ \rangle$ denotes the empty sequence.
2. $\langle x_0 \rangle$ denotes the sequence consisting of the single component x_0; it is called a *singleton* sequence.
3. If $x = \langle x_1, \ldots, x_m \rangle$ and $y = \langle y_1, \ldots, y_n \rangle$ are sequences, then
$$x \mathbin{\&} y = \langle x_1, \ldots, x_m, y_1, \ldots, y_n \rangle \tag{1.41}$$
is the *concatenation* of x and y.
4. If $x = \langle x_1, \ldots, x_n \rangle$ is a non-empty sequence, then
$$\mathit{first}(x) = x_1 \tag{1.42}$$
denotes the first element of x.
5. If $x = \langle x_1, \ldots, x_n \rangle$ is a non-empty sequence, then
$$\mathit{rest}(x) = \langle x_2, \ldots, x_n \rangle \tag{1.43}$$
is the sequence x without its first component. As a consequence, we obtain the invariant relation
$$\langle \mathit{first}(x) \rangle \mathbin{\&} \mathit{rest}(x) \equiv x \tag{1.44}$$

The introduction of these notations does not mean that they will be used in actual programs to be obeyed by real computers. In fact, it is essential that the concatenation operation is *not* used in its generality and that the handling of sequences is confined to the application of a carefully selected set of operators, which ensure a certain discipline of usage, but which are themselves defined in terms of the abstract notions of the sequence and of concatenation. A careful choice of the set of sequence operators enables implementors to find suitable and efficient representations of sequences on any given storage medium; this ensures that the associated mechanism of dynamic storage allocation can be sufficiently simple to enable the programmer to work without concern for its details.

In order to make it clear that the sequence to be introduced as a basic data structure permits only the application of a restricted set of operators that essentially allow only strictly sequential access to components, this structure is called a *sequential file* or, for short, simply *file*. In close analogy to the notations for array and set type definitions, a file type is defined by the formula

$$\textbf{type } T = \textbf{file of } T_0 \tag{1.45}$$

expressing that any file of type T consists of 0 or more components of type T_0.

EXAMPLES

type *text* = **file of** *char*
type *deck* = **file of** *card*

The essence of *sequential access* is that at any time only a single, specific component of the sequence is immediately accessible. This component is specified by a *current position* of the access mechanism. It may be changed by the file operators, usually either to the next component or to the first of the entire sequence. We formally express the file position by regarding a file x to consist of two parts, a part x_L to its left and a part x_R to its right. It is plain that the equation

$$x \equiv x_L \mathbin{\&} x_R \tag{1.46}$$

expresses an invariant relationship.

A second, most important consequence of sequential access is that the processes of constructing and of scanning a sequence are distinct and cannot be mixed in arbitrary order. Thus, a file is constructed by repeatedly appending components (at its end), and it may subsequently be inspected by a sequential scan. It is therefore customary to consider a file as being in one of two *states*: either in the state of being constructed (written) or of being scanned (read).

The advantage of strictly sequential access is particularly pronounced if files are to be allocated on secondary storage media, that is, if transfers between different media are involved. The sequential access method is the only one in which the intricacies of mechanisms required by such transfers can be successfully hidden from the programmer. In particular, it allows for the application of simple *buffering techniques* that alone guarantee optimal usage of the various resources available in a complex computer system.

There exist certain storage media in which the sequential access is indeed the only possible one. Among them are evidently all kinds of tapes. But even on magnetic drums and disks each recording track constitutes a storage facility allowing only sequential access. Strictly sequential access is the primary characteristic of every mechanically moving device and of some other ones as well.

1.11.1. Elementary File Operators

We now proceed to formulate the abstract notion of sequential access through a set of concrete *elementary file operators* available to the programmer. They are defined in terms of the notions of sequence and concatenation. There is an operator to initiate the process of file generation, one to initiate the scan, one to append a component at the tail of the sequence, and one to proceed in scanning to the next component. The latter two are here postulated in a form in which they involve an implicit, auxiliary variable that represents a buffer. We assume that such a buffer is automatically associated with every file variable x, and we denote it by $x\uparrow$. Clearly, if x is of type T, then $x\uparrow$ is of its base type T_0.

1. Constructing the empty sequence. The operation

$$rewrite(x) \tag{1.47}$$

stands for the assignment

$$x := \langle\ \rangle$$

This operation is used to overwrite the current x and to initiate the process of constructing a new sequence, and corresponds to rewinding a tape.

2. Extending a sequence. The operation

$$put(x) \tag{1.48}$$

stands for the assignment

$$x := x \ \&\ \langle x\uparrow\rangle$$

which effectively appends the value $x\uparrow$ to the sequence x.

3. Initiation of a scan. The operation

$$reset(x) \tag{1.49}$$

stands for the simultaneous assignments

$$\begin{aligned} x_L &:= \langle\ \rangle \\ x_R &:= x \\ x\uparrow &:= \mathit{first}(x) \end{aligned}$$

This operation is used to initiate the process of reading a sequence.

4. Proceeding to the next component. The operation

$$get(x) \tag{1.50}$$

stands for the simultaneous assignments

$$\begin{aligned} x_L &:= x_L \ \& \ \langle \mathit{first}(x_R)\rangle \\ x_R &:= \mathit{rest}(x_R) \\ x\uparrow &:= \mathit{first}(\mathit{rest}(x_R)) \end{aligned}$$

Note that $\mathit{first}(s)$ is defined only if $s \neq \langle\ \rangle$.

The operators *rewrite* and *reset* notably do not depend on the position of the file prior to their execution. They reposition the file in any case to its beginning.

When scanning a sequence, it is necessary to be able to recognize the end of the sequence, because otherwise the assignment

$$x\uparrow := \mathit{first}(x_R)$$

represents an undefined operation. Reaching the end of the file is evidently synonymous with the condition that the right part x_R is empty. Therefore, we introduce the predicate

$$\mathit{eof}(x) \equiv x_R = \langle\ \rangle. \tag{1.51}$$

to mean that the *e*nd *o*f the *f*ile is reached. The operation *get*(x) can therefore only be executed if the predicate *eof*(x) is false.

In principle it is possible to express all operations on files in terms of the four basic file operators. In practice, however, it is often natural to combine the operation of advancing the file position (*get* or *put*) with the access to the buffer variable. We therefore postulate two further procedures; they are expressible in terms of the basic operators. Let v be a variable and e an expression of the file component type T_0. Then

$$\begin{aligned} &\mathit{read}(x,v) \quad \text{shall be synonymous with} \\ &\quad v := x\uparrow;\ \mathit{get}(x) \end{aligned}$$

and

$$\begin{aligned} &\mathit{write}(x,e) \quad \text{shall be synonymous with} \\ &\quad x\uparrow := e;\ \mathit{put}(x) \end{aligned}$$

The advantage of the use of *read* and *write* in place of *get* and *put* lies not only in brevity, but also in conceptual simplicity. For, it is now possible to ignore the existence of the buffer variable $x\uparrow$, whose value is sometimes undefined. The buffer variable may, however, be useful for the purpose of "looking ahead."

The prerequisite conditions for the execution of the two procedures are

$$\begin{array}{lcl} \neg eof(x) & \text{for} & read(x,v) \\ eof(x) & \text{for} & write(x,e) \end{array}$$

Upon reading, the predicate $eof(x)$ becomes true as soon as the last element of the file x has been read. These considerations are neatly incorporated in two *program schemata* for the sequential construction and processing of a file x. The statements R and S and the predicate p are additional parameters of the schemata.

Writing a file x:

```
rewrite(x);
while p do
    begin R(v); write(x,v)
    end
```
(1.52)

Reading a file x:

```
reset(x);
while ¬eof(x) do
    begin read(x,v); S(v)
    end
```
(1.53)

1.11.2. Files with Substructure

In the majority of applications, large files require some kind of substructure. For instance, a book, although it may be regarded as a single sequence of characters, is subdivided into chapters or paragraphs. The purpose of the substructure is to provide some explicit points of reference, some coordinates, in order to facilitate orientation in the long sequence of information. Existing storage devices often provide some facilities for representing such points of reference (e.g., tape marks) and have the capability of locating them with greater speed than is obtained when all information in between such points is scanned.

In our framework of notation the natural way to introduce a first level of substructure is to regard such a file as a sequence of components which are themselves sequences, that is, as a file of files. Assuming that the ultimate components (or units) are of type U, the substructures are then of type

$$T' = \textbf{file of } U$$

and the entire file is of type

$$T = \textbf{file of } T'$$

It is plain that in this manner files can be constructed with a partitioning to an arbitrary depth of nesting. In general, a type T_n can be defined by the recursive relation

$$T_i = \textbf{file of } T_{i-1} \qquad i = 1 \ldots n$$

and $T_0 = U$. Such files are often called *multilevel files*, and a component of type T_i is said to be a *segment* of level i. An example of a multilevel file is a book in which the levels of segmentation correspond to chapters, sections, paragraphs, and lines. However, by far the most common case is, of course, the file with a single level of segmentation.

This single level segmented file is by no means identical to an array of files. After all, the number of segments is variable, and the file may still be extended only at its end. Remaining within the framework of our notation introduced so far, and assuming a file defined by

$$x\text{: } \textbf{file of file of } U$$

$x\uparrow$ denotes the currently accessible segment, $x\uparrow\uparrow$ the currently accessible unit component. Accordingly, $put(x\uparrow)$ and $get(x\uparrow)$ refer to a unit component, whereas $put(x)$ and $get(x)$ denote the operations of appending and proceeding to the next segment.

Segmented files are readily implemented on virtually all sequential storage devices including tapes. Their segmentation has not changed their primary characteristic of permitting only sequential access either to individual components or—by a possibly faster skipping mechanism—to segments. Other storage devices—notably magnetic *drums* and *disks*—usually contain a number of tracks, each of which represents a proper sequential device but is usually too short to accommodate an entire file. Consequently, files on disks are usually spread over several tracks and contain appropriate bookkeeping information linking the tracks. It is plain that the starting point of each track constitutes a natural segment marker, which might easily be accessed even more directly than markers on any purely sequential medium. An indexable table in primary store might, for instance, be used to address the tracks where segments begin, and to indicate the actual lengths of the segments (see Fig. 1.10).

This leads us to the so-called *indexed files* (sometimes also called *direct access files*). Actually, drums and disks are organized in a way that each track contains many physical marks at which the reading or writing may start. Therefore, it is not necessary for each segment to occupy a full track because it would result in a poor storage utilization if segments are short in comparison with the track length. The storage line between two marks is called a *physical segment* (or *sector*) in contrast to the *logical segment* which is a

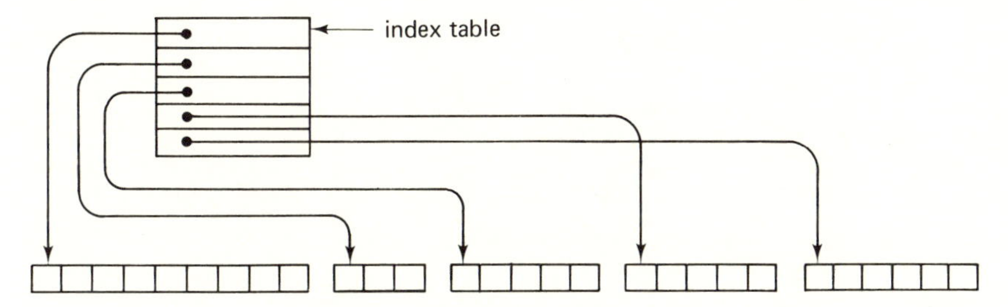

Fig. 1.10 Indexed file with five segments.

meaningful entity in the program's data structure. Clearly, each physical segment houses one logical segment at most, and each logical segment occupies at least one physical segment (even if it is empty). It should be kept in mind that even if they are called "direct access" files, the average time to locate a segment, the so-called *latency* time, is half the revolution time of the disk.

Indexed files retain the essential characteristic that writing proceeds sequentially at its end. They are therefore particularly useful in applications in which changes occur relatively infrequently. Changes are made by either extending or by recopying and updating the entire file. Inspection may occur in a much more selective and faster manner via the index points. This is the typical situation for so-called *data banks*.

Systems that allow selective rewriting of parts in the middle of a file are generally difficult and hazardous to use because the new portions of information must be of the same size as the old ones that they replace. Moreover, in applications involving large amounts of data, selective updating is not recommended because of the basic rule that upon any failure—be it caused by an erroneous program or by malfunctioning equipment—there should be a state of the data on which to fall back in order to resume and repeat the process that failed. Therefore, updating is usually made in toto, such that the old file is replaced by the new, updated copy, only after subsequent verification has established that the new file is valid. For the purpose of updating, the sequential organization is by far the best from the point of view of reliability. It is to be much preferred over more sophisticated organizations of large data sets, which may be more efficient, but often result in a total loss of data upon failure of the equipment.

1.11.3. Texts

Files whose components are of type *char* play an especially important role in computing and data processing: They constitute the interface between the computing machines and their human users. The *legible input* provided by programmers as well as the *legible output* representing the computed results are sequences of characters. This data type shall therefore be given a standard name:

type *text* = **file of** *char*

Communication between a computing process and its human inventor is ultimately established by an interface that can be represented as two text-files. One of them contains the *input* to the computing process, the other the computed results called *output*. We shall henceforth assume the existence of these two files in all programs, and their declaration as being

var *input*, *output*: *text*

With due regard to the assumption that these files represent the standard input and output media of a computer system (such as a card reader and a line printer), we will assume that the file *input* can be read only and the file *output* can be written only.

Since the two standard files are used predominantly, we postulate that if the first parameter of the procedures *read* and *write* is *not* a file variable, then *input* and *output* shall be assumed by default. Moreover, we take the liberty of providing the two standard procedures with an arbitrary number of arguments. These *notational conventions* can be summarized as follows:

read(*x*1, . . . , *xn*) stands for *read*(*input*, *x*1, . . . , *xn*)
write(*x*1, . . . , *xn*) stands for *write*(*output*, *x*1, . . . , *xn*)
read(*f*, *x*1, . . . , *xn*) stands for
  begin *read*(*f*,*x*1); . . . ; *read*(*f*,*xn*) **end**
write(*f*, *x*1, . . . , *xn*) stands for
  begin *write*(*f*,*x*1); . . . ; *write*(*f*,*xn*) **end**

Texts are typical examples of sequences displaying a substructure. The usual units of substructure are the chapters, paragraphs, and lines. An often-used method of representing this substructure in textfiles is the use of special separator characters. The blank character is the best-known example, but similar separators may be used to mark the ends of lines, paragraphs, and chapters. For instance, the widely used ISO character set (including its American version ASCII) contains several such elements, called *control characters* (see Appendix A).

In this book, we refrain from using specific separator characters and specifying a definite method of substructure representation. Instead, we think of a text as a file consisting of sequences of character sequences representing individual lines. Also, we constrain the discussion to a single level of substructure, namely, the *line*. However, instead of looking at texts as files of files of printable characters, we regard them as files of characters, and we introduce additional operators and predicates to manipulate, i.e., mark and recognize lines. Their effects may best be understood if one assumes that lines are separated by (hypothetical) separator characters (not belonging to the type *char*) and if their task is considered to be the insertion and recognition of such separators. Additional operators are the following:

writeln(f) Append a line marker to the file f.

readln(f) Skip characters on file f up to the one immediately following the next line marker.

eoln(f) A Boolean function. True, if the file position had been advanced to a marker; false otherwise. (We postulate that if *eoln*(f) is true, then $f\uparrow$ = blank.)

We are now in a position to formulate two program schemata for writing and reading texts similar to those for "writing" and "reading" other files [see (1.52) and (1.53)]. These schemata assume a textfile f and pay due attention to the generation and recognition of line structure. Let $R(x)$ be a statement assigning a value to x (of type *char*) and defining conditions p and q with meanings "this was the last character of the line" and "this was the last character of the file." Let U be a statement to be executed at the beginning of each line read, $S(x)$ a statement to be executed for each character x of the file, and V a statement executed at the end of each line.

Writing a text f.

```
rewrite(f);
while ¬q do
    begin
        while ¬p do
            begin R(x); write(f,x)
            end ;
        writeln(f)
    end
```
(1.54)

Reading a text f.

```
reset(f);
while ¬eof(f) do
    begin U;
        while ¬eoln(f) do
            begin read(f,x); S(x)
            end ;
        V; readln(f)
    end
```
(1.55)

There are cases in which the line structure of a text does not represent any particularly relevant information. Our assumption about the buffer variable's value upon encountering a line marker [see definition of *eoln*(*f*)] allows a simple program schema to be used in these situations. Note that according to the definition of *eoln* each end of a line appears as an additional blank character.

```
while ¬eof(f) do
    begin read(f,x); S(x)                    (1.56)
    end
```

In most programming languages it is customary to admit arguments of type *integer* or *real* to read and write procedures. This generalization would be straightforward if the types *integer* and *real* were defined as arrays of characters whose elements would denote the individual digits of the numbers. Languages strongly oriented towards commercial applications do indeed adhere to such definitions, and they require a representation of numbers in terms of decimal digits and of the decimal number system. The significant advantage of introducing the data types *integer* and *real* as fundamental types is that such detailed specifications may be omitted and that a computing system may use different representations of numbers which may be much more suitable to its purpose. In fact, systems oriented toward scientific calculations invariably choose a binary representation that is in most respects superior to the decimal representation.

This implies, however, that a programmer cannot assume that numbers can be read from or written onto textfiles without accompanying conversion operations. It is customary to hide these conversion operations behind read and write statements with arguments of numeric types. The professional programmer, however, is aware that such statements (so-called *I/O statements*) incorporate two distinct functions: data transfer between different storage media *and* transformations of data representation. The latter may be quite complex and time-consuming.

In the subsequent chapters of this book, read and write statements with numeric arguments will be used according to the rules of the programming language PASCAL. These rules allow for a certain kind of format specification to control the transformation process. The format specification indicates the number of desired digits in the case of write statements. This number of characters, also called "field width," is written immediately after the argument as follows:

write(*f*, *x*: *n*)

The argument x is to be written on file f; its value is converted into a sequence of (at least) n characters; if necessary, the digits are preceded by a sign and a suitable number of blanks.

Further details are unnecessary for understanding the later program examples of this book. Two examples of routines for the conversion of number representation are included here (Programs 1.3 and 1.4), however, for the sake of exhibiting the costly *complexity* of such operations which are usually assumed implicitly in write statements. The two procedures represent the conversion of real numbers from decimal representation to an arbitrary "internal" representation and vice versa. (The constants in the headings are determined by the properties of the floating-point number format of the CDC 6000 computer: 11-bit binary exponent and 48-bit mantissa. The function *expo* (*x*) denotes the exponent of *x*.)

Program 1.3 Read a real number.

```
procedure readreal (var f: text; var x: real);
   { read real number x from file f}
   {the following are system dependent constants}
   const t48 = 281474976710656;     {= 2**48}
      limit  =  56294995342131;     {= t48 div 5}
      z = 27;      {= ord('0') }
      lim 1 = 322;       { maximum exponent }
      lim 2 = -292;   { minimum exponent }
   type posint = 0 .. 323;
   var ch: char; y: real; a,i,e: integer;
      s,ss: boolean;       { signs }

   function ten(e: posint): real; { = 10**e, 0<e<322 }
      var i: integer; t: real;
   begin i := 0; t := 1.0;
      repeat if odd(e) then
         case i of
          0: t := t * 1.0E1;
          1: t := t * 1.0E2;
          2: t := t * 1.0E4;
          3: t := t * 1.0E8;
          4: t := t * 1.0E16;
          5: t := t * 1.0E32;
          6: t := t * 1.0E64;
          7: t := t * 1.0E128;
          8: t := t * 1.0E256
         end ;
         e := e div 2; i := i+1
      until e = 0;
      ten := t
   end ;
```

```
begin
   {skip leading blanks}
   while f↑=' ' do get(f);
   ch := f↑ ;
   if ch = '-' then
      begin s := true; get(f); ch := f↑
      end else
      begin s := false;
         if ch = '+' then
         begin get(f); ch := f↑
         end
      end ;
   if ¬(ch in ['0' .. '9']) then
   begin message (' DIGIT EXPECTED'); halt;
   end ;
   a := 0; e := 0;
   repeat if a < limit then a := 10*a + ord(ch)-z else e := e+1;
      get(f); ch := f↑
   until ¬(ch in ['0' .. '9']);
   if ch = '.' then
   begin { read fraction } get(f); ch := f↑;
      while ch in ['0' .. '9'] do
      begin if a < limit then
            begin a := 10*a + ord(ch)-z; e := e-1
            end ;
         get(f); ch := f↑
      end
   end ;
   if ch = 'E' then
   begin { read scale factor } get(f); ch := f↑;
      i := 0;
      if ch = '-' then
      begin ss := true; get(f); ch := f↑
      end else
      begin ss := false; if ch = '+' then
         begin get(f); ch := f↑
         end
      end ;
      while ch in ['0' .. '9'] do
      begin if i<limit then begin i := 10*i + ord(ch)-z end;
         get(f); ch := f↑
      end ;
      if ss then e := e-i else e := e+i
   end ;
```

Program 1.3 (Continued)

```
        if e < lim2 then
            begin a := 0; e := 0
            end else
        if e > lim1 then
        begin message(' NUMBER TOO LARGE '); halt end;
        { 0 < a < 2**49 }
        if a ≥ t48 then y := ((a+1) div 2) * 2.0 else y := a;
        if s then y := -y;
        if e < 0 then x := y/ten(-e) else
        if e ≠ 0 then x := y*ten(e) else x := y ;
        while (f↑ = ' ')∧(¬eof(f)) do get(f);
end {readreal}
```

Program 1.4 Write a real number.

```
procedure writereal (var f: text; x: real; n: integer);
    {write real number x with n characters in decimal flt.pt. format}
    {the following constants depend on the underlying floating-point representa-
    tion of real numbers}
    const t48 = 281474976710656; {= 2**48; 48 = size of mantissa}
          z = 27;     { ord('0') }
    type posint = 0 .. 323;     {range of decimal exponent}
    var c,d,e,e0,e1,e2,i: integer;

    function ten(e: posint): real;     { 10**e, 0<e<322 }
        var i: integer; t: real;
    begin i := 0; t := 1.0;
        repeat if odd(e) then
            case i of
              0: t := t * 1.0E1;
              1: t := t * 1.0E2;
              2: t := t * 1.0E4;
              3: t := t * 1.0E8;
              4: t := t * 1.0E16;
              5: t := t * 1.0E32;
              6: t := t * 1.0E64;
              7: t := t * 1.0E128;
              8: t := t * 1.0E256
            end ;
            e := e div 2; i := i+1
        until e = 0;
        ten := t
    end { ten } ;
```

```
begin { at least 10 characters needed: b+9.9E+999 }
  if x = 0 then
  begin repeat write(f, ' '); n := n-1
        until n ≤ 1;
        write(f, '0')
  end else
  begin
    if n ≤ 10 then n := 3 else n := n-7;
    repeat write(f, ' '); n := n-1
    until n ≤ 15;
    { 1 < n ≤ 15,  number of digits to be printed }
    begin { test sign, then obtain exponent }
      if x < 0 then
        begin write(f, '-'); x := -x
        end else write (f, ' ');
      e := expo (x);  {e = entier(log2(abs(x)))}
      if e ≥ 0 then
        begin e := e*77 div 256 +1; x := x/ten(e);
          if x ≥ 1.0 then
            begin x := x/10.0; e := e+1
            end
        end else
        begin e := (e+1)*77 div 256; x := ten(-e)*x;
          if x < 0.1 then
            begin x := 10.0*x; e := e-1
            end
        end ;
      { 0.1 ≤ x < 1.0 }
      case n of     { rounding }
         2: x := x+0.5E-2;
         3: x := x+0.5E-3;
         4: x := x+0.5E-4;
         5: x := x+0.5E-5;
         6: x := x+0.5E-6;
         7: x := x+0.5E-7;
         8: x := x+0.5E-8;
         9: x := x+0.5E-9;
        10: x := x+0.5E-10;
        11: x := x+0.5E-11;
        12: x := x+0.5E-12;
        13: x := x+0.5E-13;
        14: x := x+0.5E-14;
        15: x := x+0.5E-15
      end ;
```

Program 1.4 (Continued)

```
            if x ≥ 1.0 then
               begin x := x * 0.1; e := e+1;
               end ;
            c := trunc(x,48); {= trunc(x*(2**48))}
            c := 10*c; d := c div t48;
            write(f, chr(d+z), '.');
            for i := 2 to n do
            begin c := (c - d*t48) * 10; d := c div t48;
               write(f, chr(d+z))
            end ;
            write(f, 'E'); e := e-1;
            if e < 0 then
               begin write(f, '-'); e := -e
               end else write(f, '+');
            e1 := e * 205 div 2048; e2 := e - 10*e1;
            e0 := e1 * 205 div 2048; e1 := e1 - 10*e0;
            write(f, chr(e0+z), chr(e1+z), chr(e2+z))
         end
      end
   end {writereal}
```

Program 1.4 (Continued)

1.11.4. A File Editing Program

As an example of an application of sequential structures, we pose the following problem, which moreover serves to demonstrate a method of developing and explaining programs. This method is called *stepwise refinement* [1.4, 1.6] and will be used to explain many algorithms throughout this book.

The problem is to develop a program which edits a text x into a text y. Editing means deleting or replacing specific lines or inserting new lines. Editing is governed by a sequence of *editing instructions* represented by the standard text *input* as follows:

I, *m*.	Insertion of text after the *m*th line.
D, *m*, *n*.	Deletion of lines *m* to *n*.
R, *m*, *n*.	Replacement of lines *m* to *n*.
E.	Terminate the editing process.

Each instruction is written as a line in the standard file *input*, which we call the instruction file. *m* and *n* are decimal line numbers, and insertion texts are to follow the *I* and *R* instructions immediately. They are terminated by an empty line.

We postulate that editing instructions are issued (sequenced) with strictly increasing line numbers. This rule immediately suggests a strictly *sequential processing* of the input text *x*. It is plain that the state of the process must be characterized by the current position of *x* in terms of the number of the line currently under investigation.

Let us now assume that the editing program is to be used in an interactive fashion and that therefore the instruction file represents, for instance, the data originating at a keyboard terminal. In this mode of operation it is highly desirable that the operator receives some sort of feedback. An appropriate and useful form of feedback is the text of that line to which the last instruction caused the process to advance. We call this line the *current line*. An important consequence of the new requirement to print the current line after each instruction is that the current line must be represented by an explicit variable in which the line is buffered after reading it from *x* and before writing it onto *y*. This technique is called "lookahead." The editing program can now be formulated as follows:

```
program editor (x, y, input, output);
var lno: integer;       {number of current line}
    cl : line;          {current line}
    x,y: text;
begin read instruction;
   repeat interpret instruction;                    (1.57)
          write line;
          read instruction
   until instruction = 'E'
end.
```

We now proceed to specify the various statements in greater detail. Refining *read instruction* and *interpret instruction*, we note that an instruction generally consists of three parts: the instruction code and two parameters. We therefore introduce the three variables *code*, *m*, and *n* intended for communication between the two routines.

```
var code,ch: char;
    m,n: integer
```

Read instruction:

```
read(code,ch);
if ch = ',' then read(m,ch) else m := lno;          (1.58)
if ch = ',' then read(n) else n := m;
```

This formulation caters to the acceptance of instructions with 0, 1, or 2 parameters, substituting appropriate default values for "missing" specifications.

Interpret instruction:

```
copy;
if code = 'I' then
begin putline;
      insert;
end else
if code = 'D' then skip else
if code = 'R' then                                  (1.59)
begin insert;
      skip
end else
if code = 'E' then copyrest else Error
```

In a second step of refinement we express the statements *copy*, *insert*, and *skip* used in (1.59) in terms of operations involving single lines only, i.e., in terms of *getline* and *putline*. Their common characteristic is the repetitive structure. *Copy* serves to copy lines from x to y, starting with the current line, and terminating with the mth line. *Skip* reads lines from x without copying up to the nth line.

```
Copy:     while lno < m do
          begin putline;
                getline
          end
Skip:     while lno < n do getline
Insert:   readline;
          while noend do
             begin putline; readline                (1.60)
             end;
          getline;
Copyrest: while ¬eof(x) do
             begin putline; getline
             end;
          putline
```

In the third and last step of refinement we express the operations *getline*, *putline*, *readline*, and *writeline* in terms of operations on single characters. We note that until now all operations dealt exclusively with entire lines and that no specific assumptions were made about the detailed substructure of a line. We know that lines themselves are sequences of characters. It would be tempting to declare the variable *cl* (holding the current line) as a sequence

var *cl*: **file of** *char*

However, recall the advice that a structure with infinite cardinality should never be used if a fundamental structure (such as an array) is adequate. Indeed, we are well advised to use an array structure in the present case. This is feasible, if we limit the line length to, say, 80 characters. Hence we specify

var *cl*: **array**[1 . . 80] **of** *char*

The four routines use an index variable *i* with this array, which, in fact, is used locally and could well be declared local to each routine; moreover, it now becomes necessary to introduce a global variable *L* to denote the length of the current line.

```
Getline:   i := 0; lno := lno + 1;
           while ¬eoln(x) do
               begin i := i+1; read (x, cl[i])
               end ;
           L := i; readln(x)

Putline:   i := 0;
           while i < L do
               begin i := i+1; (write y, cl[i])
               end ;
           writeln(y)
                                                        (1.61)
Readline:  i := 0;
           while ¬eoln(input) do
               begin i := i+1; read(cl[i])
               end ;
           readln

Writeline: i := 0; write (lno);
           while i < L do
               begin i := i+1; write(cl[i])
               end ;
           writeln
```

The condition *noend* in the routine *insert* is now readily expressed as

$$L \neq 0$$

This concludes the development of this file editing program.

EXERCISES

1.1. Assume that the cardinalities of the standard types *integer*, *real*, and *char* are denoted by c_I, c_R, and c_C. What are the cardinalities of the following data types defined as examples in this chapter: *sex*, *Boolean*, *weekday*, *letter*, *digit*, *officer*, *row*, *alfa*, *complex*, *date*, *person*, *coordinate*, *charset*, *tapestatus*?

1.2. How would you represent variables of the types listed in Exercise 1.1:
(a) In the store of your computer?
(b) In FORTRAN?
(c) In your favorite programming language?

1.3. Which are the instruction sequences (on your computer) for the following:
(a) Fetch and store operations for elements of packed records and arrays?
(b) Set operations, including the test for membership?

1.4. Can the correct use of variant records be checked at run time? Can it even be verified at compile time?

1.5. What are the reasons for defining certain sets of data as sequential files instead of arrays?

1.6. Assume that you are to implement sequential files as defined in Sec. 1.11 on a computer with a very large primary store. You are allowed to impose the restriction that files will never exceed a certain length L. Hence, you can represent files in terms of arrays.

Describe a possible implementation, including the chosen data representation and procedures for the elementary file operators *get*, *put*, *reset*, and *rewrite*, which are defined by a set of axioms in Sec. 1.11.

1.7. Apply Exercise 1.6 to the case of segmented files.

1.8. Given is a railway timetable listing the daily services on several lines of a railway system. Find a representation of these data in terms of arrays, records, or files, which is suitable for lookup of arrival and departure times given a certain station and desired direction of the train.

1.9. Given a text T in the form of a file, and lists of a small number of words in the form of two arrays A and B. Assume that words are short arrays of characters of a small and fixed maximum length.

Write a program that transforms the text T into a text S by replacing each occurrence of a word A_i by its corresponding word B_i.

1.10. Which adjustments—redefinition of constants, etc.—are necessary to adapt Programs 1.3 and 1.4 to your available computer?

1.11. Write a procedure similar to Program 1.4 whose heading is

```
procedure writereal (var f: text; x: real; n,m: integer);
```

It is supposed to transform the value x into a sequence of at least n characters (to be appended to file f) representing x in decimal, fixed-point form with m digits following the decimal point. If necessary, the number is to be preceded by a suitable number of blanks and/or a sign.

1.12. Rewrite the text editor of Sec. 1.11.4 in the form of a complete program.

1.13. Compare the following three versions of the binary search with (1.17). Which of the three programs are correct? Which ones are more efficient? We assume the following variables, and a constant $N > 0$:

```
var i,j,k: integer;
    a: array[1 .. N] of T;
    x: T
```

Program A:

```
i := 1; j := N;
repeat k := (i+j) div 2;
   if a[k] < x then i := k else j := k
until (a[k]=x) ∨ (i≥j)
```

Program B:

```
i := 1; j := N;
repeat k := (i+j) div 2;
   if x ≤ a[k] then j := k-1;
   if a[k] ≤ x then i := k+1
until i > j
```

Program C:

```
i := 1; j := N;
repeat k := (i+j) div 2;
   if x < a[k] then j := k else i := k+1
until i ≥ j
```

Hint: All programs must terminate with $a[k] = x$, if such an element exists, or $a[k] \neq x$, if there exists no element with value x.

1.14. A company organizes a poll to determine the success of its products. Its products are records and tapes of hits, and the most popular hits are to be broadcast in a hit parade. The polled population is to be divided into four categories according to sex and age (say, less or equal to 20, and older than 20). Every person is asked to name five hits. Hits are identified by the numbers 1 to N (say, $N = 30$). The results of the poll are represented by a file.

```
type hit = 1 .. N;
     sex = (male, female);
     response =
        record name, firstname: alfa;
               s: sex;
               age: integer;
               choice: array [1 .. 5] of hit
        end ;
var poll: file of response
```

Hence, each file element represents a respondent and lists his name, first name, sex, age, and his five preferred hits according to priority. This file is the input to a program which is supposed to compute the following results:

1. A list of hits in the order of their popularity. Each entry consists of the hit number and the number of times it was mentioned in the poll. Hits that were never mentioned are omitted from the list.
2. Four separate lists with the names and first names of all respondents who had mentioned in first place one of the three hits most popular in their category.

The five lists are to be preceded by suitable titles.

REFERENCES

1-1. DAHL, O. J., DIJKSTRA, E. W., and HOARE, C. A. R., *Structured Programming*, (New York: Academic Press, 1972), pp. 155–65.

1-2. HOARE, C. A. R., "Notes on Data Structuring," in *Structured Programming*, Dahl, Dijkstra, and Hoare, pp. 83–174.

1-3. JENSEN, K. and WIRTH, N., "PASCAL, User Manual and Report," *Lecture Notes in Computer Science*, Vol. 18 (Berlin: Springer-Verlag, 1974).

1-4. WIRTH, N., "Program Development by Stepwise Refinement," *Comm. ACM*, **14**, No. 4 (1971), 221–27.

1-5. __________, "The Programming Language PASCAL," *Acta Informatica*, **1**, No. 1 (1971), 35–63.

1-6. __________, "On the Composition of Well-Structured Programs," *Computing Surveys*, **6**, No. 4, (1974) 247–59.

2 SORTING

2.1. INTRODUCTION

The primary purpose of this chapter is to provide an extensive set of examples illustrating the use of the data structures introduced in the preceding chapter and to show how the choice of structure for the underlying data profoundly influences the algorithms that perform a given task. Sorting is also a good example to show that such a task may be performed according to many different algorithms, each one having certain advantages and disadvantages that have to be weighted against each other in the light of the particular application.

Sorting is generally understood to be the process of re-arranging a given set of objects in a specific *order*. The purpose of sorting is to facilitate the later search for members of the sorted set. As such it is an almost universally performed, fundamental activity. Objects are sorted in telephone books, in income tax files, in tables of contents, in libraries, in dictionaries, in warehouses, and almost everywhere that stored objects have to be searched and retrieved. Even small children are taught to put their things "in order," and they are confronted with some sort of sorting long before they learn anything about arithmetic.

Hence, sorting is a relevant and essential activity, particularly in data processing. What else would be easier to sort than "data"! Nevertheless, our primary interest in sorting is devoted to the even more fundamental techniques used in the construction of algorithms. There are not many techniques that do not occur somewhere in connection with sorting algorithms. In particular, sorting is an ideal subject to demonstrate a great diversity of algorithms, all having the same purpose, many of them being optimal in some sense, and most of them having advantages over others. It is therefore an ideal subject to demonstrate the necessity of performance analysis of algorithms. The example of sorting is moreover well-suited for showing how a very significant gain in performance may be obtained by the development of sophisticated algorithms when obvious methods are readily available.

The dependence of the choice of an algorithm on the structure of the data to be processed—an ubiquitous phenomenon—is so profound in the case of sorting that sorting methods are generally classified into two categories, namely, *sorting of arrays* and *sorting of* (sequential) *files*. The two classes are often called *internal* and *external* sorting because arrays are stored in the fast, high-speed, random-access "internal" store of computers and files are conveniently located on the slower, but more spacious "external" stores based on mechanically moving devices (disks and tapes). The importance of this distinction is obvious from the example of sorting numbered cards. Structuring the cards as an array corresponds to laying them out in front of the sorter so that each card is visible and individually accessible (see Fig. 2.1).

Fig. 2.1 Array sorting.

Structuring the cards as a file, however, implies that from each pile only the card on the top is visible (see Fig. 2.2). Such a restriction will evidently have serious consequences on the sorting method to be used, but it is unavoidable if the number of cards to be laid out is larger than the available table.

Before proceeding, we introduce some terminology and notation to be used throughout this chapter. We are given items

$$a_1, a_2, \ldots, a_n$$

Fig. 2.2 File sorting.

Sorting consists of permuting these items into an order

$$a_{k_1}, a_{k_2}, \ldots, a_{k_n}$$

such that, given an *ordering function f*,

$$f(a_{k_1}) \leq f(a_{k_2}) \leq \cdots \leq f(a_{k_n}) \tag{2.1}$$

Ordinarily, the ordering function is not evaluated according to a specified rule of computation but is stored as an explicit component (field) of each item. Its value is called the *key* of the item. As a consequence, the record structure is particularly well-suited to represent the items a_i. We therefore define a type *item* to be used in all subsequent sorting algorithms:

```
type item  =  record key: integer;
                     {other components declared here}
              end
```
(2.2)

The "other components" represent relevant data about the items in the collection; the key merely assumes the purpose of identifying the items. As far as our sorting algorithms are concerned, however, the key is the *only* relevant component, and there is no need to define any particular remaining components. The choice of *integer* as the key type is somewhat arbitrary. Evidently, any type on which a total ordering relation is defined could be used just as well.

A sorting method is called *stable* if the relative order of items with equal keys remains unchanged by the sorting process. Stability of sorting is often desirable if items are already ordered (sorted) according to some secondary keys, i.e., properties not reflected by the (primary) key itself.

This chapter is *not* to be regarded as a comprehensive survey in sorting techniques. Rather, some selected, specific methods are exemplified in greater detail. For a thorough treatment of sorting, the interested reader is referred to the excellent and comprehensive compendium by D. E. Knuth [2-7] (see also Lorin [2-10]).

2.2. SORTING ARRAYS

The predominant requirement that has to be made for sorting methods on arrays is an economical use of the available store. This implies that the permutation of items which brings the items into order has to be performed *in situ* and that methods which transport items from an array a to a result array b are intrinsically of minor interest. Having thus restricted our choice of methods among the many possible solutions by the criterion of economy of storage, we proceed to a first classification according to their efficiency, i.e., their economy of time. A good *measure of efficiency* is obtained by counting the numbers C of needed *key comparisons* and M of *moves* (transpositions) *of items*. These numbers are functions of the number n of items to be sorted. Whereas good sorting algorithms require in the order of $n \cdot \log n$ comparisons, we first discuss several simple and obvious sorting techniques, called *straight methods*, all of which require in the order n^2 comparisons of keys. There are three good reasons for presenting straight methods before proceeding to the faster algorithms.

1. Straight methods are particularly well-suited for elucidating the characteristics of the major sorting principles.
2. Their programs are easy to understand and are short. Remember that programs occupy storage as well!
3. Although sophisticated methods require fewer operations, these operations are usually more complex in their details; consequently, straight methods are faster for sufficiently small n, although they must not be used for large n.

Sorting methods which sort items *in situ* can be classified into three principal categories according to their underlying method:

1. Sorting by insertion.
2. Sorting by selection.
3. Sorting by exchange.

These three principles will now be examined and compared. The programs operate on the variable a whose components are to be sorted *in situ* and refer to the data types *item* (2.2) and *index*, defined as

```
type index = 0 . . n;
var a: array[1 . . n] of item
```
(2.3)

2.2.1. Sorting by Straight Insertion

This method is widely used by card players. The items (cards) are conceptually divided into a destination sequence $a_1 \ldots a_{i-1}$ and a source sequence $a_i \ldots a_n$. In each step, starting with $i = 2$ and incrementing i by unity, the ith element of the source sequence is picked and transferred into the destination sequence by *inserting* it at the appropriate place.

Initial Keys	44	55	12	42	94	18	06	67
$i = 2$	44	55	12	42	94	18	06	67
$i = 3$	12	44	55	42	94	18	06	67
$i = 4$	12	42	44	55	94	18	06	67
$i = 5$	12	42	44	55	94	18	06	67
$i = 6$	12	18	42	44	55	94	06	67
$i = 7$	06	12	18	42	44	55	94	67
$i = 8$	06	12	18	42	44	55	67	94

Table 2.1 A Sample Process of Straight Insertion Sorting.

The process of sorting by insertion is shown in an example of eight numbers chosen at random (see Table 2.1). The algorithm of straight insertion is

```
for i := 2 to n do
  begin x := a[i];
    "insert x at the appropriate place in a1 . . . ai"
  end
```

In the process of actually finding the appropriate place, it is convenient to alternate between comparisons and moves, i.e., to let x "sift down" by comparing x with the next item a_j, and either inserting x or moving a_j to the right and proceeding to the left. We note that there are two distinct conditions that may cause the termination of the "sifting down" process:

1. An item a_j is found with a key less than the key of x.
2. The left end of the destination sequence is reached.

This typical case of a repetition with two termination conditions brings the well-known sentinel technique to our attention. It is easily applied to this case by posting a sentinel item $a_0 = x$. (Note that this must be included by extending the index range in the declaration of a to 0 . . n.) The completed algorithm is formulated in Program 2.1.

```
procedure straightinsertion;
    var i,j: index;   x: item;
begin
    for i := 2 to n do
    begin x := a[i]; a[0] := x; j := i-1;
        while x .key < a[j] .key do
            begin a[j+1] := a[j]: j := j-1;
            end ;
        a[j+1] := x
    end
end
```

Program 2.1 Sorting by Straight Insertion.

Analysis of straight insertion. The number C_i of key comparisons in the ith sift is at most $i - 1$, at least 1, and—assuming that all permutations of the n keys are equally probable—$i/2$ in the average. The number M_i of moves (assignments of items) is $C_i + 2$ (including the sentinel). Therefore, the total numbers of comparisons and moves are

$$\begin{aligned} C_{min} &= n - 1 & M_{min} &= 2(n - 1) \\ C_{ave} &= \tfrac{1}{4}(n^2 + n - 2) & M_{ave} &= \tfrac{1}{4}(n^2 + 9n - 10) \\ C_{max} &= \tfrac{1}{2}(n^2 + n) - 1 & M_{max} &= \tfrac{1}{2}(n^2 + 3n - 4) \end{aligned} \tag{2.4}$$

The least numbers occur if the items are originally in order; the worst case occurs if the items are originally in reverse order. In this sense, sorting by insertion exhibits a truly *natural behavior*. It is plain that the given algorithm also describes a *stable* sorting process: it leaves the order of items with equal keys unchanged.

The algorithm of straight insertion is easily improved by noting that the destination sequence $a_1 \ldots a_{i-1}$, in which the new item has to be inserted, is already ordered. Therefore, a faster method of determining the insertion point can be used. The obvious choice is a binary search that samples the destination sequence in the middle and continues bisecting until the insertion point is found. The modified sorting algorithm is called *binary insertion*, and is shown in Program 2.2.

```
procedure binaryinsertion;
   var i,j,l,r,m: index;  x: item;
begin
   for i := 2 to n do
   begin x := a[i]; l := 1; r := i-1;
      while l ≤ r do
      begin m := (l+r) div 2;
         if x .key < a[m] .key then r := m-1 else l := m+1
      end ;
      for j := i-1 downto l do a[j+1] := a[j];
      a[l] := x;
   end
end
```

Program 2.2. Sorting by Binary Insertion.

Analysis of binary insertion. The insertion position is found if $a_r .key \leq x .key < a_1 .key$. Thus, the search interval must in the end be 1; and this involves halving the interval of i keys $\lceil \log_2 i \rceil$ times. Thus,

$$C = \sum_{i=1}^{n} \lceil \log_2 i \rceil$$

We approximate this sum by the integral

$$\int_1^n \log x \, dx = x(\log x - c) \Big|_1^n = n(\log n - c) + c \qquad (2.5)$$

where $c = \log e = 1/\ln 2 = 1.44269\ldots$. The number of comparisons is essentially independent of the initial order of the items. However, because of the truncating character of the division involved in bisecting the search interval, the true number of comparisons needed with i items may be up to 1 higher than expected. The nature of this bias is such that insertion positions at the low end are on the average located slightly faster than those at the high end, thereby favoring those cases in which the items are originally highly out of order. In fact, the minimum number of comparisons is needed if the items are initially in reverse order and the maximum if they are already in order. Hence, this is a case of *unnatural behavior* of a sorting algorithm.

$$C \doteq n(\log n - \log e \pm 0.5)$$

Unfortunately, the improvement obtained by using a binary search method applies only to the number of comparisons but not to the number of necessary moves. In fact, since moving items, i.e., keys *and* associated information, is in general considerably more time-consuming than comparing two keys, the improvement is by no means drastic: the important term M is still of the order n^2. And, in fact, re-sorting the already sorted array takes more time than does straight insertion with sequential search! This example demonstrates that an "obvious improvement" often has much less drastic consequences than one is first inclined to estimate and that in some cases (which *do* occur) the "improvement" may actually turn out to be a deterioration. After all, sorting by insertion does not appear to be a very suitable method for digital computers: insertion of an item with the subsequent shifting of an entire row of items by a single position is uneconomical. One should expect better results from a method in which moves of items are only performed upon single items and over longer distances. This idea leads to sorting by selection.

2.2.2 Sorting by Straight Selection

This method is based on the following principle:

1. Select the item with the least key.
2. Exchange it with the first item a_1.

Then repeat these operations with the remaining $n - 1$ items, then with $n - 2$ items, until only one item—the largest—is left. This method is shown on the same eight keys as in Table 2.1.

initial keys	44	55	12	42	94	18	06	67
	06	55	12	42	94	18	44	67
	06	12	55	42	94	18	44	67
	06	12	18	42	94	55	44	67
	06	12	18	42	94	55	44	67
	06	12	18	42	44	55	94	67
	06	12	18	42	44	55	94	67
	06	12	18	42	44	55	67	94

Table 2.2 A Sample Process of Straight Selection Sorting.

The program is formulated as follows:

```
for i := 1 to n-1 do
   begin "assign the index of the least item of a[i] . . . a[n] to k";
         "exchange a_i and a_k"
   end
```

This method, called *straight selection,* is in some sense the opposite of straight insertion: Straight insertion considers in each step only the *one* next item of the *source sequence* and *all* items of the *destination array* to find the insertion point; straight selection considers *all* items of the *source array* to find the one with the least key and to deposit it as the *one* next item of the *destination sequence*. The entire program of straight selection is given in Program 2.3.

```
procedure straightselection;
    var i,j,k: index;   x: item;
begin for i := 1 to n-1 do
        begin k := i; x := a[i];
            for j := i+1 to n do
                if a[j] .key < x .key then
                begin k := j; x := a[j]
                end ;
            a[k] := a[i]; a[i] := x;
        end
end
```

Program 2.3 Sorting by Straight Selection.

Analysis of straight selection. Evidently, the number C of key comparisons is independent of the initial order of keys. In this sense, this method may be said to behave less naturally than straight insertion. We obtain

$$C = \tfrac{1}{2}(n^2 - n)$$

The number M of moves is at least

$$M_{\min} = 3(n - 1) \tag{2.6}$$

in the case of initially ordered keys and at most

$$M_{\max} = \operatorname{trunc}\left(\frac{n^2}{4}\right) + 3(n - 1)$$

if initially the keys are in reverse order. The average M_{ave} is difficult to determine in spite of the algorithm's simplicity. It depends on the number of times that k_j is found to be less than all preceding numbers $k_1 \ldots k_{j-1}$ when scanning a sequence of numbers $k_1 \ldots k_n$. This value, averaged over all $n!$ permutations of n keys is

$$H_n - 1$$

where H_n is the nth *harmonic number*

$$H_n = 1 + \frac{1}{2} + \frac{1}{3} + \cdots + \frac{1}{n} \tag{2.7}$$

(cf. Knuth, Vol. 1, pp. 95–99).

H_n can be expressed as

$$H_n = \ln n + \gamma + \frac{1}{2n} - \frac{1}{12n^2} + \cdots \tag{2.8}$$

where $\gamma = 0.577216\ldots$ is Euler's constant. For sufficiently large n, we may ignore the fractional terms and therefore approximate the average number of assignments in the ith pass as

$$F_i = \ln i + \gamma + 1$$

The average number of moves M_{ave} in a selection sort is then the sum of F_i with i ranging from 1 to n.

$$M_{\text{ave}} = \sum_{i=1}^{n} F_i = n(\gamma + 1) + \sum_{i=1}^{n} \ln i$$

By further approximating the sum of discrete terms by the integral

$$\int_1^n \ln x \, dx = x(\ln x - 1)\Big|_1^n = n \ln n - n + 1$$

we obtain an approximate value

$$M_{\text{ave}} \doteq n(\ln n + \gamma) \tag{2.9}$$

We may conclude that in general the algorithm of straight selection is to be preferred over straight insertion, although in the cases in which keys are initially sorted or almost sorted, straight insertion is still somewhat faster.

2.2.3 Sorting by Straight Exchange

The classification of a sorting method is seldom entirely clear-cut. Both previously discussed methods can also be viewed as exchange sorts. In this section, however, we present a method in which the exchange of two items is the dominant characteristic of the process. The subsequent algorithm of straight exchanging is based on the principle of comparing *and* exchanging pairs of adjacent items until all items are sorted.

As in the previous methods of straight selection, we make repeated passes over the array, each time sifting the least item of the remaining set to the left end of the array. If, for a change, we view the array to be in a vertical instead of a horizontal position, and—with the help of some imagination—the items

initial	i = 2	i = 3	i = 4	i = 5	i = 6	i = 7	i = 8
44	06	06	06	06	06	06	06
55	44	12	12	12	12	12	12
12	55	44	18	18	18	18	18
42	12	55	44	42	42	42	42
94	42	18	55	44	44	44	44
18	94	42	42	55	55	55	55
06	18	94	67	67	67	67	67
67	67	67	94	94	94	94	94

Table 2.3 A Sample of Bubblesorting.

as bubbles in a water tank with "weights" according to their keys, then each pass over the array results in the ascension of a bubble to its appropriate level of weight (see Table 2.3). This method is widely known as the *Bubblesort*. Its simplest form is shown in Program 2.4.

```
procedure bubblesort;
   var i,j: index;  x: item;
begin for i := 2 to n do
      begin for j := n downto i do
            if a[j-1].key > a[j].key then
            begin x := a[j-1]; a[j-1] := a[j]; a[j] := x
            end
      end
end {bubblesort}
```

Program 2.4 Bubblesort

This algorithm easily lends itself to some improvements. The example in Table 2.3 shows that the last three passes have no effect on the order of the items because the items are already sorted. An obvious technique for improving this algorithm is to remember whether or not any exchange had taken place during a pass. A last pass without further exchange operations is therefore necessary to determine that the algorithm may be terminated. However, this improvement may itself be improved by remembering not merely the fact that an exchange took place, but rather the position (index) of the last exchange. For example, it is plain that all pairs of adjacent items below this index k are in the desired order. Subsequent scans may therefore be terminated at this index instead of having to proceed to the predetermined lower limit i. The careful programmer will, however, notice a peculiar asymmetry: A single misplaced bubble in the "heavy" end of an otherwise sorted array will sift into order in a single pass, but a misplaced item in the "light" end will sink toward its correct position only one step in each pass. For example, the array

12 18 42 44 55 67 94 06

will be sorted by the improved Bubblesort in a single pass, but the array

94 06 12 18 42 44 55 67

will require seven passes for sorting. This unnatural asymmetry suggests a third improvement: alternating the direction of consecutive passes. We appropriately call the resulting algorithm *Shakersort*. Its behavior is illustrated in Table 2.4 by applying it to the same eight keys that were used in Table 2.3.

```
procedure shakersort;
   var j,k,l,r: index;   x: item;
begin l := 2; r := n; k := n;
   repeat
      for j := r downto l do
         if a[j-1] .key > a[j] .key then
         begin x := a[j-1]; a[j-1] := a[j]; a[j] := x;
            k := j
         end ;
      l := k+1;
      for j := l to r do
         if a[j-1] .key > a[j] .key then
         begin x := a[j-1]; a[j-1] := a[j]; a[j] := x;
            k := j
         end ;
      r := k-1;
   until l > r
end {shakersort}
```

Program 2.5 Shakersort.

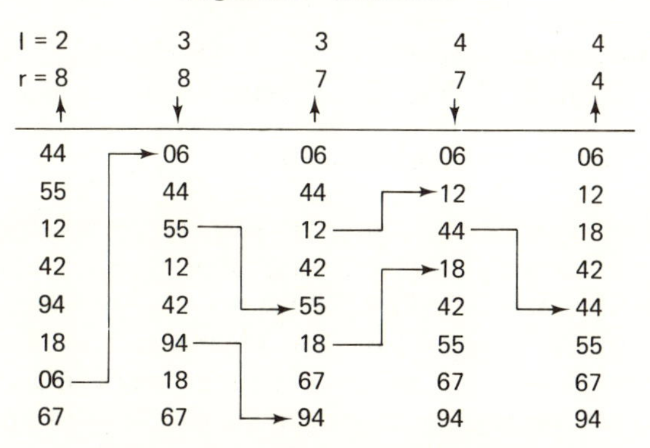

l = 2	3	3	4	4
r = 8	8	7	7	4
↑	↓	↑	↓	↑
44	06	06	06	06
55	44	44	12	12
12	55	12	44	18
42	12	42	18	42
94	42	55	42	44
18	94	18	55	55
06	18	67	67	67
67	67	94	94	94

Table 2.4 An Example of Shakersort.

Analysis of Bubblesort and Shakersort. The number of comparisons in the straight exchange algorithm is

$$C = \tfrac{1}{2}(n^2 - n) \tag{2.10}$$

and the minimum, average, and maximum numbers of moves (assignments of items) are

$$M_{\min} = 0, \qquad M_{\text{ave}} = \tfrac{3}{4}(n^2 - n), \qquad M_{\max} = \tfrac{3}{2}(n^2 - n) \tag{2.11}$$

The analysis of the improved methods, particularly that of Shakersort, is involved. The least number of comparisons is $C_{\min} = n - 1$. For the improved Bubblesort, Knuth arrives at an average number of passes proportional to $n - k_1\sqrt{n}$, and an average number of comparisons proportional to $\frac{1}{2}[n^2 - n(k_2 + \ln n)]$. But we note that all improvements mentioned above do in no way affect the number of exchanges; they only reduce the number of redundant double checks. Unfortunately, an exchange of two items is generally a much more costly operation than a comparison of keys; our clever improvements therefore have a much less profound effect than one would intuitively expect.

This analysis shows that the exchange sort and its minor improvements are inferior to both the insertion and the selection sorts; and in fact, the Bubblesort has hardly anything to recommend it except its catchy name. The Shakersort algorithm is used with advantage in those cases in which it is known that the items are already almost in order—a rare case in practice.

It can be shown that the average distance that each of the n items has to travel during a sort is $n/3$ places. This figure provides a clue in the search for improved, i.e., more effective sorting methods. All straight sorting methods essentially move each item by one position in each elementary step. Therefore, they are bound to require in the order n^2 such steps. Any improvement must be based on the principle of moving items over greater distances in single leaps.

Subsequently, three improved methods will be discussed, namely, one for each basic sorting method: insertion, selection, and exchange.

2.2.4 Insertion Sort by Diminishing Increment

A refinement of the straight insertion sort was proposed by D. L. Shell in 1959. The method is explained and demonstrated on our standard example of eight items (see Table 2.5). First, all items which are four positions apart are grouped and sorted separately. This process is called a 4-sort. In this example of eight items, each group contains exactly two items. After this first pass, the items are regrouped into groups with items two positions apart and then sorted anew. This process is called a 2-sort. Finally, in a third pass, all items are sorted in an ordinary sort or 1-sort.

One may at first wonder if the necessity of several sorting passes, each of which involves all items, will not introduce more work than it saves. However, each sorting step over a chain involves either relatively few items or the items are already quite well ordered and comparatively few re-arrangements are required.

It is obvious that the method results in an ordered array, and it is fairly obvious that each pass will profit from previous passes (since each i-sort combines two groups sorted in the preceding $2i$-sort). It is also obvious that any sequence of increments will be acceptable, as long as the last one is unity, because in the worst case the last pass will do all the work. It is, however,

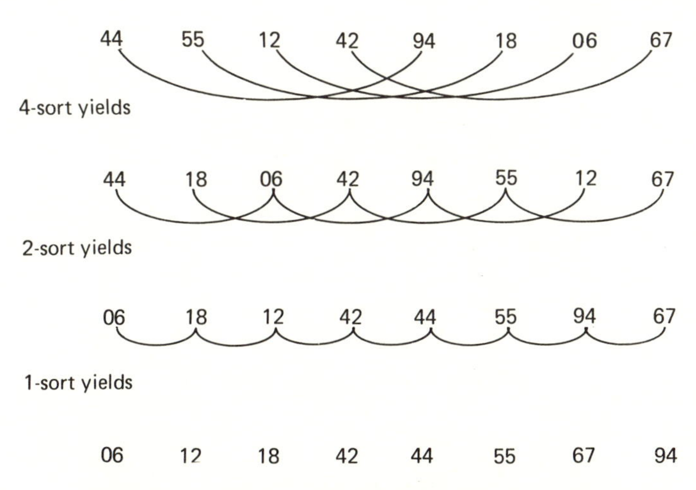

Table 2.5 An Insertion Sort with Diminishing Increments.

much less obvious that the method of diminishing increments yields even *better* results with increments other than powers of 2.

The program is therefore developed without relying on a specific sequence of increments. The t increments are denoted by

$$h_1, h_2, \ldots, h_t$$

with the conditions,

$$h_t = 1, \qquad h_{i+1} < h_i \tag{2.12}$$

Each h-sort is programmed as a straight insertion sort using the sentinel technique to provide a simple termination condition for the search of the insertion place.

It is plain that each sort needs to post its own sentinel and that the program to determine its position should be made as simple as possible. The array a therefore has to be extended not only by a single component $a[0]$, but by h_1 components, such that it is now declared as

$$a\text{: } \textbf{array}[-h_1 \,..\, n] \textbf{ of } item$$

The algorithm is described by the procedure called *Shellsort* [2.11] in Program 2.6 for $t = 4$.

Analysis of Shellsort. The analysis of this algorithm poses some very difficult mathematical problems, many of which have not yet been solved. In particular, it is not known which choice of increments yields the best results. One surprising fact, however, is that they should not be multiples of each other. This will avoid the phenomenon evident from the example given above in which each sorting pass combines two chains that before had no interaction whatsoever. It is indeed desirable that interaction between

```
procedure shellsort;
   const t = 4;
   var i,j,k,s: index;  x: item; m: 1 .. t;
      h: array [1 .. t] of integer;
begin h[1] := 9; h[2] := 5; h[3] := 3; h[4] := 1;
   for m := 1 to t do
   begin k := h[m]; s := -k; {sentinel position}
      for i := k+1 to n do
      begin x := a[i]; j := i-k;
         if s=0 then s := -k; s := s+1; a[s] := x;
         while x .key < a[j] .key do
         begin a[j+k] := a[j]; j := j-k
         end ;
         a[j+k] := x
      end
   end
end
```

Program 2.6. Shellsort.

various chains takes place as often as possible, and the following theorem holds:

If a k-sorted sequence is i-sorted, then it remains k-sorted.

Knuth [2.8] indicates evidence that a reasonable choice of increments is the sequence (written in reverse order)

$$1, 4, 13, 40, 121, \ldots$$

where $h_{k-1} = 3h_k + 1$, $h_t = 1$, and $t = \lfloor \log_3 n \rfloor - 1$. He also recommends the sequence

$$1, 3, 7, 15, 31, \ldots$$

where $h_{k-1} = 2h_k + 1$, $h_t = 1$, and $t = \lfloor \log_2 n \rfloor - 1$. For the latter choice, mathematical analysis yields an effort proportional to $n^{1.2}$ required for sorting n items with the Shellsort algorithm. Although this is a significant improvement over n^2, we will not expound further on this method, since even better algorithms are known.

2.2.5 Tree Sort

The method of sorting by straight selection is based on the repeated selection of the least key among n items, among the remaining $n - 1$ items, etc. Clearly, finding the least key among n items requires $n - 1$ comparisons, and finding it among $n - 1$ items needs $n - 2$ comparisons. So how can this selection sort possibly be improved? It can only be improved by retaining from each scan more information than just the identification of the single least item. For instance, with $n/2$ comparisons it is possible to determine the smaller key of each pair of items, with another $n/4$ comparisons the smaller of each pair of such smaller keys can be selected, and so on. Finally, with only $n - 1$ comparisons, we can construct a selection tree as shown in Fig. 2.3 and identify the root as the desired least key [2.2].

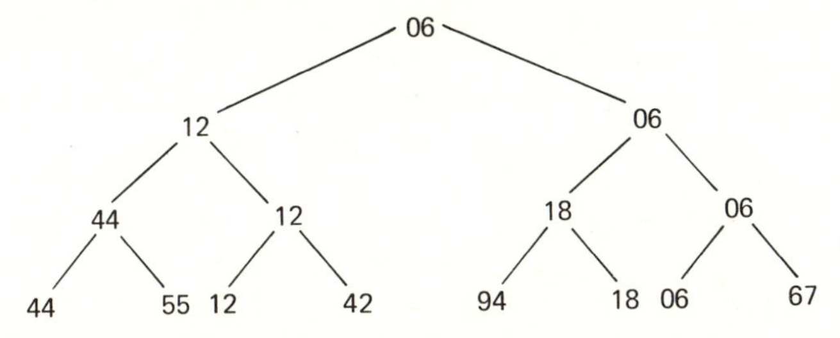

Fig. 2.3 Repeated selection between two keys.

The second step now consists of descending down along the path marked by the least key and eliminating it by successively replacing it by either an empty hole (or the key $-\infty$) at the bottom or by the item at the alternative branch at intermediate nodes (see Figs. 2.4 and 2.5). Again, the item emerging at the root of the tree has the (now second) smallest key and can be eliminated. After n such selection steps, the tree is empty (i.e., full of holes), and the sorting process is terminated. It should be noted that each of

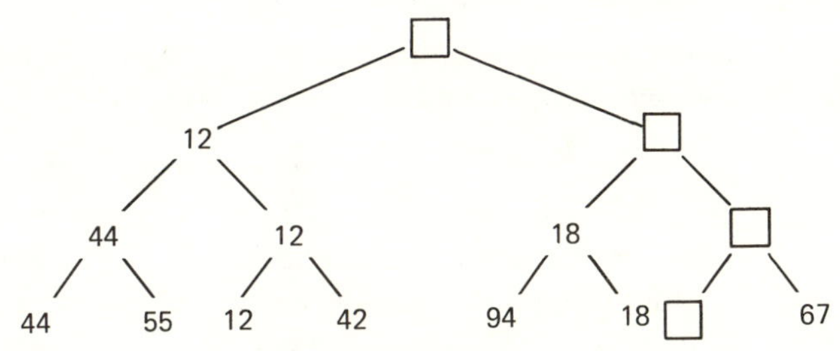

Fig. 2.4 Selecting the least key.

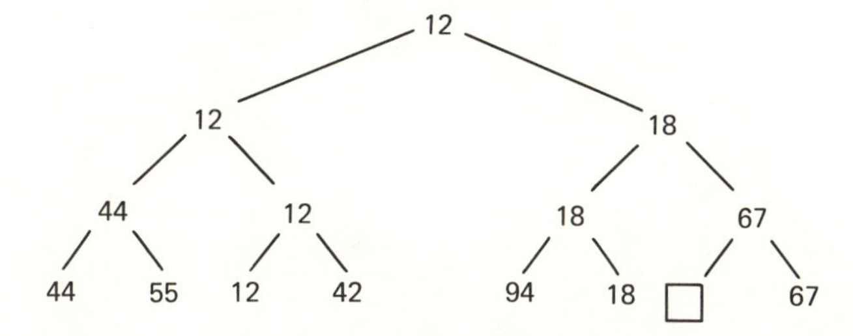

Fig. 2.5 Refilling the holes.

the n selection steps requires only $\log_2 n$ comparisons. Therefore, the total selection process requires only on the order of $n \cdot \log n$ elementary operations in addition to the n steps required by the construction of the tree. This is a very significant improvement over the straight methods requiring n^2 steps, and even over Shellsort that requires $n^{1.2}$ steps.

Naturally, the task of bookkeeping has become more elaborate, and therefore the complexity of individual steps is greater in the tree sort method; after all, in order to retain the increased amount of information gained from the initial pass, some sort of tree structure has to be created. Our next task is to find methods of organizing this information efficiently.

Of course, it would seem particularly desirable to eliminate the need for the holes ($-\infty$) that in the end populate the entire tree and are the source of many unnecessary comparisons. Moreover, a way should be found to represent the tree of n items in n units of storage, instead of in $2n - 1$ units as shown above. These goals are indeed achieved by a method called *Heapsort* by its inventor J. Williams [2-14]; it is plain that this method represents a drastic improvement over more conventional tree sorting approaches.

A *heap* is defined as a sequence of keys

$$h_l, h_{l+1}, \ldots, h_r$$

such that

$$\begin{aligned} h_i &\leq h_{2i} \\ h_i &\leq h_{2i+1} \end{aligned} \tag{2.13}$$

for all $i = l \ldots r/2$. If a binary tree is represented as an array as shown in Fig. 2.6, then it follows that the sort trees in Figs. 2.7 and 2.8 are heaps, and in particular that the element h_1 of a heap is its *least* element.

$$h_1 = \min(h_1 \ldots h_n)$$

Let us now assume that a heap with elements $h_{l+1} \ldots h_r$ is given for some values l and r, and that a new element x has to be added to form the extended heap $h_l \ldots h_r$. Take, for example, the initial heap $h_1 \ldots h_7$ shown in Fig. 2.7 and extend the heap "to the left" by an element $h_1 = 44$. A new heap is

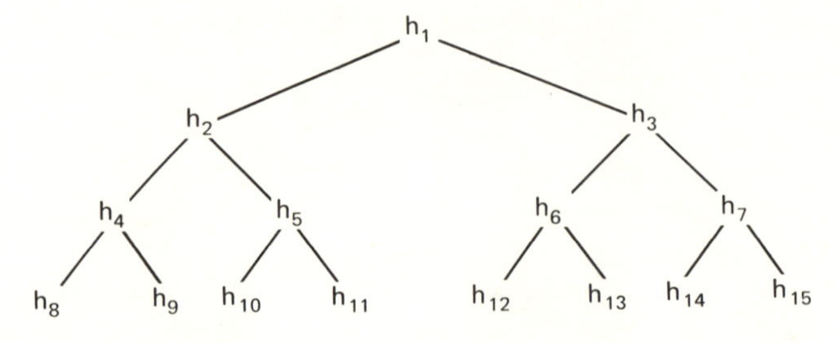

Fig. 2.6 Array h viewed as binary tree.

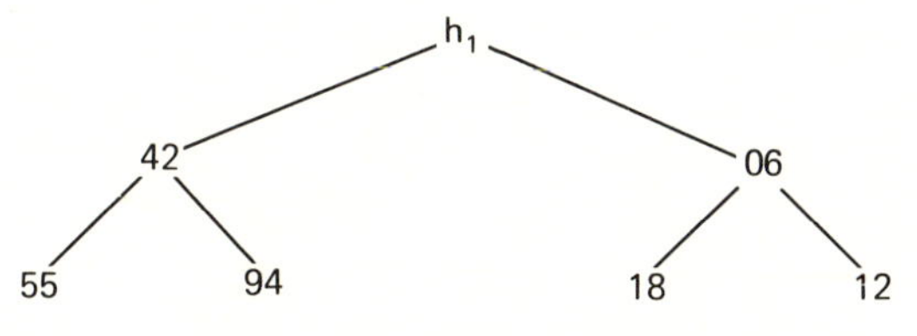

Fig. 2.7 Heap with seven elements.

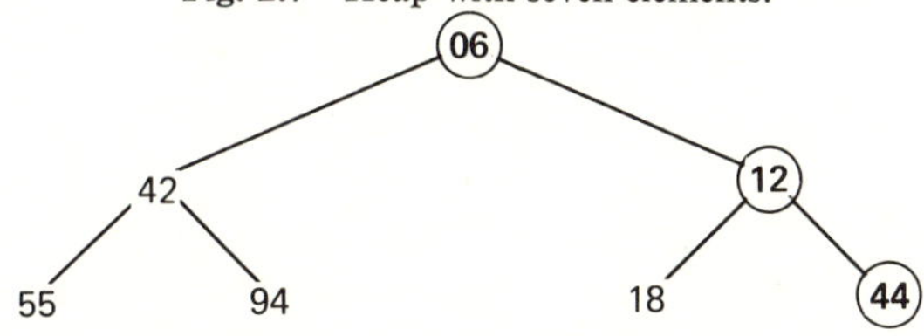

Fig. 2.8 Key 44 sifting through the heap.

obtained by first putting x on top of the tree structure and then by letting it "sift down" along the path of the smaller comparands, which at the same time move up. In the given example the value 44 is first exchanged with 06, then with 12, and thus forming the tree shown in Fig. 2.8. We now formulate this sifting algorithm as follows: i, j are the pair of indices denoting the items to be exchanged during each sift step. The reader is urged to convince himself that the proposed method of sifting actually preserves the conditions (2.13) that define a heap.

A neat way to construct a heap *in situ* was suggested by R. W. Floyd. It uses the sifting procedure shown in Program 2.7. Given is an array $h_1 \ldots h_n$; clearly, the elements $h_{n/2} \ldots h_n$ form a heap already, since no two indices i, j are such that $j = 2i$ (or $j = 2i + 1$). These elements form what may be considered as the bottom row of the associated binary tree (see Fig.

```
procedure sift(l,r: index);
   label 13;
   var i,j: index;  x: item;
begin i := l; j := 2*i; x := a[i];
   while j ≤ r do
   begin if j < r then
               if a[j] .key > a[j+1] .key then j := j+1;
         if x .key ≤ a[j] .key then goto 13;
         a[i] := a[j]; i := j; j := 2*i {sift}
   end;
13: a[i] := x
end
```

Program 2.7 Sift.

44	55	12	42	94	18	06	67
44	55	12	42	94	18	06	67
44	55	06	42	94	18	12	67
44	42	06	55	94	18	12	67
06	42	12	55	94	18	44	67

Table 2.6 Constructing a Heap.

2.6) among which no ordering relationship is required. The heap is now extended to the left where in each step a new element is included and properly positioned by a sift. This process is illustrated in Table 2.6 and yields the heap shown in Fig. 2.6. Consequently, the process of generating a heap of n elements $h_1 \ldots h_n$ *in situ* is described as follows:

```
l := (n div 2) + 1;
while l > 1 do
      begin l := l-1; sift(l,n)
      end
```

In order to obtain the elements sorted, n sift steps have now to be executed, where after each step the subsequent item may be picked off the top of the heap. Once more, there arise the questions of where to store the emerging top elements and whether or not an *in situ* sort would be possible. Of course, there is such a solution! In each step take the last component (say x) off the heap, store the top element of the heap in the now free location of x, and let x sift down into its proper position. The necessary $n - 1$ steps are illustrated on the heap of Table 2.7. The process is described with the aid of the *sift*

06	42	12	55	94	18	44	67
12	42	18	55	94	67	44	06
18	42	44	55	94	67	12	06
42	55	44	67	94	18	12	06
44	55	94	67	42	18	12	06
55	67	94	44	42	18	12	06
67	94	55	44	42	18	12	06
94	67	55	44	42	18	12	06

Table 2.7 Example of a Heapsort Process.

procedure (Program 2.7) as follows:

```
r := n;
while r > 1 do
   begin x := a[1]; a[1] := a[r]; a[r] := x;
      r := r-1; sift(1,r)
   end
```

The example of Table 2.7 shows that the resulting order is actually inverted. This, however, can easily be remedied by changing the direction of the ordering relations in the sift procedure. This results in the procedure *Heapsort* shown in Program 2.8.

```
procedure heapsort;
   var l,r: index;  x: item;
   procedure sift;
      label 13;
      var i,j: index;
   begin i := l; j := 2*i; x := a[i];
      while j ≤ r do
      begin if j < r then
               if a[j] .key < a[j+1] .key then j := j+1;
         if x .key ≥ a[j] .key then goto 13;
         a[i] := a[j]; i := j; j := 2*i
      end ;
   13: a[i] := x
   end ;
begin l := (n div 2) + 1; r := n;
   while l > 1 do
      begin l := l-1; sift
      end ;
   while r > 1 do
      begin x := a[1]; a[1] := a[r]; a[r] := x;
         r := r-1; sift
      end
end {heapsort}
```

Program 2.8 Heapsort.

Analysis of Heapsort. At first sight it is not evident that this method of sorting provides good results. After all, the large items are first sifted to the left before finally being deposited at the far right. Indeed, the procedure is not recommended for small numbers of items, such as shown in the example.

However, for large n, Heapsort is very efficient, and the larger the n, the better it is—even compared to Shellsort.

In the worst case, there are $n/2$ sift steps necessary, sifting items through $\log(n/2), \log(n/2 - 1), \ldots, \log(n - 1)$ positions, where the logarithm is taken to the base 2 and truncated to the next lower integer. Subsequently, the sorting phase takes $n - 1$ sifts, with at most $\log(n - 1), \log(n - 2), \ldots, 1$ moves. In addition, there are $n - 1$ moves for stashing the sifted item away at the right. This argument shows that Heapsort takes of the order of $n \cdot \log(n)$ steps *even in the worst case*. This excellent worst-case performance is one of the strongest qualities of Heapsort.

It is not at all clear in which case the worst (or the best) performance may be expected. But generally Heapsort seems to like initial sequences in which the items are more or less sorted in the inverse order and therefore displays an unnatural behavior. Evidently, the heap creation phase requires zero moves if the inverse order is present. For the eight items of our example, the following initial sequences result in the minimal and maximal number of moves:

$$M_{min} = 13 \text{ for the sequence}$$

94 67 44 55 12 42 18 6

$$M_{max} = 24 \text{ for the sequence}$$

18 42 12 44 6 55 67 94

The average number of moves is approximately $\frac{1}{2}n \cdot \log n$, and the deviations from this value are relatively small.

2.2.6 Partition Sort

After having discussed two advanced sorting methods based on the principles of insertion and selection, we introduce a third improved method based on the principle of exchanging. In view of the fact that Bubblesort was on the average the least effective of the three straight sorting algorithms, a relatively significant improvement factor should be expected. Still, it comes as a surprise that the improvement based on exchanges to be discussed here yields the best sorting method on arrays known so far. Its performance is so spectacular that its inventor, C. A. R..Hoare, dubbed it *Quicksort* [2.5 and 2.6].

Quicksort is based on the fact that exchanges should preferably be performed over large distances in order to be most effective. Assume that n items are given in reverse order of their keys. It is possible to sort them by performing only $n/2$ exchanges, first taking the left and the rightmost and gradually progressing inward from both sides. Naturally, this is possible only if we know that their order is exactly inverse. But something might still be learned from this example.

Let us try the following algorithm: Pick any item at random (and call it x); scan the array from the left until an item $a_i > x$ is found and then scan from the right until an item $a_j < x$ is found. Now exchange the two items and continue this "scan and swap" process until the two scans meet somewhere in the middle of the array. The result is that the array is now partitioned into a left part with keys less than x and a right part with keys greater than x. This partitioning process is now formulated in the form of a procedure in Program 2.9. Note that the relations $>$ and $<$ have been replaced by $\geq$ and $\leq$ whose negation in the **while** clause is $<$ and $>$. With this change x acts as a sentinel for both scans.

```
procedure partition;
var w,x: item;
begin i := 1; j := n;
    select at random an item x;
    repeat
        while a[i].key < x .key do i := i+1;
        while x.key < a[j] .key do j := j-1;
        if i ≤ j then
            begin w := a[i]; a[i] := a[j]; a[j] := w;
                i := i+1; j := j-1
            end
    until i > j
end
```

Program 2.9 Partition.

As an example, if the middle key 42 is selected as comparand x, then the array of keys

44 55 12 *42* 94 06 18 67

requires two exchanges to result in the partitioned array

18 06 12 |42| 94 55 44 67

the final index values are $i = 5$ and $j = 3$. Keys $a_1 \ldots a_{i-1}$ are less or equal to key $x = 42$, keys $a_{j+1} \ldots a_n$ are greater or equal to key x. Consequently, there are two partitions, namely,

$$\begin{aligned} a_k.key &\leq x.key \qquad \text{for } k = 1 \ldots i-1 \\ a_k.key &\geq x.key \qquad \text{for } k = j+1 \ldots n \end{aligned} \tag{2.14}$$

and consequently,

$$a_k.key = x.key \qquad \text{for } k = j+1 \ldots i-1$$

This algorithm is very straightforward and efficient because the essential comparands i, j, and x can be kept in fast registers throughout the scan.

However, it can also be cumbersome, as witnessed by the case with n identical keys, which results in $n/2$ exchanges. These unnecessary exchanges might easily be eliminated by changing the scanning statements to

while $a[i]$ *.key* $\leq x$ *.key* **do** $i := i+1;$
while x *.key* $\leq a[j]$ *.key* **do** $j := j-1;$

In this case, however, the choice element x, which is present as a member of the array, no longer acts as a sentinel for the two scans. The array with all identical keys would cause the scans to go beyond the bounds of the array unless more complicated termination conditions were used. The simplicity of the conditions used in Program 2.9 is well worth the extra exchanges that occur relatively rarely in the average "random" case. A slight saving, however, may be achieved by changing the clause controlling the exchange step to

$$i < j$$

instead of $i \leq j$. But this change must not be extended over the two statements

$$i := i+1; \quad j := j-1$$

which therefore require a separate conditional clause. The necessity of this clause is demonstrated by the following example with $x = 2$:

1 1 1 *2* 1 1 *1*

The first scan and exchange results in

1 1 1 1 1 1 2

and $i = 5, j = 6$. The second scan leaves the array unchanged, with $i = 7$, $j = 6$. Had the exchange not been subjected to the condition $i \leq j$, an erroneous exchange of a_6 and a_7 would have been executed.

Confidence in the correctness of the partition algorithm can be gained by verifying that the two assertions (2.14) are invariants of the **repeat** statement. Initially, with $i = 1$ and $j = n$, they are trivially true, and upon exit with $i > j$, they imply the desired result.

We now have to recall that our goal is not only to find partitions of the original array of items, but also to sort it. However, it is only a small step from partitioning to sorting: after partitioning the array, apply the same process to both partitions, then to the partitions of the partitions, and so on, until every partition consists of a single item only. This recipe is described by Program 2.10.

Procedure *sort* activates itself recursively. Such use of recursion in algorithms is a very powerful tool and will be discussed further in Chap. 3. In some programming languages of older provenience, recursion is disallowed for certain technical reasons. We will now show how this same algorithm can be expressed as a non-recursive procedure. Obviously, the solution

```
procedure quicksort;
    procedure sort (l,r: index);
        var i,j: index;  x,w: item;
    begin i := l; j := r;
        x := a[(l+r) div 2];
        repeat
            while a[i] .key < x .key do i := i+1;
            while x .key < a[j] .key do j := j-1;
            if i <= j then
            begin w := a[i]; a[i] := a[j]; a[j] := w;
                i := i+1; j := j-1
            end
        until i > j;
        if l < j then sort(l,j);
        if i < r then sort(i,r)
    end ;
begin sort(1,n)
end {quicksort}
```

Program 2.10 Quicksort.

is to express recursion as an iteration, whereby a certain amount of additional bookkeeping operations become necessary.

The key to an iterative solution lies in maintaining a list of partitioning requests that have yet to be performed. After each step, two partitioning tasks arise. Only one of them can be attacked directly by the subsequent iteration; the other one is stacked away on that list. It is, of course, essential that the list of requests is obeyed in a specific sequence, namely, in reverse sequence. This implies that the first request listed is the last one to be obeyed, and vice versa; the list is treated as a pulsating stack. In the following non-recursive version of Quicksort, each request is represented simply by a left and a right index specifying the bounds of the partition to be further partitioned. Thus, we introduce an array variable called *stack* and an index s designating its most recent entry (see Program 2.11). The appropriate choice of the stack size m will be discussed during the analysis of Quicksort.

Analysis of Quicksort. In order to analyze the performance of Quicksort, we need to investigate the behavior of the partitioning process first. After having selected a bound x, it sweeps the entire array. Hence, exactly n comparisons are performed. The number of exchanges can be determined by the following probabilistic argument.

Assume that the data set to be partitioned consists of the n keys $1 \ldots n$, and that we have selected x as the bound. After the partitioning process, x will occupy the xth position in the array. The number of exchanges required

```
procedure quicksort 1;
    const m = 12;
    var i,j,l,r: index;
        x,w: item;
        s: 0 .. m;
        stack: array [1 .. m] of
                record l,r: index end;
begin s := 1; stack[1] .l := 1; stack[1] .r := n;
    repeat {take top request from stack}
        l := stack[s] .l; r := stack[s] .r; s := s-1;
        repeat {split a[l] ... a[r]}
            i := l; j := r; x := a[(l+r) div 2];
            repeat
                while a[i] .key < x .key do i := i+1;
                while x .key < a[j] .key do j := j-1;
                if i ≤ j then
                begin w := a[i]; a[i] := a[j]; a[j] := w;
                    i := i+1; j := j-1
                end
            until i > j;
            if i < r then
            begin {stack request to sort right partition}
                s := s+1; stack[s] .l := i; stack[s] .r := r
            end ;
            r := j
        until l ≥ r
    until s = 0
end {quicksort 1}
```

Program 2.11 Non-recursive Version of Quicksort.

is equal to the number of elements in the left partition $(x - 1)$ times the probability of a key having been interchanged. A key is exchanged, if it is not less than the bound x. This probability is $(n - x + 1)/n$. The expected number of exchanges is obtained by summation over all possible choices of the bound and dividing by n.

$$M = \frac{1}{n} \sum_{x=1}^{n} \frac{n-x}{n} \cdot (n - x + 1) = \frac{n}{6} - \frac{1}{6n} \tag{2.15}$$

Hence, the expected number of exchanges is approximately $n/6$.

Assuming that we are very lucky and always happen to select the median as the bound, then each partitioning process splits the array in two halves, and the number of necessary passes to sort is $\log n$. The resulting total number of comparisons is then $n \cdot \log n$, and the total number of exchanges is $n/6 \cdot \log n$.

Of course, one cannot expect to hit the median all the time. In fact, the chance of doing so is only $1/n$. Surprisingly, however, the average performance of Quicksort is inferior to the optimal case by a factor of only $2 \cdot \ln 2$ if the bound is chosen at random.

But Quicksort does have its pitfalls. First of all, it performs moderately well for small values of n, as do all advanced methods. Its advantage over the other advanced methods lies in the ease with which a straight sorting method can be incorporated to handle small partitions. This is particularly advantageous when considering the recursive version of the program.

Still, there remains the question of the worst case. How does Quicksort perform then? The answer is unfortunately disappointing and it unveils the one weakness of Quicksort (which in those cases becomes Slowsort). Consider, for instance, the unlucky case in which each time the largest value of a partition happens to be picked as comparand x. Then each step splits a segment of n items into a left partition with $n - 1$ and a right partition with one single item. The result is that n instead of only $\log n$ splits become necessary and that the worst-case performance is of the order n^2.

Apparently, the crucial step is the selection of the comparand x. In our example program it is chosen as the middle element. Note that one might almost as well select either the first or the last element $a[l]$ or $a[r]$. In these cases, the worst case is the initially sorted array; Quicksort then shows a definite dislike for the trivial job and a preference for disordered arrays. In choosing the middle element, the strange characteristic of Quicksort is less obvious because the initially sorted array becomes the optimal case! In fact, the average performance is slightly better if the middle element is selected. Hoare suggests that the choice of x be made "at random" or by selecting it as the median of a small sample of, say, three keys [2.12 and 2.13]. Such a judicious choice hardly influences the average performance of Quicksort, but it improves the worst-case performance considerably. It becomes evident that sorting on the basis of Quicksort is somewhat like a gamble in which one should be aware of how much one may afford to lose if bad luck were to strike.

There is one important lesson to be learned from this experience; it concerns the programmer directly. What are the consequences of the worst case behavior mentioned above to the performance of Program 2.11? We have realized that each split results in a right partition of only a single element; the request to sort this partition is stacked for later execution. Consequently, the maximum number of requests, and therefore the total required stack size, is n. This is, of course, totally unacceptable. (Note that we fare no better—and, in fact, even worse—with the recursive version because a system allowing recursive activation of procedures will have to store the values of local variables and parameters of all procedure activations automatically, and it will use an implicit stack for this purpose.) The remedy lies in stacking the sort request for the longer partition and in continuing directly with the further partitioning of the smaller sections. In this case, the size of the stack m can be limited to $m = \log_2 n$.

The change necessary to Program 2.11 is localized in the section setting up new requests. It now reads

```
if j-l < r-i then
begin if i < r then
      begin {stack request for sorting right partition}
            s := s+1; stack[s] .l := i; stack[s] .r := r
      end;
      r := j  {continue sorting left partition}
end else                                                    (2.16)
begin if l < j then
      begin {stack request for sorting left partition}
            s := s+1; stack[s] .l := l; stack[s] .r := j
      end;
      l := i  {continue sorting right partition}
end
```

2.2.7. Finding the Median

The median of n items is defined as that item which is less than (or equal to) half of the n items and which is larger than (or equal to) the other half of the n items. For example, the median of

16 12 99 95 18 87 10

is 18.

The problem of finding the median is customarily connected with that of sorting because one sure method of determining the median is to sort the n items and then to pick the item in the middle. But partitioning by Program 2.9 yields a potentially much faster way of finding the median. The method to be displayed easily generalizes to the problem of finding the kth smallest of n items. Finding the median represents the special case $k = n/2$.

The algorithm invented by C. A. R. Hoare [2-4] functions as follows. First, the partitioning operation of Quicksort is applied with $l = 1$ and $r = n$ and with $a[k]$ selected as splitting value (bound) x. The resulting index values i and j are such that

1. $a[h] \leq x$ for all $h < i$
2. $a[h] \geq x$ or all $h > j$ (2.17)
3. $i > j$

There are three possible cases that may arise:

1. The splitting value x was too small; as a result, the limit between the two partitions is below the desired value k. The partitioning process has to be repeated upon the elements $a[i] \ldots a[r]$ (see Fig. 2.9).

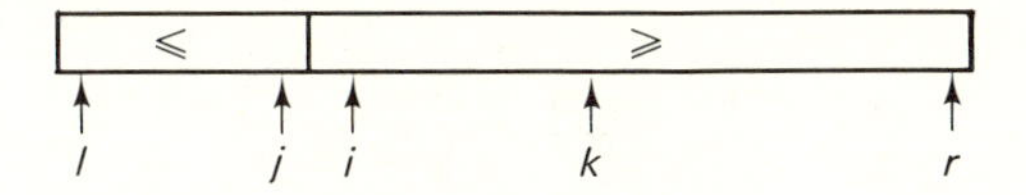

Fig. 2.9 Bound too small.

2. The chosen bound x was too large. The splitting operation has to be repeated on the partition $a[l] \ldots a[j]$ (see Fig. 2.10).

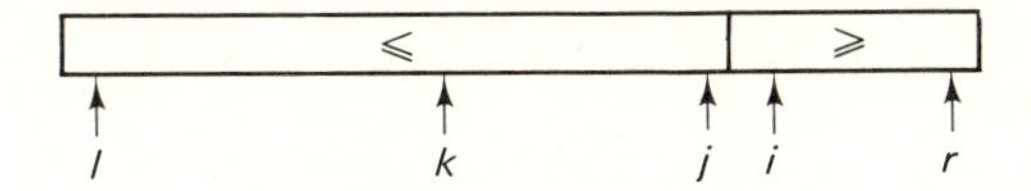

Fig. 2.10 Bound too large.

3. $j < k < i$: the element $a[k]$ splits the array into two partitions in the specified proportions and therefore is the desired quantile (see Fig. 2.11).

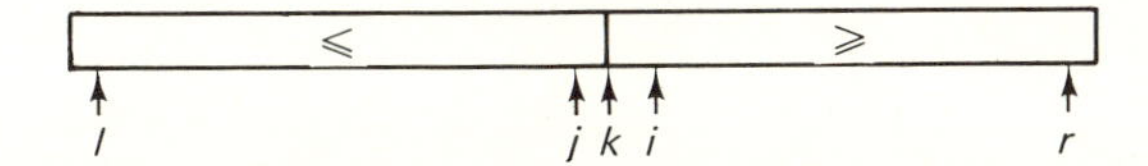

Fig. 2.11 Correct bound.

The splitting process has to be repeated until case 3. arises. This iteration is expressed by the following piece of program:

```
l := 1; r := n;
while l < r do
   begin x := a[k];
      partition(a[l] . . . a[r]);                    (2.18)
      if j < k then l := i;
      if k < i then r := j
   end
```

For a formal proof of the correctness of this algorithm, the reader is referred to the original article by Hoare. The entire program *Find* is readily derived from this.

```
procedure find (k; integer);
    var l,r,i,j,w,x: integer;
begin l := 1; r := n;
    while l < r do
    begin x := a[k]; i := l; j := r;
        repeat {split}
            while a[i] < x do i := i+1;
            while x < a[j] do j := j-1;
            if i <= j then
                begin w := a[i]; a[i] := a[j]; a[j] := w;
                    i := i+1; j := j-1
                end
        until i > j;
        if j < k then l := i;
        if k < i then r := j
    end
end {find}
```

Program 2.12 Find the kth element.

If we assume that on the average each split halves the size of the partition in which the desired quantile lies, then the number of necessary comparisons is

$$n + \frac{n}{2} + \frac{n}{4} + \cdots + 1 \doteq 2n \tag{2.19}$$

i.e., it is of order n. This explains the power of the program Find for finding medians and similar quantiles, and it explains its superiority over the straightforward method of sorting the entire set of candidates before selecting the kth (where the best is of order $n \cdot \log n$). In the worst case, however, each partitioning step reduces the size of the set of candidates only by 1, resulting in a required number of comparisons of order n^2. Again, there is hardly any advantage in using this algorithm if the number of elements is small, say, fewer than 10.

2.2.8. A Comparison of Array Sorting Methods

To conclude this parade of sorting methods, we shall try to compare their effectiveness. If n denotes the number of items to be sorted, C and M shall again stand for the number of required key comparisons and item moves, respectively. Closed analytical formulas can be given for all three straight sorting methods. They are tabulated in Table 2.8. The column indicators min, ave, max, specify the respective minima, maxima, and expected values averaged over all $n!$ permutations of n items.

		Min	Ave	Max
Straight Insertion	$C=$ $M=$	$n-1$ $2(n-1)$	$(n^2+n-2)/4$ $(n^2-9n-10)/4$	$(n^2-n)/2-1$ $(n^2+3n-4)/2$
Straight Selection	$C=$ $M=$	$(n^2-n)/2$ $3(n-1)$	$(n^2-n)/2$ $n(\ln n+0.57)$	$(n^2-n)/2$ $n^2/4+3(n-1)$
Straight Exchange (Bubblesort)	$C=$ $M=$	$(n^2-n)/2$ 0	$(n^2-n)/2$ $(n^2-n)*0.75$	$(n^2-n)/2$ $(n^2-n)*1.5$

Table 2.8. Comparison of Straight Sorting Methods.

No reasonably simple accurate formulas are available on the advanced methods. The essential facts are that the computational effort needed is $c_1 \cdot n^{1.2}$ in the case of Shellsort and is $c_i \cdot n \cdot \log(n)$ in the cases of Heapsort and Quicksort.

These formulas merely provide a rough measure of performance as functions of n, and they allow the classification of sorting algorithms into primitive, straight methods (n^2) and advanced or "logarithmic" methods ($n \cdot \log n$). For practical purposes, however, it is helpful to have some experimental data available that shed some light on the coefficients c_i which further distinguish the various methods. Moreover, the formulas do not take into account the computational effort expended on operations other than key comparisons and item moves, such as loop control, etc. Clearly, these factors depend to some degree on individual systems, but an example of experimentally obtained data is nevetheless informative. Table 2.9 shows the times (in milliseconds) consumed by the sorting methods previously discussed, as executed by the PASCAL system on a CDC 6400 computer. The three col-

	Ordered		Random		Inversely Ordered	
Straight insertion	12	23	366	1444	704	2836
Binary insertion	56	125	373	1327	662	2490
Straight selection	489	1907	509	1956	695	2675
Bubblesort	540	2165	1026	4054	1492	5931
Bubblesort with flag	5	8	1104	4270	1645	6542
Shakersort	5	9	961	3642	1619	6520
Shellsort	58	116	127	349	157	492
Heapsort	116	253	110	241	104	226
Quicksort	31	69	60	146	37	79
Mergesort*	99	234	102	242	99	232

**See Sect. 2.3.1.*

Table 2.9. Execution Times of Sort Programs.

umns contain the times used to sort the already ordered array, a random permutation, and the inversely ordered array. The left figure in each column is for 256 items, the right one for 512 items. The data clearly separate the n^2 methods from the $n \cdot \log n$ methods. Noteworthy are the following points:

1. The improvement of binary insertion over straight insertion is marginal indeed, and even negative in the case of an already existing order.
2. Bubblesort is definitely the worst sorting method among all compared. Its improved version "Shakersort" is still worse than straight insertion and straight selection (except in the pathological case of sorting a sorted array).
3. Quicksort beats Heapsort by a factor of 2 to 3. It sorts the inversely ordered array with speed practically identical to the one which is already sorted.

It must be added that the data were gathered by sorting items consisting of a key only, without associated data. This is not a very realistic assumption; Table 2.10 shows the influence of enlarging the size of the items. In the example chosen the associated data occupy seven times the storage space of the key. The left figure in each column displays the time needed for sorting records without associated data; the right figure relates to sorting with associated data; $n = 256$.

	Ordered		Random		Inversely Ordered	
Straight insertion	12	46	366	1129	704	2150
Binary insertion	56	76	373	1105	662	2070
Straight selection	489	547	509	607	695	1430
Bubblesort	540	610	1026	3212	1492	5599
Bubblesort with flag	5	5	1104	3237	1645	5762
Shakersort	5	5	961	3071	1619	5757
Shellsort	58	186	127	373	157	435
Heapsort	116	264	110	246	104	227
Quicksort	31	55	60	137	37	75
Mergesort*	99	196	102	195	99	187

**See Sect. 2.3.1.*

Table 2.10. Execution Times of Sort Programs. (Keys with Associated Data).

The following details should be noted:

1. Straight selection has gained significantly and now emerges as the best of the straight methods.
2. Bubblesort is still the worst method by a large margin (it has even lost ground!), and only its "improvement" called Shakersort is slightly worse in the case of the inversely ordered array.

3. Quicksort has even strengthened its position as the quickest method and appears as the best array sorter by far.

2.3. SORTING SEQUENTIAL FILES

2.3.1. Straight Merging

Unfortunately, the sorting algorithms presented in the preceding chapter are inapplicable if the amount of data to be sorted does not fit into a computer's main store, but if it is, for instance, represented on a peripheral and sequential storage device such as a tape. In this case we describe the data as a (sequential) file whose characteristic is that at each moment one and only one component is directly accessible. This is a severe restriction compared to the possibilities offered by the array structure, and therefore different sorting techniques have to be used. The most important one is sorting by *merging*. Merging (or collating) means combining two (or more) ordered sequences into a single, ordered sequence by repeated selection among the currently accessible components. Merging is a much simpler operation than sorting, and it is used as an auxiliary operation in the more complex process of sequential sorting. One way of sorting on the basis of merging, called *straight merging*, is the following:

1. Split the sequence a into two halves, called b and c.
2. Merge b and c by combining single items into ordered pairs.
3. Call the merged sequence a, and repeat steps 1 and 2, this time merging ordered pairs into ordered quadruples.
4. Repeat the previous steps, merging quadruples into octets, and continue doing this, each time doubling the lengths of the merged subsequences, until the entire sequence is ordered.

As an example, consider the sequence

44 55 12 42 94 18 06 67

In step 1, the split results in the sequences

44 55 12 42

94 18 06 67

The merging of single components (which are ordered sequences of length 1), into ordered pairs yields

44 94 ' 18 55 ' 06 12 ' 42 67

Splitting again in the middle and merging ordered pairs yields

06 12 44 94 ' 18 42 55 67

A third split and merge operation finally produces the desired result

06 12 18 42 44 55 67 94

Each operation that treats the entire set of data once is called a *phase*, and the smallest subprocess which by repetition constitutes the sort process is called a *pass* or a *stage*. In the above example the sort took three passes, each pass consisting of a splitting phase and a merging phase. In order to perform the sort, three tapes are needed; the process is therefore called a *three-tape merge*.

Actually, the splitting phases do not contribute to the sort since they do in no way permute the items; in a sense they are unproductive, although they constitute half of all copying operations. They can be eliminated altogether by combining the split and the merge phases. Instead of merging into a single sequence, the output of the merge process is immediately redistributed onto two tapes, which constitute the sources of the subsequent pass. In contrast to the previous two-phase merge sort, this method is called a *single-phase merge* or a *balanced merge*. It is evidently superior because only half as many copying operations are necessary; the price for this advantage is a fourth tape.

We shall develop a merge program in detail and initially let the data be represented as an array which, however, is scanned in *strictly sequential* fashion. A later version of a merge sort will then be based on the file structure, allowing a comparison of the two programs and demonstrating the strong dependence of the form of a program on the underlying representation of its data.

A single array may easily be used in place of two files if it is regarded as a double-ended sequence. Instead of merging from two source files, we may pick items off the two ends of the array. Thus, the general form of the combined merge-split phase can be illustrated as shown in Fig. 2.12. The destination of the merged items is switched after each ordered pair in the first pass, after each ordered quadruple in the second pass, etc., thus evenly filling the two destination sequences, represented by the two ends of a single array. After each pass, the two arrays interchange their roles, the source becomes the new destination, and vice versa.

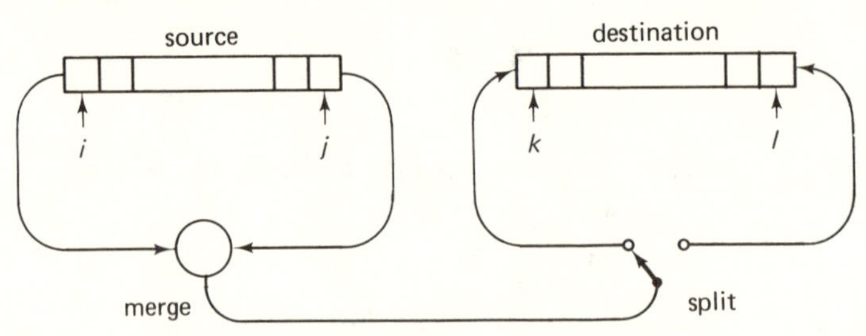

Fig. 2.12 Straight mergesort with two arrays.

A further simplification of the program can be achieved by joining the two conceptually distinct arrays into a single array of doubled size. Thus, the data will be represented by

$$a \;:\; \mathbf{array}[1\,..\,2*n]\ \mathbf{of}\ item \tag{2.20}$$

and we let the indices i and j denote the two source items, whereas k and l designate the two destinations (see Fig. 2.12). The initial data are, of course, the items $a_1 \ldots a_n$. Clearly, a Boolean variable *up* is needed to denote the direction of the data flow; $up = true$ shall mean that in the current pass components $a_1 \ldots a_n$ will be moved "up" to the variables $a_{n+1} \ldots a_{2n}$, whereas $up = false$ will indicate that $a_{n+1} \ldots a_{2n}$ will be transferred "down" into $a_1 \ldots a_n$. The value of *up* strictly alternates between consecutive passes. And, finally, a variable p is introduced to denote the length of the subsequences to be merged. Its value is initially 1, and it is doubled before each successive pass. To simplify matters somewhat, we shall assume that n is always a power of 2. Thus, the first version of the straight merge program assumes the following form:

```
procedure mergesort;
    var i,j,k,l: index;
        up: Boolean; p: integer;
begin up := true; p := 1;
    repeat {initialize indices}
        if up then
            begin i := 1; j := n; k := n+1; l := 2*n
            end else                                        (2.21)
            begin k := 1; l := n; i := n+1; j := 2*n
            end;
        "merge p-tuples from i- and j-sequences into
        k- and l-sequences";
        up := ¬up; p := 2*p
    until p = n
end
```

In the next development step we refine the statement expressed (within quotes) in natural language. Evidently, this merge pass involving n items is itself a sequence of merges of subsequences, i.e., of p-tuples. Between every such partial merge the destination is switched from the lower to the upper end of the destination array, or vice versa, to guarantee equal distribution onto both destinations. If the destination of the merged items is the lower end of the destination array, then the destination index is k, and k is incremented by 1 after each move of an item. If they are to be moved to the upper end of the destination array, the destination index is l, and l is to be decremented by 1 after each move. In order to simplify the actual merge statement, we choose the destination to be designated by k at all times, switching the values of the variables k and l after each p-tuple merge, and denote the increment to be used at all times by h, where h is either 1 or -1. These design discussions lead to the following refinement:

```
h := 1; m := n;  {m = no. of items to be merged}
repeat q := p; r := p; m := m-2*p;
        "merge q items from i with r items from j,
        destination index is k with increment h";          (2.22)
    h := -h;
    exchange k and l
until m = 0
```

In the further refinement step the actual merge statement is to be formulated. Here we have to keep in mind that the tail of the one subsequence which is left non-empty after the merge has to be appended to the output sequence by simple copying operations.

```
while (q≠0) ∧ (r≠0) do
begin {select an item from i or j}
   if a[i] .key < a[j] .key then
      begin "move an item from i to k, advance i and k"; q := q-1
      end else
      begin "move an item from j to k; advance j and k"; r := r-1
      end
end;
"copy tail of i-sequence";
"copy tail of j-sequence"                                  (2.23)
```

After this further refinement of the tail copying operations, the program is laid out in complete detail. Before writing it out in full, we wish to eliminate the restriction that n be a power of 2. Which parts of the algorithm are affected by this relaxation of constraints? We easily convince ourselves that the best way to cope with the more general situation is to adhere to the old method as long as possible. In this example this means that we continue merging p-tuples until the remainders of the source sequences are of length less than p. The one and only one part that is influenced are the statements that determine the values of q and r, the lengths of the sequences to be merged. The following four statements replace the three statements

$$q := p;\ r := p;\ m := m-2*p$$

and, as the reader should convince himself, they represent an effective implementation of the strategy specified above; note that m denotes the total number of items in the two source sequences which remain to be merged:

```
if m ≥ p then q := p else q := m; m := m-q;
if m ≥ p then r := p else r := m; m := m-r;
```

In addition, in order to guarantee termination of the program, the condition $p = n$, which controls the "outer" repetition, must be changed to $p \geq n$. After these modifications, we may now proceed to describe the entire algorithm in terms of a complete program (see Program 2.13).

```
procedure mergesort;
    var i,j,k,l,t: index;
        h,m,p,q,r: integer; up: boolean;
    {note that a has indices 1 . . . 2*n}
begin up := true; p := 1;
    repeat h := 1; m := n;
        if up then
        begin i := 1; j := n; k := n+1; l := 2*n
        end else
        begin k := 1; l := n; i := n+1; j := 2*n
        end ;
        repeat {merge a run from i and j to k}
            {q = length of i-run, r = length of j-run}
            if m ≥ p then q := p else q := m; m := m−q;
            if m ≥ p then r := p else r := m; m := m−r;
            while (q≠0) ∧ (r≠0) do
            begin {merge}
                if a[i] .key < a[j] .key then
                begin a[k] := a[i]; k := k+h; i := i+1; q := q−1
                end else
                begin a[k] := a[j]; k := k+h; j := j−1; r := r−1
                end
            end ;
            {copy tail of j-run}
            while r ≠ 0 do
                begin a[k] := a[j]; k := k+h; j := j−1; r := r−1
                end ;
            {copy tail of i-run}
            while q ≠ 0 do
                begin a[k] := a[i]; k := k+h; i := i+1; q := q−1
                end ;
            h := −h; t := k; k := l; l := t
        until m = 0;
        up := ¬up; p := 2*p
    until p ≥ n;
    if ¬up then
    for i := 1 to n do a[i] := a[i+n]
end {mergesort}
```

Program 2.13 Straight Mergesort.

Analysis of Mergesort. Since each pass doubles p, and since the sort is terminated as soon as $p \geq n$, it involves $\lceil \log_2 n \rceil$ passes. Each pass, by definition, copies the entire set of n items exactly once. As a consequence, the total number of moves is exactly

$$M = n \cdot \lceil \log n \rceil \tag{2.24}$$

The number C of key comparisons is even less than M since no comparisons are involved in the tail copying operations. However, since the mergesort technique is usually applied in connection with the use of peripheral storage devices, the computational effort involved in the move operations dominates the effort of comparisons often by several orders of magnitude. The detailed analysis of the number of comparisons is therefore of little practical interest.

The merge sort algorithm apparently compares well with even the advanced sorting techniques discussed in the previous chapter. However, the administrative overhead for the manipulation of indices is relatively high, and the decisive disadvantage is the need for storage of $2n$ items. This is the reason why sorting by merging is rarely used upon arrays, i.e., upon data located in main store. Figures comparing the real time behavior of this Mergesort algorithm appear in the last lines of Tables 2.9 and 2.10. They compare favorably with Heapsort but unfavorably with Quicksort.

2.3.2. Natural Merging

In straight merging no advantage is gained when the data are initially already partially sorted. The length of all merged subsequences in the kth pass is (less than or) equal to 2^k, independent of whether longer subsequences are already ordered and could as well be merged. In fact, any two ordered subsequences of lengths m and n might be merged directly into a single sequence of $m + n$ items. A mergesort which at any time merges the two longest possible subsequences is called a *natural merge sort*.

An ordered subsequence is often called a *string*. However, since the word string is even more frequently used to describe sequences of characters, we will follow Knuth in our terminology and use the word *run* instead of *string* when referring to ordered subsequences. We call a subsequence $a_i \ldots a_j$ such that

$$\begin{aligned} &a_k \leq a_{k+1} \qquad \text{for} \quad k = i \ldots j-1 \\ &a_{i-1} > a_i \\ &a_j > a_{j+1} \end{aligned} \tag{2.25}$$

a *maximal run* or, for short, a run. A natural merge sort, therefore, merges (maximal) runs instead of sequences of fixed, predetermined length. Runs have the property that if two sequences of n runs are merged, a single sequence of exactly n runs emerges. Therefore, the total number of runs is halved in each pass, and the number of required moves of items is in the worst case $n \cdot \lceil \log_2 n \rceil$, but in the average case it is even less. The expected number of comparisons, however, is much larger because in addition to the comparisons necessary for the selection of items, further comparisons are needed between consecutive items of each file in order to determine the end of each run.

Our next programming exercise develops a natural merge algorithm in the same stepwise fashion that was used to explain the straight merging algorithm. It employs the sequential file structure instead of the array, and it represents an unbalanced, two-phase, three-tape merge sort. We assume that the initial sequence of items is given as the file *c*, on which the sorted output will appear. (Naturally, in actual data processing applications, the initial data are first copied from the original tape onto file *c* for reasons of safety.) The two auxiliary tapes are *a* and *b*. Each pass consists of a distribution phase that distributes runs equally from *c* onto *a* and *b* and a merge phase that merges runs from *a* and *b* onto *c*. This process is illustrated in Fig. 2.13.

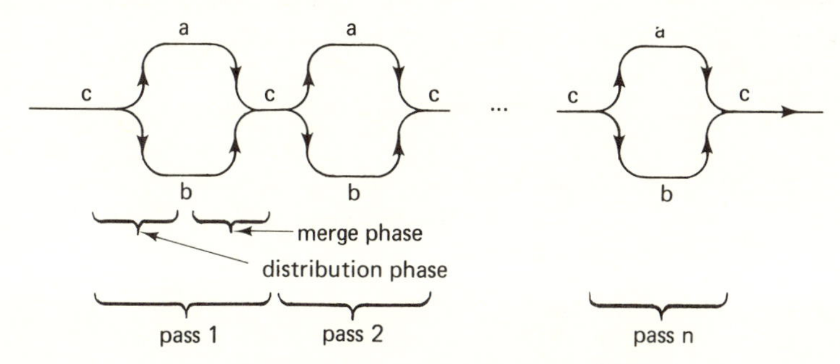

Fig. 2.13 Sort phases and sort passes.

17	31′	5	59′	13	41	43	67′	11	23	29	47′	3	7	71′	2	19	57′	37	61
5	17	31	59′	11	13	23	29	41	43	47	67′	2	3	7	19	57	71′	37	61
5	11	13	17	23	29	31	41	43	47	59	67′	2	3	7	19	37	57	61	71
2	3	5	7	11	13	17	19	23	29	31	37	41	43	47	57	59	61	67	71

Table 2.11. Example of a Natural Mergesort.

As an example, Table 2.11 shows the file *c* in its original state (line 1) and after each pass (lines 2-4) in a natural merge sort involving 20 numbers. Note that only three passes are needed. The sort terminates as soon as the number of runs on *c* is 1. (We assume that there exists at least one non-empty run on the initial file.) We therefore let a variable *l* be used for counting the number of runs merged onto file *c*. If we define the global entities

```
type tape = file of item;
var c: tape
```
(2.26)

then the program can be formulated as follows:

```
procedure naturalmerge;
    var l: integer;
        a,b: tape;
begin
    repeat rewrite(a); rewrite(b); reset(c);
        distribute;                                        (2.27)
        reset(a); reset(b); rewrite(c);
        l := 0; merge
    until l = 1
end
```

The two phases clearly emerge as two distinct statements. They are now to be refined, i.e., expressed in more detail. The refined descriptions can either be directly substituted or they may be described as procedures, and the abbreviated statements must be regarded as procedure calls. This time we choose the latter method and define

```
procedure distribute; · {from c to a and b}
begin
    repeat copyrun(c,a);                                   (2.28)
        if ¬eof(c) then copyrun(c,b)
    until eof(c)
end
```

and

```
procedure merge;
begin  {from a and b to c}
    repeat mergerun; l := l+1                              (2.29)
    until eof(b);
    if ¬eof(a) then
        begin copyrun(a,c); l := l+1
        end
end
```

This method of distribution supposedly results in either equal numbers of runs on files *a* and *b* or in file *a* containing one run more than *b*. Since corresponding pairs of runs are merged, a leftover run may still be on file *a*, which simply has to be copied. The procedures *merge* and *distribute* are formulated in terms of subordinate procedures *mergerun* and *copyrun* with obvious tasks. These procedures are now explained in further detail; they require the introduction of a global Boolean variable *eor* that specifies whether or not the *e*nd *o*f the *r*un has been reached.

```
procedure copyrun(var x,y: tape);
begin  {copy one run from x to y}                              (2.30)
    repeat copy(x,y) until eor
end

procedure mergerun;
begin  {merge a run from a and b to c}
    repeat if a↑.key < b↑.key then
        begin copy(a,c);
            if eor then copyrun(b,c)
        end else                                               (2.31)
        begin copy(b,c);
            if eor then copyrun(a,c)
        end
    until eor
end
```

The comparison and selection process of keys in merging a run terminates as soon as one of the two runs is exhausted. After this, the other run (which is not exhausted yet) has to be transferred to the resulting run by merely copying its tail. This is done by a call of the procedure *copyrun*.

The two procedures are defined in terms of the subordinate procedure *copy*, which transfers one item from a source file x to a destination file y and determines whether or not the end of a run has been reached. It is readily expressed in terms of a *read* and a *write* statement. In order to determine the end of a run, the key of the last item read (copied) must be retained for comparison with its successor. This "lookahead" is achieved by inspecting the file buffer variable $x\uparrow$.

```
procedure copy(var x,y: tape);
    var buf: item;
begin read(x, buf); write(y, buf);                             (2.32)
    if eof(x) then eor := true else eor := buf.key > x↑.key
end
```

This terminates the development of the natural merging sort procedure. Regrettably, the program is incorrect, as the careful reader may have noticed. The program is incorrect in the sense that it does not sort properly in some cases. Consider, for example, the following sequence of input data:

3 2 5 11 7 13 19 17 23 31 29 37 43 41 47 59 57 61 71 67

By distributing consecutive runs alternately onto the files a and b, we obtain

$$a = 3\ '\ 7\quad 13\quad 19\ '\ 29\quad 37\quad 43\ '\ 57\quad 61\quad 71\ '$$

$$b = 2\quad 5\quad 11\ '\ 17\quad 23\quad 31\ '\ 41\quad 47\quad 59\ '\ 67$$

These sequences are readily merged into a single run, whereafter the sort terminates successfully. The example, although it does not lead to an erroneous behavior of the program, makes us aware that mere distribution of runs onto several files may result in a number of output runs that is less than the number of input runs. This is because the first item of the $i + 2$nd run may be larger than the last item of the ith run, thereby causing the two runs to merge automatically into a single run.

Although the procedure *distribute* supposedly outputs runs equally onto the two files, the important consequence is that the actual number of resulting runs on a and b may significantly differ. Our merge procedure, however, will only merge pairs of runs and terminate as soon as file b is read, thereby losing the tail of one of the files. Consider the following input data which are sorted (and truncated) in two subsequent passes:

17	19	13	57	23	29	11	59	31	37	7	61	41	43	5	67	47	71	2	3
13	17	19	23	29	31	37	41	43	47	57	71	11	59						
11	13	17	19	23	29	31	37	41	43	47	57	59	71						

Table 2.12 Incorrect Result of Mergesort Program.

The example of this programming mistake is typical for many programming situations. The mistake is caused by an oversight of one of the possible consequences of a presumably simple operation. It is also typical in the sense that several ways of remedying the mistake are open and that one of them has to be chosen. Often there exist two possibilities that differ in a very important, fundamental way:

1. We recognize that the operation of distribution is incorrectly programmed and does not satisfy the requirement that the number of runs are equal (or differ by at most 1). We stick to the original scheme of operation and correct the faulty procedure accordingly.
2. We recognize that the correction of the faulty part involves far-reaching modifications, and we try to find ways in which other parts of the algorithm may be changed to accommodate the currently incorrect part.

In general, the first path seems to be the safer, cleaner one, the more honest way, providing a fair degree of immunity from later consequences of overlooked, intricate side effects. It is, therefore, the way toward a solution that is generally (and rightly) recommended.

It is to be pointed out, however, that the second possibility should sometimes not be entirely ignored. It is for this reason that we further elaborate on this example and illustrate a fix by modification of the merge procedure rather than the distribution procedure, which is primarily at fault.

This implies that we leave the distribution scheme untouched and renounce the condition that runs are equally distributed. This may result in a less than optimal performance. However, the worst-case performance remains unchanged, and, moreover, the case of highly unequal distribution is statistically *very unlikely*. Efficiency considerations are therefore no serious argument against this solution.

If the condition of equal distribution of runs no longer exists, then the merge procedure has to be changed so that, after reaching the end of one file, the *entire* tail of the remaining file is copied instead of at most one run.

This change is straightforward and is very simple in comparison with any change in the distribution scheme. (The reader is urged to convince himself of the truth of this claim.) The revised version of the merge algorithm is included in the complete Program 2.14.

Program 2.14 Natural Mergesort.

```
program mergesort (input, output);
{3-tape, 2-phase natural merge sort}
type item = record key: integer
                {other fields defined here}
            end ;
   tape = file of item;
var c: tape; n: integer; buf: item;
procedure list (var f: tape);
   var x: item;
begin reset(f);
   while ¬eof(f) do
      begin read(f,x); write(output, x.key)
      end ;
   writeln
end {list} ;
procedure naturalmerge;
   var l: integer;  {no. of runs merged}
      eor: boolean; {end -of -run indicator}
      a,b: tape;
   procedure copy(var x,y: tape);
      var buf: item;
   begin read(x, buf); write(y,buf);
      if eof(x) then eor := true else eor := buf.key > x↑.key
   end ;
   procedure copyrun (var x,y: tape);
   begin {copy one run from x to y}
      repeat copy(x,y) until eor
   end ;
```

```
    procedure distribute;
    begin {from c to a and b}
        repeat copyrun (c,a);
            if ¬eof(c) then copyrun (c,b)
        until eof(c)
    end ;
    procedure mergerun;
    begin {from a and b to c}
        repeat
            if a↑.key ≤ b↑.key then
            begin copy (a,c);
                if eor then copyrun (b,c)
            end else
            begin copy (b,c);
                if eor then copyrun (a,c)
            end
        until eor
    end ;
    procedure merge;
    begin {from a and b to c}
        while ¬eof(a) ∧ ¬eof(b) do
        begin mergerun; l := l+1
        end;
        while ¬eof(a) do
        begin copyrun (a,c); l := l+1
        end;
        while ¬eof(b) do
        begin copyrun (b,c); l := l+1
        end ;
        list (c)
    end ;
begin
    repeat rewrite(a); rewrite(b); reset(c);
        distribute;
        reset(a); reset(b); rewrite(c);
        l := 0; merge;
    until l = 1
end ;
begin {main program; read input sequence ending with 0}
    rewrite(c); read(buf.key);
    repeat write(c.buf); read(buf.key)
    until buf .key = 0;
    list (c);
    naturalmerge;
    list(c)
end .
```

Pro ram 2.14 (Continued)

2.3.3. Balanced Multiway Merging

The effort involved in a sequential sort is proportional to the number of required passes since, by definition, every pass involves the copying of the entire set of data. One way to reduce this number is to distribute runs onto more than two files. Merging r runs which are equally distributed on N tapes results in a sequence of r/N runs. A second pass reduces their number to r/N^2, a third pass to r/N^3, and after k passes there are r/N^k runs left. The total number of passes required to sort n items by *N-way merging* is therefore $k = \lceil \log_N n \rceil$. Since each pass requires n copy operations, the total number of copy operations is in the worst case

$$M = n \cdot \lceil \log_N n \rceil$$

As the next programming exercise, we will develop a sort program based on multiway merging. In order to contrast further the program from the previous natural two-phase merging procedure, we shall formulate the multiway merge as a single phase, balanced mergesort. This implies that in each pass there are an equal number of input and output files onto which consecutive runs are alternately distributed. Using N files, the algorithm will therefore be based on $N/2$-way merging, assuming that N is even. Following the previously adopted strategy, we will not bother to detect the automatic merging of two consecutive runs distributed onto the same tape. Consequently, we are forced to design the merge program without assuming strictly equal numbers of runs on the input tapes.

In this program we encounter for the first time a natural application of a data structure consisting of an array of files. As a matter of fact, it is surpising how strongly the following program differs from the previous one because of the change from two-way to multiway merging. The change is primarily a result of the circumstance that the merge process can no longer simply be terminated after one of the input runs is exhausted. Instead, a list of inputs which are still active, i.e., not yet exhausted, must be kept. Another complication stems from the need to switch the groups of input and output tapes after each pass.

We start out by defining, in addition to the two familiar types *item* and *tape*, a type

$$tapeno = 1 \,..\, N \tag{2.33}$$

Obviously, tape numbers are used to index the array of files of items. Let us then assume that the initial sequence of items is given as a variable

$$f0 : tape \tag{2.34}$$

and that for the sorting process N tapes are available, where N is even

$$f\text{: } \textbf{array } [tapeno] \textbf{ of } tape \tag{2.35}$$

A recommended technique of approaching the problem of tape switching is to introduce a tape index map. Instead of directly addressing a tape by its index i, it is addressed via a map t, i.e., instead of each

$$f[i] \quad \text{we write} \quad f[t[i]]$$

where the map is defined as

$$t: \textbf{array } [tapeno] \textbf{ of } tapeno \tag{2.36}$$

If initially $t[i] = i$ for all i, then a switch consists in merely exchanging the pairs of map components

$$t[1] \longleftrightarrow t[nh + 1]$$
$$t[2] \longleftrightarrow t[nh + 2]$$
$$\cdots$$
$$t[nh] \longleftrightarrow t[n]$$

where $nh = n/2$. Consequently, we may always consider

$$f[t[1]], \ldots, f[t[nh]]$$

as input tapes, and we may always consider

$$f[t[nh + 1]], \ldots, f[t[n]]$$

as output tapes. (Subsequently, we will simply call $f[t[j]]$ "tape j" within comments.) The algorithm can now be formulated initially as follows:

```
procedure tapemergesort;
     var i,j: tapeno;
          l: integer; {no. of runs distributed}
          t: array [tapeno] of tapeno;
begin {distribute initial runs to t[1] . . . t[nh]}                    (2.37)
     j := nh; l := 0;
     repeat if j < nh then j := j+1 else j := 1;
               "copy one run from f0 to tape j";
               l := l+1
     until eof(f0);
     for i := 1 to n do t[i] := i;
     repeat {merge from t[1] . . . t[nh] to t[nh+1] . . . t[n]}
               "reset input tapes";
               l := 0;
               j := nh+1; {j = index of output tape}
               repeat l := l+1;
                         "merge a run from inputs to t[j]"
                    if j < n then j := j+1 else j := nh+1
               until "all inputs exhausted";
               "switch tapes"
     until l = 1;
     {sorted tape is t[1]}
end
```

First, we refine the copy operation used in the initial distribution of runs; we again introduce an auxiliary variable to buffer the last item read:

buf: *item*

and replace "copy one run from $f0$ to tape j" by the statement

$$\begin{array}{l} \textbf{repeat}\ read(f0,\ buf); \\ \quad write(f[j],\ buf) \\ \textbf{until}\ (buf.key > f0\uparrow.key) \lor eof(f0) \end{array} \tag{2.38}$$

Copying a run terminates when either the first item of the next run is encountered ($buf.key > f0\uparrow.key$) or when the end of the entire input file is reached ($eof(f0)$).

In the actual sort algorithm there remain the statements

1. Reset input tapes
2. Merge a run from inputs to $t[j]$
3. Switch tapes

and the predicate

4. All inputs exhausted

to be specified in more detail. First, we must accurately identify the current input files. Notably, the number of "active" input files may be less than $n/2$. In fact, there can be at most as many sources as there are runs; the sort terminates as soon as there is one single file left. This leaves open the possibility that at the initiation of the last sort pass there are fewer than nh runs. We therefore introduce a variable, say $k1$, to denote the actual number of input files used. We incorporate the initialization of $k1$ in the statement "reset input tapes" as follows:

$$\begin{array}{l} \textbf{if}\ l < nh\ \textbf{then}\ k1 := l\ \textbf{else}\ k1 := nh; \\ \textbf{for}\ i := 1\ \textbf{to}\ k1\ \textbf{do}\ reset(f[t[i]]); \end{array}$$

Naturally, statement (2) is to decrement $k1$ whenever an input source ceases. Hence, predicate (4) may easily be expressed by the relation

$$k1 = 0$$

Statement (2) is more difficult to refine; it consists of the repeated selection of the least key among the available sources and its subsequent transport to the destination, i.e., the current output tape. The process is complicated again by the necessity of determining the end of each run. The end of a run may be reached because (1) the subsequent key is less than the current key or (2) the end of the source file is reached. In the latter case the tape is eliminated by decrementing $k1$; in the former case the run is closed by excluding the file from further selection of items, but only until the creation of the current output run is completed. This makes it obvious that a second variable, say $k2$, is needed to denote the number of source tapes actually available for the selection of the next item. This value is initially set equal to $k1$ and is decremented whenever a run terminates because of condition (1).

Unfortunately, the introduction of $k2$ is not enough; knowledge of the number of tapes does not suffice. We need to know exactly *which* tapes are still in the game. An obvious solution is to use an array with Boolean components indicating the availability of the tapes. We choose, however, a different method which leads to a more efficient selection procedure which, after all, is the most frequently repeated part of the entire algorithm. Instead of using a Boolean array, a second tape map, say *ta*, is introduced. This map is used in place of *t* such that *ta*[1] . . . *ta*[*k*2] are the indices of the *t*apes *a*vailable. Thus statement (2) can be formulated as follows:

```
k2 := k1;
repeat "select the minimal key, let ta[mx] be its tape number";
       read(f[ta[mx]], buf);                                        (2.39)
       write(f[t[j]], buf);
       if eof(f[ta[mx]]) then "eliminate tape" else
       if buf.key > f[ta[mx]]↑.key then "close run"
until k2 = 0
```

Since the number of tape units available in any computer installation is usually fairly small, the selection algorithm to be specified in further detail in the next refinement step may as well be a straightforward linear search. The statement "eliminate tape" implies a decrease of $k1$ as well as $k2$ and implies a re-assignment of indices in the map *ta*. The statement "close run" merely decrements $k2$ and re-arranges components of *ta* accordingly. The details are shown in Program 2.15, which is a last refinement of (2.37)

Program 2.15 Balanced Mergesort.

```
program balancedmerge (output);
{balanced n-way tape merge sort}
const n = 6; nh = 3;        {no. of tapes}
type item = record
                key: integer
            end ;
   tape = file of item;
   tapeno = 1 .. n;
var leng, rand: integer;        {used to generate file}
   eot: boolean;                {end of tape}
   buf: item;
   f0: tape; {f0 is the input tape with random numbers}
   f: array [1 .. n] of tape;
```

```
procedure list(var f: tape; n: tapeno);
   var z: integer;
begin writeln('TAPE', n:2); z := 0;
   while ¬eof(f) do
   begin read(f, buf); write(output, buf.key: 5); z := z+1;
      if z = 25 then
         begin writeln(output); z := 0;
         end
   end ;
   if z ≠ 0 then writeln (output); reset(f)
end {list} ;

procedure tapemergesort;
   var i,j,mx,tx: tapeno;
      k1,k2,l; integer;
      x, min: integer;
      t, ta: array [tapeno] of tapeno;
begin {distribute initial runs to t[1] . . . t[nh]}
   for i := 1 to nh do rewrite(f[i]);
   j := nh; l := 0;
   repeat if j < nh then j := j+1 else j := 1;
      {copy one run from f0 to tape j}
      l := l+1;
      repeat read(f0, buf); write(f[j], buf)
      until (buf .key > f0↑ .key) ∨ eof(f0)
   until eof(f0);
   for i := 1 to n do t[i] := i;
   repeat {merge from t[1] . . . t[nh] to t[nh+1] . . . t[n]}
      if l < nh then k1 := l else k1 := nh;
      {k1 = no. of input tapes in this phase}
      for i := 1 to k1 do
         begin reset(f[t[i]]); list(f[t[i]], t[i]); ta[i] := t[i]
         end ;
      l := 0;  {l = number of runs merged}
      j := nh+1;     {j = index of output tape}
      repeat {merge a run from t[1] . . . t[k1] to t[j]}
         k2 := k1; l := l+1;  {k2 = no. of active input tapes}
         repeat {select minimal element}
            i := 1; mx := 1; min := f[ta[1]]↑.key;
            while i < k2 do
            begin i := i+1; x := f[ta[i]]↑.key;
                  if x < min then
                  begin min := x; mx := i
                  end
            end ;
```

Program 2.15 (Continued)

```
                {ta[mx] has minimal element, move it to t[j]}
                read(f[ta[mx]], buf); eot := eof(f[ta[mx]]);
                write(f[t[j]], buf);
                if eot then
                begin rewrite(f[ta[mx]]); {eliminate tape}
                   ta[mx] := ta[k2]; ta[k2] := ta[k1];
                   k1 := k1-1; k2 := k2-1
                end else
                if buf .key > f[ta[mx]]↑ .key then
                begin tx := ta[mx]; ta[mx] := ta[k2]; ta[k2] := tx;
                   k2 := k2-1
                end
             until k2 = 0;
             if j < n then j := j+1 else j := nh+1
          until k1 = 0;
          for i := 1 to nh do
             begin tx := t[i]; t[i] := t[i+nh]; t[i+nh] := tx
             end
       until l = 1;
       reset(f[t[1]]); list(f[t[1]], t[1]);  {sorted output is on t[1]}
    end {tapemergesort} ;

    begin {generate random file f0}
       leng := 200; rand := 7789; rewrite(f0);
       repeat rand := (131071*rand) mod 2147483647;
          buf .key := rand div 2147484; write(f0, buf); leng := leng - 1
       until leng = 0;
       reset(f0); list(f0, 1);
       tapemergesort
    end .
```

Program 2.15 (Continued)

through (2.39). Note that tapes are rewound by the procedure *rewrite* as soon as their last run has been read. The statement "switch tapes" is elaborated according to explanations given earlier.

2.3.4. Polyphase Sort

We have now discussed the necessary techniques and have acquired the proper background to investigate and program yet another sorting algorithm whose performance is superior to the balanced sort. We have seen that balanced merging eliminates the pure copying operations necessary when the distribution and the merging operations are united into a single phase.

The question arises whether or not the given tapes could still be better utilized. This is indeed the case; the key to this next improvement lies in abandoning the rigid notion of strict passes, i.e., to use the tapes in a more sophisticated way than by always having $N/2$ source tapes and as many destination tapes and exchanging source and destination tapes at the end of each distinct pass. Instead, the notion of a pass becomes diffuse. The method was invented by R. L. Gilstad [2-3] and christened *Polyphase Sort.*

It is first illustrated by an example using three tapes. At any time, items are merged from two tapes onto the third tape. Whenever one of the source tapes is exhausted, it immediately becomes the destination tape of the merge operations from the non-exhausted tape and the previous destination tape.

As we know that n runs on each input tape are transformed into n runs on the output tape, we need to list only the number of runs present on each tape (instead of specifying actual keys). In Fig. 2.14 we assume that initially the two input tapes $f1$ and $f2$ contain 13 and 8 runs, respectively. Thus, in the first "pass" 8 runs are merged from $f1$ and $f2$ to $f3$, in the second "pass" the remaining 5 runs are merged from $f3$ and $f1$ onto $f2$, etc. In the end, $f1$ is the sorted file.

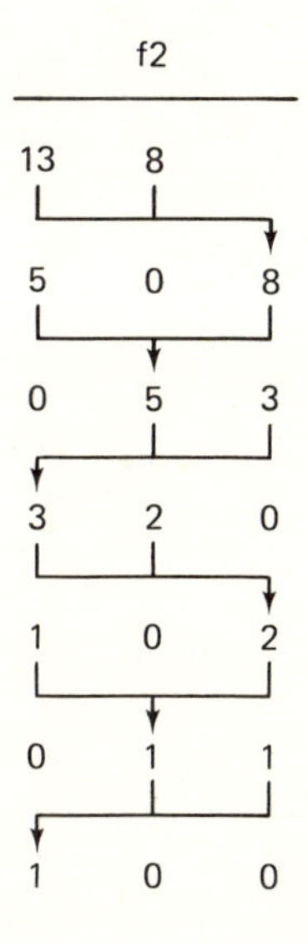

Fig. 2.14 Polyphase mergesort of 21 runs with three tapes.

A second example shows the Polyphase method with 6 tapes. Let there initially be 16 runs on $f1$, 15 on $f2$, 14 on $f3$, 12 on $f4$, and 8 on $f5$; in the first partial pass, 8 runs are merged onto $f6$; In the end, $f2$ contains the sorted set of items (see Fig. 2.15).

Polyphase is more efficient than balanced merge because—given N tapes—it always operates with an $N-1$-way merge instead of an $N/2$-way merge. As the number of required passes is approximately $\log_N n$, n being the num-

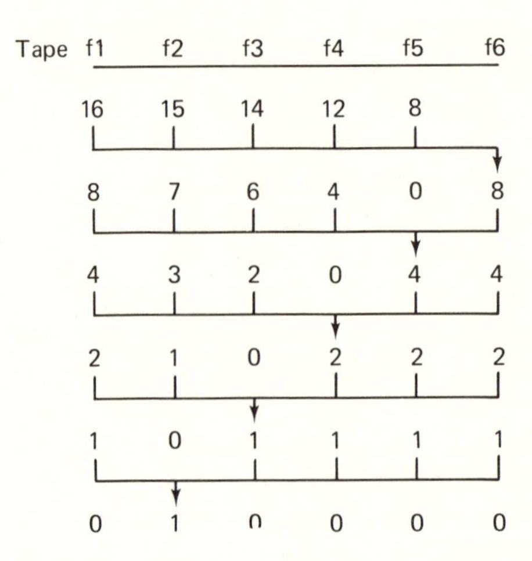

Fig. 2.15 Polyphase mergesort of 65 runs with six tapes.

ber of items to be sorted and N being the degree of the merge operations, Polyphase promises a significant improvement over balanced merge.

Of course, the distribution of initial runs was carefully chosen in the above examples. In order to find out which initial distributions of runs lead to a proper functioning, we work backward, starting with the final distribution (last line in Fig. 2.15). Rewriting the tables of the two examples and rotating each row by one position with respect to the prior row yields Tables 2.13 and 2.14 for six passes and for three and six tapes, respectively.

l	$a_1^{(l)}$	$a_2^{(l)}$	$\sum a_i^{(l)}$
0	1	0	1
1	1	1	2
2	2	1	3
3	3	2	5
4	5	3	8
5	8	5	13
6	13	8	21

Table 2.13 Perfect Distribution of Runs on Two Tapes.

l	$a_1^{(l)}$	$a_2^{(l)}$	$a_3^{(l)}$	$a_4^{(l)}$	$a_5^{(l)}$	$\sum a_i^{(l)}$
0	1	0	0	0	0	1
1	1	1	1	1	1	5
2	2	2	2	2	1	9
3	4	4	4	3	2	17
4	8	8	7	6	4	33
5	16	15	14	12	8	65

Table 2.14 Perfect Distribution of Runs on Five Tapes.

From Table 2.13 we can deduce the relations

$$\left.\begin{aligned} a_2^{(l+1)} &= a_1^{(l)} \\ a_1^{(l+1)} &= a_1^{(l)} + a_2^{(l)} \end{aligned}\right\} \quad \text{for } l > 0 \tag{2.40}$$

and $a_1^{(0)} = 1$, $a_2^{(0)} = 0$. Setting $a_1^{(i)} = f_i$, we obtain

$$\begin{aligned} f_{i+1} &= f_i + f_{i-1}, \qquad \text{for } i \geq 1 \\ f_1 &= 1 \\ f_0 &= 0 \end{aligned} \tag{2.41}$$

These are the recursive rules (or recurrence relations) defining the so-called *Fibonacci* numbers:

$$0, 1, 1, 2, 3, 5, 8, 13, 21, 34, 55, \ldots$$

Each Fibonacci number is the sum of its two predecessors. As a consequence, the numbers of initial runs on the two tapes must be two consecutive Fibonacci numbers in order to make Polyphase with three tapes work properly.

How about the second example (Table 2.14) with six tapes? The formation rules are easily derived as

$$\begin{aligned} a_5^{(l+1)} &= a_1^{(l)} \\ a_4^{(l+1)} &= a_1^{(l)} + a_5^{(l)} = a_1^{(l)} + a_1^{(l+1)} \\ a_3^{(l+1)} &= a_1^{(l)} + a_4^{(l)} = a_1^{(l)} + a_1^{(l-1)} + a_1^{(l-2)} \\ a_2^{(l+1)} &= a_1^{(l)} + a_3^{(l)} = a_1^{(l)} + a_1^{(l-1)} + a_1^{(l-2)} + a_1^{(l-3)} \\ a_1^{(l+1)} &= a_1^{(l)} + a_2^{(l)} = a_1^{(l)} + a_1^{(l-1)} + a_1^{(l-2)} + a_1^{(l-3)} + a_1^{(l-4)} \end{aligned} \tag{2.42}$$

Substituting f_i for $a_1^{(i)}$ yields

$$\begin{aligned} f_{i+1} &= f_i + f_{i-1} + f_{i-2} + f_{i-3} + f_{i-4}, \qquad \text{for } i \geq 4 \\ f_4 &= 1 \\ f_i &= 0, \qquad \text{for } i < 4 \end{aligned} \tag{2.43}$$

These numbers are the so-called Fibonacci numbers of order 4. In general, the *Fibonacci numbers of order p* are defined as follows:

$$\begin{aligned} f_{i+1}^{(p)} &= f_i^{(p)} + f_{i-1}^{(p)} + \cdots + f_{i-p}^{(p)}, \qquad \text{for } i \geq p \\ f_p^{(p)} &= 1 \\ f_i^{(p)} &= 0, \qquad \text{for } 0 \leq i < p \end{aligned} \tag{2.44}$$

Note that the ordinary Fibonacci numbers are those of order 1.

We have now seen that the initial numbers of runs for a perfect Polyphase Sort with n tapes are the sums of any $n-1, n-2, \ldots, 1$ (see Table 2.15) consecutive Fibonacci numbers of order $n-2$. This apparently implies

l \ n	3	4	5	6	7	8
1	2	3	4	5	6	7
2	3	5	7	9	11	13
3	5	9	13	17	21	25
4	8	17	25	33	41	49
5	13	31	49	65	81	97
6	21	57	94	129	161	193
7	34	105	181	253	321	385
8	55	193	349	497	636	769
9	89	355	673	977	1261	1531
10	144	653	1297	1921	2501	3049
11	233	1201	2500	3777	4961	6073
12	377	2209	4819	7425	9841	12097
13	610	4063	9289	14597	19521	24097
14	987	7473	17905	28697	38721	48001
15	1597	13745	34513	56417	76806	95617
16	2584	25281	66526	110913	152351	190465
17	4181	46499	128233	218049	302201	379399
18	6765	85525	247177	428673	599441	755749
19	10946	157305	476449	842749	1189041	1505425
20	17711	289329	918385	1656801	2358561	2998753

Table 2.15 Numbers of Runs Allowing for Perfect Distribution.

that this method is only applicable to inputs whose number of runs is the sum of $n - 1$ such Fibonacci sums. The important question thus arises: What is to be done when the number of initial runs is not such an ideal sum? The answer is simple (and typical for such situations): We simulate the existence of hypothetical empty runs, such that the sum of real and hypothetical runs is a perfect sum. The empty runs are called *dummy runs.* But this is not really a satisfactory answer because it immediately raises the further and more difficult question: How do we recognize dummy runs during merging? Before answering this question we must first investigate the prior problem of initial run distribution and decide upon a rule for the distribution of actual and dummy runs onto the $n - 1$ tapes.

In order to find an appropriate rule for distribution, however, we must know how actual and dummy runs are merged. Clearly, the selection of a dummy run from tape i means precisely that tape i is ignored during this merge, resulting in a merge from fewer than $n - 1$ sources. Merging of a dummy run from all $n - 1$ source tapes implies no actual merge operation, but instead the recording of the resulting dummy run on the output tape. From this we conclude that dummy runs should be distributed to the $n - 1$

tapes as uniformly as possible since we are interested in active merges from as many source tapes as possible.

Let us forget dummy runs for a moment and consider the problem of distributing an *unknown* number of runs onto $n - 1$ tapes. It is plain that the Fibonacci numbers of order $n - 2$ specifying the desired numbers of runs on each tape can be generated while the distribution progresses. Assuming $n = 6$, for example, and referring to Table 2.14, we start by distributing runs as indicated by the row with index $l = 1$ (1, 1, 1, 1, 1); if there are more runs available, we proceed to the second row (2, 2, 2, 2, 1); if the source is still unexhausted, the distribution proceeds according to the third row (4, 4, 4, 3, 2), and so on. We shall call the row index *level*. Evidently, the larger the number of runs, the higher will be the level of Fibonacci numbers which, incidentally, is equal to the number of merge passes or tape switchings necessary for the subsequent sort.

The distribution algorithm can now be formulated in a first version as follows:

1. Let the distribution goal be the Fibonacci numbers of order $n - 2$, level 1.
2. Distribute according to the set goal.
3. If the goal is reached, compute the next level of Fibonacci numbers; the difference between them and those on the former level constitutes the new distribution goal. Return to step 2. If the goal cannot be reached because the source is exhausted, terminate the distribution process.

The rules for calculating the next level of Fibonacci numbers are contained in their definition (2.44). We can thus concentrate our attention on step 2, where, with a given goal, the subsequent runs are to be distributed one after the other onto the $n - 1$ tapes. It is here where the dummy runs have to re-appear in our considerations.

Let us assume that when raising the level, we record the next goal by the differences d_i for $i = 1 \ldots n - 1$, where d_i denotes the number of runs to be put onto tape i in this step. We can now assume that we immediately put d_i dummy runs onto tape i and then regard the subsequent distribution as the *replacement* of dummy runs by actual runs, each time recording a replacement by subtracting 1 from d_i. Thus, the d_i's will indicate the number of dummy runs on tape i when the source becomes empty.

It is not known which algorithm will yield the optimal distribution, but the following has proved to be a very good method. It is called *horizontal distribution* (cf. Knuth, vol 3. p. 270), a term that can be understood by regarding the runs as being piled up in the form of silos, as shown in Fig. 2.16 for $n = 6$, level 5 (cf. Table 2.14).

In order to reach an equal distribution of remaining dummy runs as quickly as possible, their replacement by actual runs reduces the size of the piles by picking off dummy runs on horizontal levels proceeding from left to right. In this way, the runs are distributed onto the tapes as indicated by their sequence numbers in Fig. 2.16.

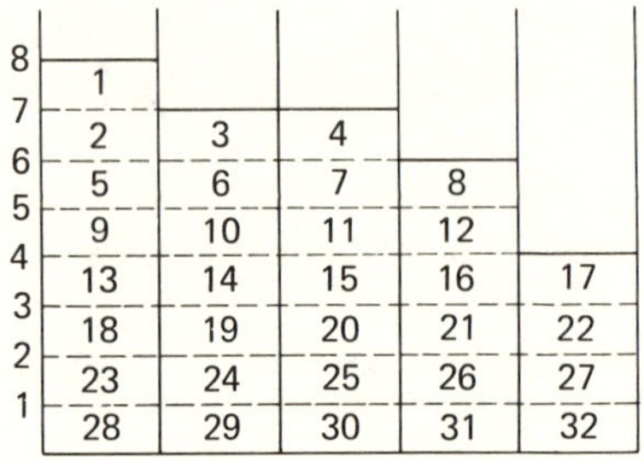

8	1				
7	2	3	4		
6	5	6	7	8	
5	9	10	11	12	
4	13	14	15	16	17
3	18	19	20	21	22
2	23	24	25	26	27
1	28	29	30	31	32

Fig. 2.16 "Horizontal distribution" of runs.

We are now in a position to describe the algorithm in the form of a procedure called *selecttape*, which is called each time a run has been copied and a new tape is to be selected for the next run. We assume the existence of a variable j denoting the index of the current destination tape. a_i and d_i denote the ideal and dummy distribution numbers for tape i.

j: *tapeno*;
a,d: **array** [*tapeno*] **of** *index*; (2.45)
level: *integer*

These variables are initialized with the following values:

$$a_i = 1, \quad d_i = 1 \quad \text{for } i = 1 \ldots n-1$$
$$a_n = 0, \quad d_n = 0 \quad \text{(dummy)}$$
$$j = 1$$
$$level = 1$$

Note that *selecttape* is to compute the next row of Table 2.14, i.e., the values $a_1^{(l)} \ldots a_{n-1}^{(l)}$, each time that the level is increased. The "next goal," i.e., the differences $d_i = a_i^{(l)} - a_i^{(l-1)}$ are also computed at that time. The indicated algorithm relies on the fact that the resulting d_i's decrease with increasing index (decreasing stair in Fig. 2.16). (Note that the exception is the transition from level 0 to level 1; this algorithm must therefore be used starting at level 1). *Selecttape* ends by decreasing d_j by 1; this operation stands for the replacement of a dummy run on tape j by an actual run.

```
procedure selecttape;
    var i: tapeno; z: integer;
begin
    if d[j] < d[j+1] then j := j+1 else
    begin if d[j] = 0 then
        begin level := level + 1; z := a[1];
            for i := 1 to n-1 do                                    (2.46)
            begin d[i] := z+a[i+1]-a[i]; a[i] := z+a[i+1]
            end
        end ;
        j := 1
    end ;
    d[j] := d[j] -1
end
```

Assuming the availability of a routine to copy a run from the source $f0$ onto $f[j]$, we can formulate the initial distribution phase as follows (always assuming that the source contains at least one run):

```
repeat selecttape; copyrun                                          (2.47)
until eof(f0)
```

Here, however, we must pause for a moment to recall the effect encountered in distributing runs in the previously discussed natural merge algorithm: The fact that two runs consecutively arriving at the same destination may turn out to constitute a single run causes the assumed numbers of runs to be incorrect. By devising the sort algorithm such that its correctness does not depend on the number of runs, this side effect can safely be ignored. In the Polyphase Sort, however, we are particularly concerned about keeping track of the *exact* numbers of runs on each tape. Consequently, we cannot afford to overlook the effect of such a coincidental merge.

An additional complication of the distribution algorithm therefore cannot be avoided. It becomes necessary to retain the keys of the last item of the last run on each tape. For this purpose, we introduce a variable

last: **array** [*tapeno*] **of** *integer*

A next attempt to describe the distribution algorithm could be

```
repeat selecttape;
    if last[j] ≤ f0↑ .key then
        "continue old run";                                         (2.48)
    copyrun; last[j] := f0↑ .key
until eof(f0)
```

The obvious mistake lies in forgetting that *last*[*j*] has only obtained a (defined) value after copying the first run! A correct solution first distributes one run onto each of the $n - 1$ tapes without inspection of *last*[*j*]. The remaining runs are distributed according to (2.49).

```
while ¬eof(f0) do
begin selecttape;
   if last[j] ≤ f0↑ .key then
   begin {continue old run}                              (2.49)
      copyrun;
      if eof(f0) then d[j] := d[j] + 1 else copyrun
   end
   else copyrun
end
```

Here, the assignment to *last*[*j*] is assumed to be included in the procedure *copyrun*.

Now we are finally in a position to tackle the main polyphase merge sort algorithm. Its principal structure is similar to the main part of the n-way merge program: an outer loop whose body merges runs until the sources are exhausted, an inner loop whose body merges a single run from each source tape, and an innermost loop whose body selects the initial key and transmits the involved item to the target file. The principal differences are the following:

1. Instead of $n/2$, there is only one output tape in each pass.
2. Instead of switching $n/2$ input and $n/2$ output tapes after each pass, the tapes are *rotated*. This is achieved by using a tape index mapping t.
3. The number of input tapes varies from run to run; at the start of each run, it is determined from the counts d_i of dummy runs. If $d_i > 0$ for all i, then $n - 1$ dummy runs are pseudo-merged into one dummy run by merely incrementing the count d_n of the output tape. Otherwise, one run is merged from all tapes with $d_i = 0$, and d_i is decremented for all other tapes, indicating that one dummy run was taken off. We denote the number of input tapes involved in a merge by k.
4. It is impossible to derive termination of a phase by the end-of-file status of the n — 1st tape because more merges might be necessary involving dummy runs from that tape. Instead, the theoretically necessary number of runs is determined from the coefficients a_i. The coefficients $a_i^{(l)}$ were computed during the distribution phase; they can now be recomputed "backward."

The main part of the Polyphase Sort can now be formulated according to these rules, assuming that all $n - 1$ tapes with initial runs are reset and that the tape map is initially set to $t_i = i$.

```
repeat {merge from t[1] ... t[n-1] to t[n]}
   z := a[n-1]; d[n] := 0; rewrite(f[t[n]]);
   repeat k := 0;  {merge one run}
      {determine no. k of active input tapes}
      for i := 1 to n-1 do                                    (2.50)
      if d[i] > 0 then d[i] := d[i] -1 else
         begin k := k+1; ta[k] := t[i]
         end ;
      if k = 0 then d[n] := d[n] + 1 else
         "merge one real run from t[1] ... t[k]";
      z := z-1
   until z = 0;
   reset(f[t[n]]);
   "rotate tapes in map t; compute a[i] on next level";
   rewrite(f[t[n]]); level := level - 1
until level = 0;
{sorted output is on t[1]}
```

The actual merge operation is almost identical with the program in the n-way merge sort, the only difference being that the tape elimination algorithm is somewhat simpler. The rotation of the tape index map and the corresponding counts d_i (and the down-level recomputation of the coefficients a_i) is straightforward and can be inspected in detail from Program 2.16 which represents the Polyphase algorithm in its entirety.

Program 2.16 Polyphasesort.

```
program polysort (output);
{polyphase sort with n tapes}
const n = 6;              {no. of tapes}
type item = record
               key: integer
            end ;
   tape = file of item;
   tapeno = 1 .. n;
var leng, rand: integer;        {used to generate file}
   eot: boolean;
   buf: item;
   f0: tape;  {f0 is the input tape with random numbers}
   f: array [1 .. n] of tape;
```

```
procedure list (var f: tape; n: tapeno);
   var z: integer;
begin z := 0;
   writeln ('TAPE', n: 2);
   while ¬eof(f) do
   begin read(f, buf); write(output, buf.key: 5); z := z+1;
      if z = 25 then
         begin writeln (output); z := 0
         end
   end ;
   if z ≠ 0 then writeln (output); reset(f)
end {list} ;

procedure polyphasesort;
   var i,j,mx,tn: tapeno;
      k, level: integer;
      a, d: array [tapeno] of integer;
         {a[j] = ideal number of runs on tape j}
         {d[j] = number of dummy runs on tape j}
      dn,x,min,z: integer;
      last: array [tapeno] of integer;
         {last[j] = key of tail item on tape j}
      t,ta: array [tapeno] of tapeno;
         {mappings of tape numbers}

   procedure selecttape;
      var i: tapeno; z: integer;
   begin
      if d[j] < d[j+1] then j := j+1 else
      begin if d[j] = 0 then
            begin level := level + 1; z := a[1];
               for i := 1 to n-1 do
               begin d[i] := z + a[i+1] - a[i]; a[i] := z + a[i+1]
               end
            end ;
            j := 1
      end ;
      d[j] := d[j] -1
   end ;

   procedure copyrun;
   begin {copy one run from f0 to tape j}
      repeat read(f0, buf); write(f[j], buf);
      until eof(f0) ∨ (buf .key > f0↑ .key);
      last[j] := buf .key
   end ;
```

Program 2.16 (Continued)

```
begin {distribute initial runs}
   for i := 1 to n-1 do
      begin a[i] := 1; d[i] := 1: rewrite(f[i])
      end ;
   level := 1; j := 1; a[n] := 0; d[n] := 0;
   repeat selecttape; copyrun
   until eof(f0) ∨ (j=n-1);
   while ¬eof(f) do
   begin selecttape;
      if last[j] ≤ f0↑ .key then
      begin {continue old run}
         copyrun;
         if eof(f0) then d[j] := d[j] + 1 else copyrun
      end
      else copyrun
   end ;
   for i := 1 to n-1 do reset(f[i]);
   for i := 1 to n do t[i] := i;
   repeat {merge from t[1] . . . t[n-1] to t[n]}
      z := a[n-1]; d[n] := 0; rewrite(f[t[n]]);
      repeat k := 0;      {merge one run}
         for i := 1 to n-1 do
         if d[i] > 0 then d[i] := d[i]-1 else
            begin k := k+1; ta[k] := t[i]
            end ;
         if k = 0 then d[n] := d[n] + 1 else
         begin {merge one real run from t[1] . . . t[k]}
            repeat i := 1; mx := 1;
               min := f[ta[1]]↑ .key;
               while i < k do
               begin i := i+1; x := f[ta[i]]↑ .key;
                  if x < min then
                  begin min := x; mx := i
                  end
               end ;
               {ta[mx] contains minimal element; move it to t[n]}
               read(f[ta[mx]], buf); eot := eof(f[ta[mx]]);
               write(f[t[n]], buf);
               if (buf .key > f[ta[mx]]↑.key) ∨ eot then
               begin {drop this tape}
                  ta[mx] := ta[k]; k := k-1
               end
            until k = 0
         end ;
```

Program 2.16 (Continued)

```
            z := z-1
        until z = 0;
        reset(f[t[n]]); list(f[t[n]], t[n]);  {rotate tapes}
        tn := t[n]; dn := d[n]; z := a[n-1];
        for i := n downto 2 do
            begin t[i] := t[i-1]; d[i] := d[i-1]; a[i] := a[i-1] - z
            end ;
        t[1] := tn; d[1] := dn; a[1] := z;
        {sorted output is on t[1]}
        list(f[t[1]], t[1]); level := level - 1
    until level = 0;
end {polyphasesort} ;
begin {generate random file}
    leng := 200; rand := 7789;
    repeat rand := (131071*rand) mod 2147483647;
        buf .key := rand div 2147484; write(f0, buf); leng := leng - 1
    until leng = 0;
    reset(f0); list(f0, 1);
    polyphasesort
end .
```

Program 2.16 (Continued)

2.3.5. Distribution of Initial Runs

We were led to the sophisticated sequential sorting programs because the simpler methods operating on arrays rely on the availability of a random-access store sufficiently large to hold the entire set of data to be sorted. Very often such a store is unavailable; instead, sufficiently large sequential storage devices such as tapes must be used. We note that the sequential sorting methods developed so far need practically no primary store whatsoever, except for the file buffers and, of course, the program itself. However, it is a fact that even small computers include some random access, primary store that is almost always larger than what is needed by the programs developed here. Failing to make optimal use of it cannot be justified.

The solution lies in *combining* array and file sorting techniques. In particular, an adapted array sort may be used in the distribution phase of initial runs with the effect that these runs do already have a length l of approximately the size of the available primary data store. It is plain that in the subsequent merge passes no additional array sorts could improve the performance because the runs involved are steadily growing in length, and thus they always remain larger than the available main store. As a result, we may fortunately concentrate our attention on improving the algorithm that generates initial runs.

Naturally, we immediately concentrate our search on the logarithmic array sorting methods. The most suitable of them is the tree sort or Heapsort method (see Sect. 2.2.5). The heap may be regarded as a tunnel through which all file components must pass, some quicker and some more slowly. The least key is readily picked off the top of the heap, and its replacement is a very efficient process. The action of funnelling a component from the input tape $f0$ through a full "heap tunnel" h onto an output tape $f[j]$ may be described simply as follows:

$$\begin{aligned}&write(f[j], h[1]);\\&read(f0, h[1]);\\&sift(1, n)\end{aligned} \qquad (2.51)$$

"Sift" is the process described in Sect. 2.2.5 for sifting the newly inserted component $h[1]$ down into its proper place. Note that $h[1]$ is the *least* item on the heap. An example is shown in Fig. 2.17.

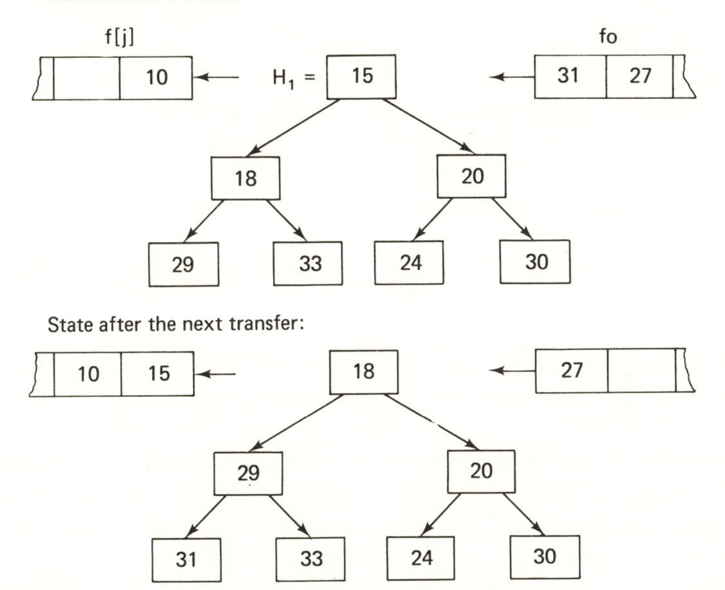

Fig. 2.17 Sifting a key through a heap.

The program eventually becomes considerably more complex because

1. The heap h is initially empty and must first be filled.
2. Toward the end, the heap is only partially filled, and it ultimately becomes empty.
3. We must keep track of the beginning of new runs in order to change the output tape index j at the right time.

Before proceeding, let us formally declare the variables that are evidently involved in the game:

```
var f0: tape;
    f : array [tapeno] of tape;
    h : array [1 .. m] of item;                    (2.52)
    l,r: integer
```

m is the size of the heap *h*. We use the constant *mh* to denote $m/2$; *l* and *r* are indices on *h*. The funnelling process can then be divided into five distinct parts.

1. Read the first *mh* items from *f*0 and put them into the upper half of the heap where no ordering among the keys is prescribed.
2. Read another *mh* items and put them into the lower half of the heap, sifting each item into its appropriate position (build heap).
3. Set *l* to *m* and repeat the following step for all remaining items on *f*0: Feed $h[1]$ to the appropriate output tape. If its key is less or equal to the key of the next item on the input tape, then this next item belongs to the same run and can be sifted into proper position. Otherwise, reduce the size of the heap and place the new item into a second, "upper" heap which is built up to contain the next run. We indicate the borderline between the two heaps with the index *l*. Thus, the "lower" or current heap consists of the items $h[1] \ldots h[l]$, the "upper" or next heap of $h[l+1] \ldots h[m]$. If $l = 0$, then change the output tape and reset *l* to *m*.
4. Now the source is exhausted. First, set *r* to *m*; then flush the lower part terminating the current run, and at the same time build up the upper part and gradually relocate it into positions $h[l+1] \ldots h[r]$.
5. The last run is generated from the remaining items in the heap.

We are now in a position to describe the five stages in detail as a complete program, calling a procedure *selecttape* whenever the end of a run is detected and some action to alter the index of the output tape has to be invoked. In Program 2.17 a dummy routine is used instead; it merely counts the number of runs generated. All elements are written onto tape *f*1.

If we now try to integrate this program with, for instance, the Polyphase Sort, we encounter a serious difficulty. It arises from the following circumstances: The sort program consists in its initial part of a fairly complicated routine for tape switching and relies on the availability of a procedure *copyrun* which delivers exactly one run to the selected tape. The Heapsort program, on the other hand, is a complex routine relying on the availability of a closed procedure *selecttape* which simply selects a new tape. There would be no problem if in one (or both) of the programs the desired procedure would be called at a single place only; but instead, they are called at several places in both programs.

This situation is best reflected by the use of a so-called *coroutine*; it is suitable in those cases in which several processes coexist. The most typical representative is the combination of a process which produces a stream of information in distinct entities and a process which consumes this stream. This producer-consumer relationship can be expressed in terms of two coroutines. One of them may well be the main program itself.

The coroutine may be considered a procedure or subroutine that contains one or more breakpoints. If such a breakpoint is encountered, then control

Program 2.17 Distribution of Initial Runs Through a Heap.

```
program distribute(f0,f1,output);
{initial distribution of runs by heap sort}
const m = 30; mh = 15; {size of heap}
type item = record
               key: integer
            end ;
   tape = file of item;
   index = 0 .. m;
var l,r: index;
   f0,f1: tape;
   count: integer;      {counter of runs}
   h: array [1 .. m] of item; {heap}

procedure selecttape;
begin count := count + 1;
   {dummy; count number of distributed runs}
end {selecttape} ;

procedure sift(l,r: index);
   label 13;
   var i,j: integer: x: item;
begin i := l; j := 2*i; x := h[i];
   while j ≤ r do
   begin if j < r then
            if h[j] .key > h[j+1] .key then j := j+1;
         if x .key ≤ h[j] .key then goto 13;
         h[i] := h[j]; i := j; j := 2*i
   end ;
13: h[i] := x
end ;
```

```
begin {create initial runs by heapsort}
   count := 0; reset(f0); rewrite(f1);
   selecttape;
  {step 1: fill upper half of heap h}
   l := m;
   repeat read(f0, h[l]); l := l-1
   until l = mh;
  {step 2: fill lower half of heap h}
   repeat read(f0, h[l]); sift(l,m); l := l-1
   until l = 0;
  {step 3: pass runs through full heap}
   l := m;
   while ¬eof(f0) do
   begin write(f1, h[1]);
      if h[1] .key ≤ f0↑ .key then
      begin {new record belongs to same run}
         read(f0, h[1]); sift(1,l);
      end else
      begin {new record belongs to next run}
         h[1] := h[l]; sift(1,l-1);
         read(f0, h[l]); if l ≤ mh then sift(l,m); l := l-1;
         if l = 0 then
         begin {heap is full; start new run}
            l := m; selecttape;
         end
      end
   end ;
  {step 4: flush lower part of heap}
   r := m;
   repeat write(f1, h[1]);
      h[1] := h[l]; sift(1,l-1);
      h[l] := h[r]; r := r-1;
      if l ≤ mh then sift(l,r); l := l-1
   until l = 0;
  {step 5: flush upper part of heap. generate last run}
   selecttape;
   while r > 0 do
   begin write(f1,h[1]);
      h[1] := h[r]; sift(1,r); r := r-1
   end ;
   writeln (count)
end .
```

Program 2.17 (Continued)

returns to the program which had called the coroutine. Whenever the coroutine is called again, execution is resumed at that breakpoint. In our example we might consider the Polyphase Sort as the main program, calling upon *copyrun* which is formulated as a coroutine. It consists of the main body of Program 2.17 in which each call of *selecttape* now represents a breakpoint. The test for end of file would then have to be replaced systematically by a test of whether or not the coroutine had reached its endpoint. A logically sound formulation would be *eoc*(*copyrun*) in place of *eof*($f0$).

Analysis and conclusions. What performance can be expected from a Polyphase Sort with initial distribution of runs by a Heapsort? We first discuss the improvement to be expected by introducing the heap.

In a sequence with randomly distributed keys the expected average length of runs is 2. What is this length after the sequence has been funnelled through a heap of size m? One is inclined to say m, but, fortunately, the actual result of probabilistic analysis is much better, namely, $2m$ (see Knuth, vol. 3, p. 254). Therefore, the expected improvement factor is m.

An estimate of the performance of Polyphase can be gathered from Table 2.15, indicating the maximal number of initial runs that can be sorted in a given number of partial passes (levels) with a given number n of tapes. As an example, with $n = 6$ tapes, and a heap of size $m = 100$, a file with up to 165,680,100 initial runs can be sorted within 20 partial passes. This is a remarkable performance.

Reviewing again the combination of Polyphase Sort and Heapsort, one cannot help but be amazed at the complexity of this program. After all, it performs the same easily defined task of re-ordering a set of items as is done by any of the short programs based on the straight array sorting principles. The moral of the entire chapter may be taken as an exhibition of the following:

1. The intimate connection between algorithm and underlying data structure and particularly the influence of the latter on the former.
2. The sophistication by which the performance of a program can be improved, even when the available structure for its data (sequence instead of array) is rather ill-suited for the task.

EXERCISES

2.1. Which of the algorithms given by Programs 2.1 through 2.6, 2.8, 2.10, and 2.13 are stable sorting methods?

2.2. Would Program 2.2 still work correctly if $l \leq r$ were replaced by $l < r$ in the while clause? Would it still be correct if the statements $r := m-1$ and

$l := m+1$ were simplified to $r := m$ and $l := m$? If not, find sets of values $a_1 \ldots a_n$ upon which the altered program would fail.

2.3. Program and measure the execution time of the three straight sorting methods on your computer and find weights by which the factors C and M have to be multiplied to yield real time estimates.

2.4. Test the Heapsort Program 2.8 with various random input sequences and determine the average number of times that the statement **goto** 13 is executed. Since this number is relatively small, the following question becomes of interest: Is there a way of extracting the test

$$x.key \geq a[j].key$$

from the while loop?

2.5. Consider the following "obvious" version of the Partition Program 2.9:

```
i := 1; j := n;
x := a[(n+1) div 2].key;
repeat
    while a[i].key < x do i := i+1;
    while x < a[j].key do j := j-1;
    w := a[i]; a[i] := a[j]; a[j] := w
until i > j
```

Find sets of values $a_1 \ldots a_n$ for which this version fails.

2.6. Write a program that combines the Quicksort and Bubblesort algorithms as follows: Use Quicksort to obtain (unsorted) partitions of length m $(1 \leq m \leq n)$; then use Bubblesort to complete the task. Note that the latter may sweep over the entire array of n elements, hence, minimizing the "bookkeeping" effort. Find that value of m which minimizes the total sort time.

Note: Clearly, the optimum value of m will be quite small. It may therefore pay to let the Bubblesort sweep exactly $m - 1$ times over the array instead of including a last pass establishing the fact that no further exchange is necessary.

2.7. Perform the same experiment as in Exercise 6 with a straight selection sort instead of a Bubblesort. Naturally, the selection sort cannot sweep over the whole array; therefore, the expected amount of index handling is somewhat greater.

2.8. Write a recursive Quicksort algorithm according to the recipe that the sorting of the shorter partition should be tackled before the sorting of the longer partition. Perform the former task by an iterative statement, the latter by a recursive call. (Hence, your sort procedure will contain one recursive call instead of two in Program 2.10 and none in Program 2.11.)

2.9. Find a permutation of the keys $1, 2, \ldots, n$ for which Quicksort displays its worst (best) behavior ($n = 5, 6, 8$).

2.10. Construct a natural merge program similar to the straight merge Program 2.13, operating on a double length array from both ends inward; compare its performance with that of Program 2.13.

2.11. Note that in a (two-way) natural merge we do not blindly select the least value among the available keys. Instead, upon encountering the end of a run, the tail of the other run is simply copied onto the output sequence. For example, merging of

$$2,\ 4,\ 5,\ 1,\ 2,\ \ldots$$
$$3,\ 6,\ 8,\ 9,\ 7,\ \ldots$$

results in the sequence

$$2,\ 3,\ 4,\ 5,\ 6,\ 8,\ 9,\ 1,\ 2,\ \ldots$$

instead of

$$2,\ 3,\ 4,\ 5,\ 1,\ 2,\ 6,\ 8,\ 9,\ \ldots$$

which seems to be better ordered. What is the reason for this strategy?

2.12. What purpose does the variable *ta* in Program 2.15 serve? Under which circumstances is the statement

begin *rewrite* (*f*[*ta*[*mx*]]); . . .

executed, and when the statement

begin *tx* := *ta*[*mx*]; . . . ?

2.13. Why do we need the variable *last* in the Polyphase Sort Program 2.16, but not in Program 2.15?

2.14. A sorting method similar to the Polyphase is the so-called *Cascade merge sort* [2.1 and 2.9]. It uses a different merge pattern. Given, for instance, six tapes $T1, \ldots, T6$, the cascade merge, also starting with a "perfect distribution" of runs on $T1 \ldots T5$, performs a five-way merge from $T1 \ldots T5$ onto $T6$ until $T5$ is empty, then (without involving $T6$) a four-way merge onto $T5$, then a three-way merge onto $T4$, a two-way merge onto $T3$, and finally a copy operation from $T1$ onto $T2$. The next pass operates in the same way starting with a five-way merge to $T1$, and so on. Although this scheme seems to be inferior to Polyphase because at times it chooses to leave some tapes idle and because it involves simple copy operations, it surprisingly is superior to Polyphase for (very) large files and for six or more tapes. Write a well-structured program for the Cascade merge principle.

REFERENCES

2-1. BETZ, B. K. and CARTER, *ACM National Conf.*, **14**, (1959), Paper 14.

2-2. FLOYD, R. W., "Treesort" (Algorithms 113 and 243), *Comm. ACM*, **5**, No. 8 (1962), 434, and *Comm. ACM*, **7**, No. 12 (1964), 701.

2-3. GILSTAD, R. L., "Polyphase Merge Sorting—An Advanced Technique," *Proc. AFIPS Eastern Jt. Comp. Conf.*, **18**, (1960), 143–48.

2-4. HOARE, C. A. R., "Proof of a Program: FIND," *Comm. ACM*, **13**, No. 1 (1970), 39–45.

2-5. _____, "Proof of a Recursive Program: Quicksort," *Comp. J.*, **14**, No. 4 (1971), 391–95.

2-6. ___________, "Quicksort," *Comp. J.*, **5**, No. 1 (1962), 10–15.

2-7. KNUTH, D. E., *The Art of Computer Programming*, Vol. 3 (Reading, Mass.: Addison-Wesley, 1973).

2-8. _____, *The Art of Computer Programming*, **3**, pp. 86–95.

2-9. _____, *The Art of Computer Programming*, **3**, p. 289.

2-10. LORIN, H., "A Guided Bibliography to Sorting," *IBM Syst. J.*, **10**, No. 3 (1971), 244-54.

2-11. SHELL, D. L., "A Highspeed Sorting Procedure," *Comm. ACM*, **2**, No. 7 (1959), 30–32.

2-12. SINGLETON, R. C., "An Efficient Algorithm for Sorting with Minimal Storage" (Algorithm 347), *Comm. ACM*, **12**, No. 3 (1969), 185.

2-13. VAN EMDEN, M. H., "Increasing the Efficiency of Quicksort" (Algorithm 402), *Comm. ACM*, **13**, No. 9 (1970), 563–66, 693.

2-14. WILLIAMS, J. W. J., "Heapsort" (Algorithm 232), *Comm. ACM*, **7**, No. 6 (1964), 347–48.

3 RECURSIVE ALGORITHMS

3.1. INTRODUCTION

An object is said to be *recursive* if it partially consists or is defined in terms of itself. Recursion is encountered not only in mathematics, but also in daily life. Who has never seen an advertising picture which contains itself?

Fig. 3.1 A recursive picture.

Recursion is a particularly powerful means in mathematical definitions. A few familiar examples are those of natural numbers, tree structures, and of certain functions:

1. Natural numbers:
   (a) 1 is a natural number.
   (b) the successor of a natural number is a natural number.
2. Tree structures
   (a) o is a tree (called the empty tree).
   (b) If t_1 and t_2 are trees, then

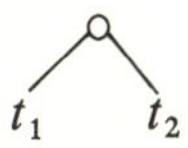

   is a tree (drawn upside down).
3. The factorial function $n!$ (for non-negative integers):
   (a) $0! = 1$
   (b) If $n > 0$, then $n! = n \cdot (n - 1)!$

The power of recursion evidently lies in the possibility of defining an infinite set of objects by a finite statement. In the same manner, an infinite number of computations can be described by a finite recursive program, even if this program contains no explicit repetitions. Recursive algorithms, however, are primarily appropriate when the problem to be solved, or the function to be computed, or the data structure to be processed are already defined in recursive terms. In general, a recursive program can be expressed as a composition $\mathcal{P}$ of base statements S_i (not containing P) and P itself.

$$P \equiv \mathcal{P}[S_i, P] \tag{3.1}$$

The necessary and sufficient tool for expressing programs recursively is the *procedure* or subroutine, for it allows a statement to be given a name by which this statement may be invoked. If a procedure P contains an explicit reference to itself, then it is said to be *directly recursive*; if P contains a reference to another procedure Q which contains a (direct or indirect) reference to P, then P is said to be *indirectly recursive*. The use of recursion may therefore not be immediately apparent from the program text.

It is common to associate a set of local objects with a procedure, i.e., a set of variables, constants, types, and procedures which are defined locally to this procedure and have no existence or meaning outside this procedure. Each time such a procedure is activated recursively, a new set of local, bound variables is created. Although they have the same names as their corresponding elements in the set local to the previous instance of the procedure, their values are distinct, and any conflict in naming is avoided by the rules of scope of identifiers: the identifiers always refer to the most recently created set of variables. The same rule holds for procedure parameters which by definition are bound to the procedure.

Like repetitive statements, recursive procedures introduce the possibility of non-terminating computations, and thereby also the necessity of consider-

ing the problem of *termination*. A fundamental requirement is evidently that the recursive call of a procedure P is subjected to a condition B, which at some time becomes non-satisfied. The scheme for recursive algorithms may therefore be expressed more precisely as

$$P \equiv \textbf{if } B \textbf{ then } \mathcal{P}[S_i, P] \tag{3.2}$$

or

$$P \equiv \mathcal{P}[S_i, \textbf{if } B \textbf{ then } P] \tag{3.3}$$

The basic technique of demonstrating that a repetition terminates is to define a function $f(x)$ (x is the set of variables in the program), such that $f(x) \leq 0$ implies the terminating condition (of the while or repeat clause), and to prove that $f(x)$ decreases during each repetition. In the same manner, termination of a recursive program can be proved by showing that each execution of P decreases $f(x)$. A particularly evident way to ensure termination is to associate a (value) parameter, say n, with P and to recursively call P with $n - 1$ as parameter value. Replacement of the condition B by $n > 0$ then guarantees termination. This may be expressed by the following program schemata:

$$P(n) \equiv \textbf{if } n > 0 \textbf{ then } \mathcal{P}[S_i, P(n-1)] \tag{3.4}$$

$$P(n) \equiv \mathcal{P}[S_i, \textbf{if } n > 0 \textbf{ then } P(n-1)] \tag{3.5}$$

In practical applications it is mandatory to show that the ultimate depth of recursion is not only finite, but that it is actually small. The reason is that upon each recursive activation of a procedure P some amount of storage is required to accommodate its variables. In addition to these local bound variables, the current state of the computation must be recorded in order to be retrievable when the new activation of P is terminated, and the old one has to be resumed. We have already encountered this situation in the development of the procedure Quicksort in Chap. 2. It was discovered that by "naively" composing the program out of a statement that splits the n items into two partitions and of two recursive calls sorting the two partitions, the depth of recursion may in the worst case approach n. By a clever re-assessment of the situation, it was possible to limit this depth to $\log n$. The difference between n and $\log n$ is sufficient to convert a case highly inappropriate for recursion into one in which recursion is perfectly practical.

3.2. WHEN NOT TO USE RECURSION

Recursive algorithms are particularly appropriate when the underlying problem or the data to be treated are defined in recursive terms. This does not mean, however, that such recursive definitions guarantee that a recursive algorithm is the best way to solve the problem. In fact, the explanation of the concept of recursive algorithm by such inappropriate examples has been a chief cause of creating widespread apprehension and antipathy toward the use of recursion in programming, and of equating recursion with inefficiency. This happened in addition to the fact that the widespread programming language FORTRAN forbids the recursive use of subroutines, thus preventing the invention of recursive solutions even when they are appropriate.

Programs in which the use of algorithmic recursion is to be avoided can be characterized by a schema which exhibits the pattern of their composition. The schema is that of (3.6) and, equivalently, of (3.7).

$$P \equiv \textbf{if } B \textbf{ then } (S\ ;\ P) \tag{3.6}$$

$$P \equiv (S;\ \textbf{if } B \textbf{ then } P) \tag{3.7}$$

These schemata are natural in those cases in which values are to be computed that are defined in terms of simple recurrence relations. Let us look at the well-known example of the factorial numbers $f_i = i!$:

$$\begin{aligned} i &= 0, 1, 2, 3, \ \ 4, \quad 5, \ldots \\ f_i &= 1, 1, 2, 6, 24, 120, \ldots \end{aligned} \tag{3.8}$$

The "zeroth" number is explicitly defined as $f_0 = 1$, whereas the subsequent numbers are usually defined—recursively—in terms of their predecessor:

$$f_{i+1} = (i+1) \cdot f_i \tag{3.9}$$

This formula suggests a recursive algorithm to proceed to the nth factorial number. If we introduce the two variables I and F to denote the values i and f_i at the ith level of recursion, we find the computation necessary to proceed to the next numbers in the sequences (3.8) to be

$$I := I + 1; \qquad F := I*F \tag{3.10}$$

and, substituting (3.10) for S in (3.6), we obtain the recursive program

$$\begin{aligned} &P \equiv \textbf{if } I < n \textbf{ then } (I := I + 1;\ F := I*F;\ P) \\ &I := 0; \quad F := 1; \quad P \end{aligned} \tag{3.11}$$

The first line of (3.11) is expressed in terms of our conventional programming notation as

```
procedure P;
begin if I < n then
      begin I := I + 1; F := I*F; P
      end
end                                              (3.12)
```

A more frequently used, but essentially equivalent, form is the one given in (3.13). P is replaced by a so-called function procedure, i.e., a procedure with which a resulting value is explicitly associated, and which therefore may be used directly as a constituent of expressions. The variable F thereby becomes superfluous; and the role of I is taken by the explicit procedure parameter.

```
function F(I: integer): integer;
begin if I > 0 then F := I*F(I - 1)
                else F := 1
end
```
(3.13)

It is quite plain that in this case recursion can be replaced by simple iteration, namely, by the program

```
I := 0;  F := 1;
while I < n do
   begin I := I + 1; F := I*F
   end
```
(3.14)

In general, programs corresponding to the general schemata (3.6) or (3.7) should be transcribed into one according to schema (3.15)

$$P \equiv (x := x_0;\ \textbf{while}\ B\ \textbf{do}\ S) \qquad (3.15)$$

There are also more complicated recursive composition schemes that can and should be translated into an iterative form. An example is the computation of the Fibonacci numbers which are defined by the recurrence relation

$$\text{fib}_{n+1} = \text{fib}_n + \text{fib}_{n-1} \qquad \text{for } n > 0 \qquad (3.16)$$

and $\text{fib}_1 = 1$, $\text{fib}_0 = 0$. A direct, naive approach leads to the program

```
function Fib(n: integer): integer;
begin if n = 0 then Fib := 0 else
      if n = 1 then Fib := 1 else
      Fib := Fib(n-1) + Fib(n-2)
end
```
(3.17)

Computation of fib_n by a call *Fib*(n) causes this function procedure to be activated recursively. How often? We notice that each call with $n > 1$ leads to 2 further calls, i.e., the total number of calls grows exponentially (see Fig. 3.2). Such a program is clearly impractical.

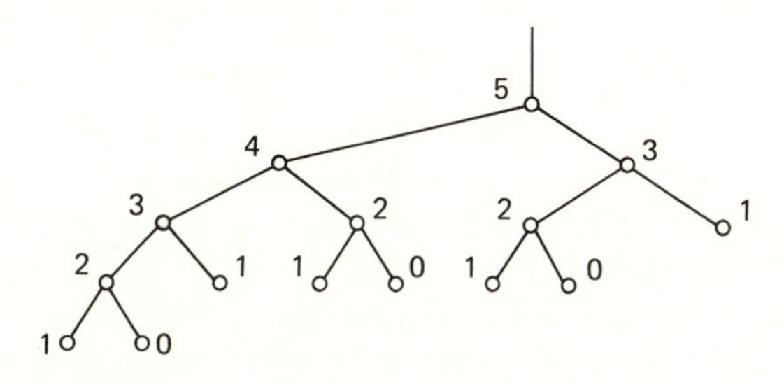

Fig. 3.2 The 15 calls of *Fib*(n) for $n = 5$.

However, it is plain that the Fibonacci numbers can be computed by an iterative scheme that avoids the recomputation of the same values by use of auxiliary variables such that $x = \text{fib}_i$ and $y = \text{fib}_{i-1}$.

```
{compute x = fib_n for n > 0}
i := 1; x := 1; y := 0;
while i < n do
    begin z := x; i := i + 1;
          x := x + y; y := z
    end                                   (3.18)
```

(Note that the three assignments to x, y, z may be expressed by merely two assignments without need for the auxiliary variable z: $x := x + y$; $y := x - y$).

Thus, the lesson to be drawn is to avoid the use of recursion when there is an *obvious* solution by iteration.

This, however, should not lead to shying away from recursion at any price. There are many good applications of recursion, as the following paragraphs and chapters will demonstrate. The fact that implementions of recursive procedures on essentially non-recursive machines exist proves that for practical purposes every recursive program can be transformed into a purely iterative one. This, however, involves the explicit handling of a recursion stack, and these operations will often obscure the essence of a program to such an extent that it becomes most difficult to comprehend. The lesson is that algorithms which by their nature are recursive rather than iterative should be formulated as recursive procedures. In order to appreciate this point, the reader is referred to Programs 2.10 and 2.11 for a comparison.

The remaining part of this chapter is devoted to the development of some recursive programs in situations in which recursion is justifiably appropriate. Also Chaps. 4 and 5 make great use of recursion in cases in which the underlying data structures let the choice of recursive solutions appear obvious and natural.

3.3. TWO EXAMPLES OF RECURSIVE PROGRAMS

The attractive graphic pattern shown in Fig. 3.5 consists of the superposition of five curves. These curves follow a regular pattern and suggest that they might be drawn by a plotter under control of a computer. Our goal is to discover the recursion schema, according to which the plotting program might be constructed. Inspection reveals that three of the superimposed curves have the shapes shown in Fig. 3.3; we denote them by H_1, H_2, and H_3. The figures show that H_{i+1} is obtained by the composition of four instances of H_i of half size and appropriate rotation and by tying together

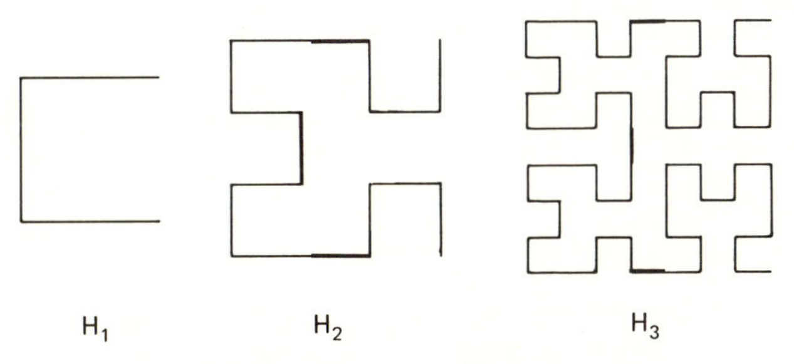

Fig. 3.3 Hilbert curves of order 1, 2, and 3.

the four H_i's by three connecting lines. Notice that H_1 may be considered as consisting of four instances of an empty H_0 connected by three straight lines. H_i is called the *Hilbert curve* of order i after its inventor D. Hilbert (1891).

Let us assume that our basic tools for plotting are two coordinate variables x and y, a procedure *setplot* (setting the pen to coordinates x and y), and a procedure *plot* (moving the drawing pen from its present position to the position indicated by x and y).

Since each curve H_i consists of four half-sized copies of H_{i-1}, it is natural to express the procedure to draw H_i as a composition of four parts, each of them drawing H_{i-1} in appropriate size and rotation. If we denote the four parts by A, B, C, and D and the routines drawing the interconnecting lines by arrows pointing in the corresponding direction, then the following *recursion scheme* emerges (see Fig. 3.3).

$$\begin{array}{ll} A: & D \leftarrow A \downarrow A \rightarrow B \\ B: & C \uparrow B \rightarrow B \downarrow A \\ C: & B \rightarrow C \uparrow C \leftarrow D \\ D: & A \downarrow D \leftarrow D \uparrow C \end{array} \tag{3.19}$$

If the length of the unit line is denoted by h, the procedure corresponding to the scheme A is readily expressed by using recursive activations of analogously designed procedures B and D and of itself.

```
procedure A(i: integer);
begin if i > 0 then
      begin D(i-1); x := x-h; plot;
            A(i-1); y := y-h; plot;
            A(i-1); x := x+h; plot;          (3.20)
            B(i-1)
      end
end
```

This procedure is initiated by the main program once for every Hilbert curve to be superimposed. The main program determines the initial point of the curve, i.e., the initial values of x and y, and the unit increment h. $h0$ denotes the full width of the page and must satisfy $h0 = 2^k$ for some $k \geq n$ (see Fig. 3.4). The entire program draws the n Hilbert curves $H_1 \ldots H_n$ (see Program 3.1 and Fig. 3.5).

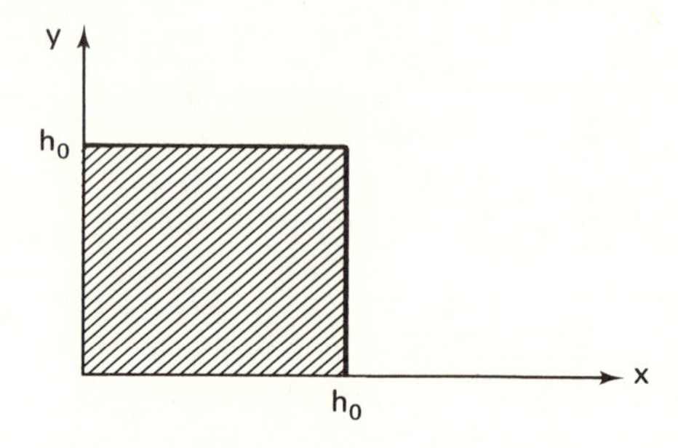

Fig. 3.4 The unit frame.

Program 3.1 Hilbert Curves.

```
program Hilbert(pf,output);
{plot Hilbert curves of orders 1 to n}
const n = 4; h0 = 512;
var i,h,x,y,x0,y0: integer;
   pf: file of integer;      {plot file}
procedure A(i: integer);
begin if i > 0 then
      begin D(i-1); x := x-h; plot;
            A(i-1); y := y-h; plot;
            A(i-1); x := x+h; plot;
            B(i-1)
      end
end ;
procedure B(i: integer);
begin if i > 0 then
      begin C(i-1); y := y+h; plot;
            B(i-1); x := x+h; plot;
            B(i-1); y := y-h; plot;
            A(i-1)
      end
end ;
procedure C(i: integer);
begin if i > 0 then
      begin B(i-1); x := x+h; plot;
            C(i-1); y := y+h; plot;
            C(i-1); x := x-h; plot;
            D(i-1)
      end
end ;
```

```
procedure D(i: integer);
begin if i > 0 then
      begin A(i-1); y := y-h; plot;
            D(i-1); x := x-h; plot;
            D(i-1); y := y+h; plot;
            C(i-1)
      end
end ;
begin startplot;
   i := 0; h := h0; x0 := h div 2; y0 := x0;
   repeat {plot Hilbert curve of order i}
      i := i+1; h := h div 2;
      x0 := x0 + (h div 2); y0 := y0 + (h div 2);
      x := x0; y := y0; setplot;
      A(i)
   until i = n;
   endplot
end .
```

Program 3.1 (Continued)

Fig. 3.5 Hilbert curves $H_1 \ldots H_5$.

A similar but slightly more complex and aesthetically more sophisticated example is shown in Fig. 3.7. This pattern is again obtained by superimposing several curves, two of which are shown in Fig. 3.6. S_i is called the *Sierpinski* curve of order i. What is the recursion scheme? One is tempted to single out

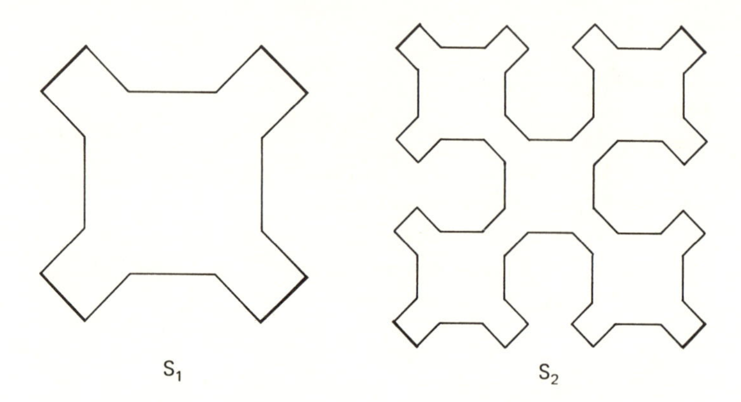

Fig. 3.6 Sierpinski curves of orders 1 and 2.

Fig. 3.7 Sierpinski curves $S_1 \ldots S_4$.

the leaf S_1 as a basic building block, possibly with one edge left off. But this does not lead to a solution. The principal difference between Sierpinski curves and Hilbert curves is that Sierpinski curves are closed (without crossovers). This implies that the basic recursion scheme must be an open curve and that the four parts are connected by links not belonging to the recursion pattern itself. Indeed, these links consist of the four straight lines in the outermost four "corners," drawn in boldface in Fig. 3.6. They may be regarded as belonging to a *non-empty* initial curve S_0, which is a square standing on one corner.

Now the recursion schema is readily established. The four constituent patterns are again denoted by *A*, *B*, *C*, and *D*, and the connecting lines are drawn explicitly. Notice that the four recursion patterns are indeed identical except for 90° rotations.

The base pattern of the Sierpinski curves is

$$S\colon A \searrow B \swarrow C \nwarrow D \nearrow \tag{3.21}$$

and the recursion patterns are

$$\begin{aligned} A&\colon A \searrow B \Rightarrow D \nearrow A \\ B&\colon B \swarrow C \Downarrow A \searrow B \\ C&\colon C \nwarrow D \Leftarrow B \swarrow C \\ D&\colon D \nearrow A \Uparrow C \nwarrow D \end{aligned} \tag{3.22}$$

(Double arrows denote lines of double unit length.)

If we use the same primitives for plot operations as in the Hilbert curve example, the above recursion scheme is transformed without difficulties into a (directly and indirectly) recursive algorithm.

```
procedure A(i: integer);
begin if i > 0 then
      begin A(i-1); x := x+h; y := y-h; plot;
            B(i-1); x := x+2*h;  plot;
            D(i-1); x := x+h; y := y+h; plot;          (3.23)
            A(i-1)
      end
end
```

This procedure is derived from the first line of the recursion scheme (3.22). Procedures corresponding to the patterns *B*, *C*, and *D* are derived analogously. The main program is composed according to the pattern (3.21). Its task is to set the initial values for the drawing coordinates and to determine the unit line length *h* according to the size of the paper, as shown in Program 3.2. The result of executing this program with $n = 4$ is shown in Fig. 3.7. Note that S_0 is not drawn.

The elegance of the use of recursion in these examples is obvious and convincing. The correctness of the programs can readily be deduced from their structure and composition patterns. Moreover, the use of the explicit level parameter *i* according to schema (3.5) guarantees termination since the

Program 3.2 Sierpinski Curves.

```
program Sierpinski (pf,output);
{plot Sierpinski curves of orders 1 to n}
const n = 4; h0 = 512;
var i,h,x,y,x0,y0: integer;
   pf: file of integer;      {plot file}
procedure A(i: integer);
begin if i > 0 then
      begin A(i-1); x := x+h; y := y-h; plot;
            B(i-1); x := x + 2*h; plot;
            D(i-1); x := x+h; y := y+h; plot;
            A(i-1)
      end
end ;
procedure B(i: integer);
begin if i > 0 then
      begin B(i-1); x := x-h; y := y-h; plot;
            C(i-1); y := y - 2*h; plot;
            A(i-1); x := x+h; y := y-h; plot;
            B(i-1)
      end
end ;
procedure C(i: integer);
begin if i > 0 then
      begin C(i-1); x := x-h; y := y+h; plot;
            D(i-1); x := x - 2*h; plot;
            B(i-1); x := x-h; y := y-h; plot;
            C(i-1)
      end
end ;
procedure D(i: integer);
begin if i > 0 then
      begin D(i-1); x := x+h; y := y+h; plot;
            A(i-1); y := y + 2*h; plot;
            C(i-1); x := x-h; y := y+h; plot;
            D(i-1)
      end
end ;
```

```
begin startplot;
   i := 0; h := h0 div 4; x0 := 2*h; y0 := 3*h;
   repeat i := i+1; x0 := x0-h;
      h := h div 2; y0 := y0+h;
      x := x0; y := y0; setplot;
      A(i); x := x+h; y := y-h; plot;
      B(i); x := x-h; y := y-h; plot;
      C(i); x := x-h; y := y+h; plot;
      D(i); x := x+h; y := y+h; plot;
   until i = n;
   endplot
end .
```

Program 3.2 (Continued)

depth of recursion cannot become greater than n. In contrast to this recursive formulation, equivalent programs that avoid the explicit use of recursion are extremely cumbersome and obscure. The reader is urged to convince himself of this claim by trying to understand the programs shown in [3-3].

3.4. BACKTRACKING ALGORITHMS

A particularly intriguing programming endeavor is the subject of "general problem solving." The task is to determine algorithms for finding solutions to specific problems not by following a fixed rule of computation, but by trial and error. The common pattern is to decompose the trial-and-error process into partial tasks. Often these tasks are most naturally expressed in recursive terms and consist of the exploration of a finite number of subtasks. We may generally view the entire process as a trial or search process that gradually builds up and scans (prunes) a tree of subtasks. In many problems this search tree grows very rapidly, usually exponentially, depending on a given parameter. The search effort increases accordingly. Frequently, the search tree can be pruned by the use of heuristics only, thereby reducing computation to tolerable bounds.

It is not our aim to discuss general heuristic rules in this text. Rather, the general principle of breaking up such problem-solving tasks into subtasks and the application of recursion is to be the subject of this chapter. We start out by demonstrating the underlying technique by using an example, namely, the well-known *knight's tour*.

Given is a $n \times n$ board with n^2 fields. A knight—being allowed to move according to the rules of chess—is placed on the field with initial coordinates x_0, y_0. The problem is to find a covering of the entire board, if there exists one, i.e., to compute a tour of $n^2 - 1$ moves such that every field of the board is visited exactly once.

The obvious way to reduce the problem of covering n^2 fields is to consider the problem of either performing a next move or finding out that none is possible. Let us therefore define an algorithm trying to perform a next move. A first approach is shown in (3.24).

```
procedure try next move;
begin initialize selection of moves;
    repeat select next candidate from list of next moves;
        if acceptable then
        begin record move;
            if board not full then
            begin try next move;                                    (3.24)
                if not successful then erase previous recording
            end
        end
    until (move was successful) ∨ (no more candidates)
end
```

If we wish to be more precise in describing this algorithm, we are forced to make some decisions on data representation. An obvious step is to represent the board by a matrix, say h. Let us also introduce a type to denote index values:

```
type index = 1 . . n;                                               (3.25)
var h: array [index, index] of integer
```

The decision to represent each field of the board by an integer instead of a Boolean value denoting occupation is because we wish to keep track of the history of successive board occupations. The following convention is an obvious choice:

$h[x, y] = 0$: field $\langle x, y \rangle$ has not been visited
$h[x, y] = i$: field $\langle x, y \rangle$ has been visited in the ith move $(1 \leq i \leq n^2)$ (3.26)

The next decision concerns the choice of appropriate parameters. They are to determine the starting conditions for the next move and also to report on its success. The former task is adequately solved by specifying the coordinates x, y from which the move is to be made and by specifying the number i of the move (for recording purposes). The latter task requires a Boolean result parameter: $q = true$ denotes success; $q = false$ failure.

Which statements can now be refined on the basis of these decisions? Certainly "board not full" can be expressed as "$i < n^2$." Moreover, if we introduce two local variables u and v to stand for the coordinates of possible move destinations determined according to the jump pattern of knights, then the predicate "acceptable" can be expressed as the logical combination of the conditions that the new field lies on the board, i.e., $1 \leq u \leq n$ and

$1 \leq v \leq n$, and that it has not been visited previously, i.e., $h[u, v] = 0$. The operation of recording the legal move is expressed by the assignment $h[u, v] := i$, and the cancellation of this recording as $h[u, v] := 0$. If a local variable $q1$ is introduced and used as the result parameter in the recursive call of this algorithm, then $q1$ may be substituted for "move was successful." Thereby we arrive at the formulation shown in (3.27).

```
procedure try (i: integer; x,y: index; var q: boolean);
var u,v: integer; q1: boolean;
begin initialize selection for moves;
      repeat let u,v be the coordinates of the next move defined
         by the rules of chess;
         if (1≤u≤n) ∧ (1≤v≤n) ∧ (h[u,v]=0) then
         begin h[u,v] := i;
               if i < sqr(n) then                                    (3.27)
               begin try(i+1,u,v,q1);
                  if ¬q1 then h[u,v] := 0
               end else q1 := true
         end
      until q1 ∨ (no more candidates);
      q := q1
end
```

Just one more refinement step will lead us to a program expressed fully in terms of our basic programming notation. We should note that so far the program was developed completely independently of the laws governing the jumps of the knight. This delaying of considerations of particularities of the problem was quite deliberate. But now is the time to take them into account.

Given a starting coordinate pair $\langle x, y \rangle$, there are eight potential candidates for coordinates $\langle u, v \rangle$ of the destination. They are numbered 1 to 8 in Fig. 3.8.

A simple method of obtaining u, v from x, y is by addition of the coordinate differences stored in either an array of difference pairs or in two arrays

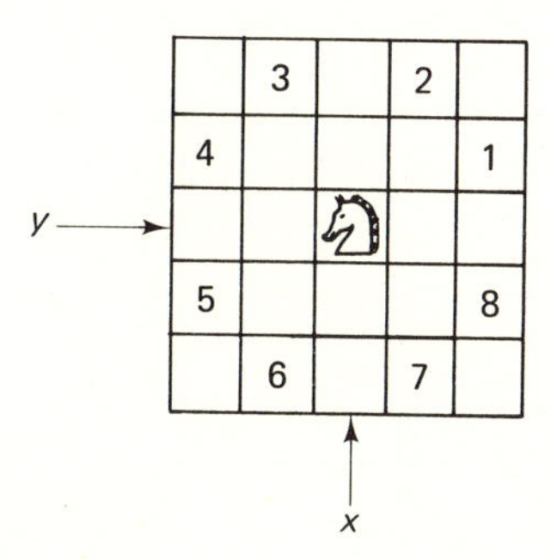

Fig. 3.8 The eight possible moves of a knight.

```
program knightstour (output);
const n = 5; nsq = 25;
type index = 1 .. n;
var i,j: index;
    q: boolean;
    s: set of index;
    a,b: array [1. .8] of integer;
    h: array [index, index] of integer;
procedure try (i: integer; x,y: index; var q: boolean);
    var k,u,v: integer; q1: boolean;
begin k := 0;
    repeat k := k+1; q1 := false;
        u := x + a[k]; v := y + b[k];
        if (u in s)∧(v in s) then
        if h[u,v] = 0 then
        begin h[u,v] := i;
            if i < nsq then
              begin try (i+1,u,v,q1);
                 if ¬q1 then h[u,v] := 0
              end else q1 := true
        end
    until q1 ∨ (k=8);
    q := q1
end {try} ;
begin s := [1,2,3,4,5];
    a[1] :=  2; b[1] :=  1;
    a[2] :=  1; b[2] :=  2;
    a[3] := −1; b[3] :=  2;
    a[4] := −2; b[4] :=  1;
    a[5] := −2; b[5] := −1;
    a[6] := −1; b[6] := −2;
    a[7] :=  1; b[7] := −2;
    a[8] :=  2; b[8] := −1;
    for i := 1 to n do
        for j := 1 to n do h[i,j] := 0;
    h[1,1] := 1; try(2,1,1,q);
    if q then
        for i := 1 to n do
        begin for j := 1 to n do write(h[i,j]:5);
            writeln
        end
    else writeln(' NO SOLUTION ')
end .
```

Program 3.3 Knight's Tour.

of single differences. Let these arrays be denoted by a and b, appropriately initialized. Then an index k may be used to number the "next candidate." The details are shown in Program 3.3. The recursive procedure is initiated by a call with the coordinates x_0, y_0 of that field as parameters, from which the tour is to start. This field must be given the value 1; all others are to be marked free.

$$H[x_0, y_0] := 1; \qquad \text{try}\ (2, x_0, y_0, q)$$

One further detail must not be overlooked: A variable $h[u, v]$ does exist only if both u and v lie within the array bounds $1 \ldots n$. Consequently, the expression in (3.27), substituted for "acceptable" in (3.24), is valid only if its first two constituent terms are true. A proper reformulation is shown in Program 3.3 in which, furthermore, the double relation $1 \le u \le n$ is replaced by the expression u **in** $[1, 2, \ldots, n]$, which for sufficiently small n is possibly more efficient (see Sect. 1.10.3). Table 3.1 indicates solutions obtained with initial positions $\langle 1, 1\rangle$, $\langle 3, 3\rangle$ for $n = 5$ and $\langle 1, 1\rangle$ for $n = 6$.

It is possible to replace the result parameter q and the local variable $q1$ by a global variable, thereby simplifying the program somewhat.

1	6	15	10	21
14	9	20	5	16
19	2	7	22	11
8	13	24	17	4
25	18	3	12	23

23	10	15	4	25
16	5	24	9	14
11	22	1	18	3
6	17	20	13	8
21	12	7	2	19

1	16	7	26	11	14
34	25	12	15	6	27
17	2	33	8	13	10
32	35	24	21	28	5
23	18	3	30	9	20
36	31	22	19	4	29

Table 3.1 Three Knights' Tours.

What abstractions can now be made from this example? Which pattern does it exhibit that is typical for this kind of "problem-solving" algorithms? What does it teach us? The characteristic feature is that steps toward the total solution are attempted and recorded that may later be taken back and erased in the recordings when it is discovered that the step does not possibly lead to the total solution, that the step leads into a "dead-end street." This action is called *backtracking*. The general pattern (3.28) is derived from (3.24), assuming that the number of potential candidates in each step is finite.

```
procedure try;
begin intialize selection of candidates;
    repeat select next;
        if acceptable then
        begin record it;
            if solution incomplete then
            begin try next step;                                (3.28)
                if not successful then cancel recording
            end
        end
    until successful ∨ no more candidates
end
```

Actual programs may, of course, assume various derivative forms of schema (3.28). A frequently encountered pattern uses an explicit level parameter indicating the depth of recursion and allowing for a simple termination condition.

If, moreover, at each step the number of candidates to be investigated is fixed, say m, then the derived schema (3.29) applies; it is to be invoked by the statement "try(1)".

```
procedure try(i: integer);
    var k: integer;
begin k := 0;
    repeat k := k+1; select k-th candidate;
        if acceptable then
        begin record it;
            if i < n then                                       (3.29)
            begin try(i+1);
                if not successful then cancel recording
            end
        end
    until successful ∨ (k = m)
end
```

The remainder of this chapter is devoted to the treatment of three more examples. They display various incarnations of the abstract schema (3.29) and are included as further illustrations of the appropriate use of recursion.

3.5. THE EIGHT QUEENS PROBLEM

The problem of the eight queens is a well-known example of the use of trial-and-error methods and of backtracking algorithms. It was investigated by C. F. Gauss in 1850, but he did not completely solve it. This should not surprise anyone. After all, the characteristic property of these problems is that they defy analytic solution. Instead, they require large amounts of exacting labor, patience, and accuracy. Such algorithms have therefore gained relevance almost exclusively through the automatic computer, which possesses these properties to a much higher degree than people, and even geniuses, do.

The eight queens problem is stated as follows (see also [3-4]): Eight queens are to be placed on a chess board in such a way that no queen checks against any other queen.

Using the schema of Eq. (3.29) as a template, we readily obtain the following crude version of a solution:

```
procedure try(i: integer);
begin
    initialize selection of positions for i-th queen;
    repeat make next selection;
        if safe then
        begin setqueen;
            if i < 8 then                                        (3.30)
            begin try(i+1);
                if not successful then remove queen
            end
        end
    until successful ∨ no more positions
end
```

In order to proceed, it is necessary to make some commitments concerning the data representation. Since we know from the rules of chess that a queen checks all other figures lying in either the same column, row, or diagonals on the board, we infer that each column may contain one and only one queen, and that the choice of a position for the ith queen may be restricted to the ith column. The parameter i therefore becomes the column index, and the selection process for positions then ranges over the eight possible values for a row index j.

There remains the question of representing the eight queens on the board. An obvious choice would again be a square matrix to represent the board, but a little inspection reveals that such a representation would lead to fairly cumbersome operations for checking the availability of positions. This is highly undesirable since it is the most frequently executed operation. We should therefore choose a data representation which makes checking as simple as possible. The best recipe is to represent as directly as possible that information which is truly relevant and most often used. In our case this is not the position of the queens, but whether or not a queen has already been placed along each row and diagonals. (We already know that exactly one is placed in each column k for $1 \leq k \leq i$). This leads to the following choice of variables:

```
var x : array [1 .. 8] of integer;
    a : array [1 .. 8] of Boolean;
    b : array [b1 .. b2] of Boolean;                    (3.31)
    c : array [c1 .. c2] of Boolean;
```

where

$x[i]$ denotes the position of the queen in the ith column;
$a[j]$ means no queen lies in the jth row;
$b[k]$ means no queen occupies the kth ↙-diagonal;
$c[k]$ means no queen sits on the kth ↘-diagonal.

The choice for index bounds $b1, b2, c1, c2$ is dictated by the way that indices of b and c are computed; we note that in a ↙-diagonal all fields have the same sums of their coordinates i and j, and that in a ↘-diagonal the coordinate differences $i - j$ are constant. The appropriate solution is shown in Program 3.4.

Given these data, the statement *setqueen* is elaborated to

$$x[i] := j;\; a[j] := false;\; b[i+j] := false;\; c[i-j] := false \qquad (3.32)$$

The statement *removequeen* is refined into

$$a[j] := true;\quad b[i+j] := true;\quad c[i-j] := true \qquad (3.33)$$

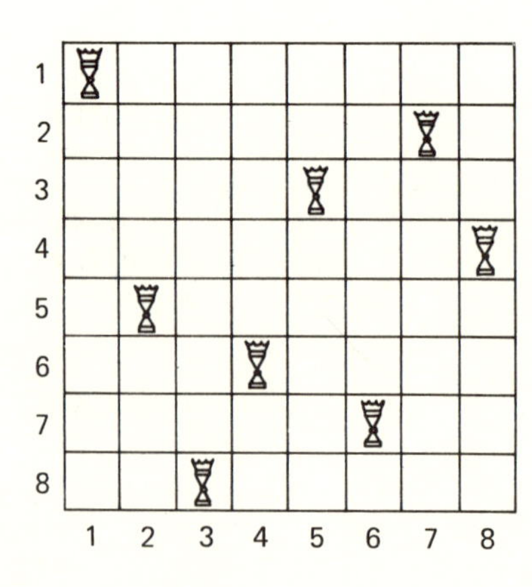

Fig. 3.9 A solution to the eight queens problem.

and the condition *safe* is fulfilled if the field $\langle i, j \rangle$ lies in a row and in diagonals which are still free (represented by *true*). Hence, it can be expressed by the logical expression

$$a[j] \wedge b[i+j] \wedge c[i-j] \tag{3.34}$$

This completes the development of this algorithm, which is shown in full as Program 3.4. The computed solution is $x = (1, 5, 8, 6, 3, 7, 2, 4)$ and is shown in Fig. 3.9.

```
program eightqueen1(output);
{find one solution to eight queens problem}
var i: integer; q: boolean;
    a: array [ 1 ..  8] of boolean;
    b: array [ 2 .. 16] of boolean;
    c: array [-7 ..  7] of boolean;
    x: array [ 1 ..  8] of integer;

procedure try(i: integer; var q: boolean);
    var j: integer;
begin j := 0;
    repeat j := j+1; q := false;
        if a[j] ∧b[i+j] ∧c[i-j] then
        begin x[i] := j;
            a[j] := false; b[i+j] := false; c[i-j] := false;
            if i < 8 then
            begin try (i+1,q);
                if ¬q then
                begin a[j] := true; b[i+j] := true; c[i-j] := true
                end
            end else q := true
        end
    until q ∨ (j=8)
end {try} ;

begin
    for i :=  1 to  8 do a[i] := true;
    for i :=  2 to 16 do b[i] := true;
    for i := -7 to  7 do c[i] := true;
    try (1,q);
    if q then
        for i := 1 to 8 do write (x[i]: 4);
    writeln
end .
```

Program 3.4 Eight Queens.

Before we abandon the context of the chess board, the eight queens example is to serve as an illustration of an important extension of the trial-and-error algorithm. The extension is—in general terms—to find not only one, but all solutions to a posed problem.

The extension is easily accommodated. We are to recall the fact that the generation of candidates must progress in a systematic manner which guarantees that no candidate is generated more than once. This property of the algorithm corresponds to a search of the candidate tree in a systematic fashion in which every node is visited exactly once. It allows—once a solution is found and duly recorded—merely to proceed to the next candidate delivered by the systematic selection process. The general schema is derived from (3.29) and shown in (3.35)

```
procedure try(i: integer);
    var k: integer;
begin
    for k := 1 to m do
    begin select k-th candidate;
        if acceptable then
        begin record it;                                        (3.35)
            if i < n then try(i+1) else print solution;
            cancel recording
        end
    end
end
```

Note that because of the simplification of the termination condition of the selection process to the single term $k = m$, the repeat statement is appropriately replaced by a for statement. It comes as a surprise that the search for *all* possible solutions is realized by a simpler program than the search for a single solution.

The extended algorithm to determine all 92 solutions of the eight queens problem is shown in Program 3.5. Actually, there are only 12 significantly differing solutions; our program does not recognize symmetries. The 12

Program 3.5 Eight Queens.

```
program eightqueens (output);
var i: integer;
    a: array [ 1 .. 8] of boolean;
    b: array [ 2 .. 16] of boolean;
    c: array [-7 .. 7] of boolean;
    x: array [ 1 .. 8] of integer;
```

```
procedure print;
   var k: integer;
begin for k := 1 to 8 do write(x[k]: 4);
   writeln
end {print} ;

procedure try(i: integer);
   var j: integer;
begin
   for j := 1 to 8 do
      if a[j] ∧ b[i+j] ∧ c[i-j] then
      begin x[i] := j;
         a[j] := false; b[i+j] := false; c[i-j] := false;
         if i < 8 then try(i+1) else print;
         a[j] := true; b[i+j] := true; c[i-j] := true
      end
end {try} ;

begin
   for i :=  1 to  8 do a[i] := true;
   for i :=  2 to 16 do b[i] := true;
   for i := -7 to  7 do c[i] := true;
   try (1)
end .
```

Program 3.5 (Continued)

solutions generated first are listed in Table 3.2. The numbers N to the right indicate the frequency of execution of the test for safe fields. Its average over all 92 solutions is 161.

x_1	x_2	x_3	x_4	x_5	x_6	x_7	x_8	N
1	5	8	6	3	7	2	4	876
1	6	8	3	7	4	2	5	264
1	7	4	6	8	2	5	3	200
1	7	5	8	2	4	6	3	136
2	4	6	8	3	1	7	5	504
2	5	7	1	3	8	6	4	400
2	5	7	4	1	8	6	3	72
2	6	1	7	4	8	3	5	280
2	6	8	3	1	4	7	5	240
2	7	3	6	8	5	1	4	264
2	7	5	8	1	4	6	3	160
2	8	6	1	3	5	7	4	336

Table 3.2 Twelve Solutions to the Eight Queens Problem.

3.6. THE STABLE MARRIAGE PROBLEM

Assume that two disjoint sets A and B of equal cardinality n are given. Find a set of n pairs $\langle a, b\rangle$ such that a **in** A and b **in** B satisfy some constraints. Many different criteria for such pairs exist; one of them is the rule called the "stable marriage rule."

Assume that A is a set of men and B is a set of women. Each man and each woman has stated distinct preferences for their partners. If the n couples are chosen such that there exists a man and a woman who are not married, but who would both prefer each other to their actual marriage partners, then the assignment is said to be unstable. If no such pair exists, it is called *stable.*

This situation characterizes many related problems in which assignments have to be made according to preferences, such as, for example, the choice of a school by students, the choice of recruits by different branches of the armed services, etc. The example of marriages is particularly intuitive; note, however, that the stated list of preferences is invariant and does not change after a particular assignment has been made. This rule simplifies the problem, but it also represents a distortion of reality (called abstraction).

One way to search for a solution is to try pairing off members of the two sets one after the other until the two sets are exhausted. Setting out to find *all* stable assignments, we can readily sketch a solution by using the program schema (3.35) as a template. Let *try*(m) denote the algorithm to find a partner for man m, and let this search proceed in the order of the man's list of stated preferences. The first version based on these assumptions is (3.36)

```
procedure try(m: man);
    var r: rank;
begin
    for r := 1 to n do
    begin pick the r-th preference of man m;
        if acceptable then
        begin record the marriage;                         (3.36)
            if m is not last man then try(succ(m))
                else record the stable set;
            cancel the marriage
        end
    end
end
```

Again, we have arrived at the point where we cannot proceed without first making some decisions about data representation. We introduce three scalar types, and, for reasons of simplicity, let their values be the integers 1 to n. Although the three types are formally identical, their designation by distinct names contributes significantly to clarity. In particular, it can more easily be made evident what a variable stands for.

$$\begin{aligned} \textbf{type } &man &= 1\,..\,n;\\ &woman &= 1\,..\,n;\\ &rank &= 1\,..\,n \end{aligned} \qquad (3.37)$$

The initial data are represented by two matrices that indicate the men's and women's preferences.

$$\begin{aligned} \textbf{var } wmr&: \textbf{array } [man, rank] \textbf{ of } woman\\ mwr&: \textbf{array } [woman, rank] \textbf{ of } man \end{aligned} \qquad (3.38)$$

Accordingly, $wmr[m]$ denotes the preference list of man m, i.e., $wmr[m][r] = wmr[m, r]$ is the woman who occupies the rth rank in the list of man m. Similarly, $mwr[w]$ is the preference list of woman w, and $mwr[w, r]$ is her rth choice.

Rank	1	2	3	4	5	6	7	8
Man 1 selects woman	7	2	6	5	1	3	8	4
2	4	3	2	6	8	1	7	5
3	3	2	4	1	8	5	7	6
4	3	8	4	2	5	6	7	1
5	8	3	4	5	6	1	7	2
6	8	7	5	2	4	3	1	6
7	2	4	6	3	1	7	5	8
8	6	1	4	2	7	5	3	8
Woman 1 selects man	4	6	2	5	8	1	3	7
2	8	5	3	1	6	7	4	2
3	6	8	1	2	3	4	7	5
4	3	2	4	7	6	8	5	1
5	6	3	1	4	5	7	2	8
6	2	1	3	8	7	4	6	5
7	3	5	7	2	4	1	8	6
8	7	2	8	4	5	6	3	1

Table 3.3 Sample Input Data for the Stable Marriage Program.

The result is represented by an array of women x, such that $x[m]$ denotes the partner of man m. In order to maintain symmetry—also called "equal rights"—between men and women, an additional array y is introduced, such that $y[w]$ denotes the partner of woman w.

$$\begin{aligned} \textbf{var } x&: \textbf{array } [man] \textbf{ of } woman;\\ y&: \textbf{array } [woman] \textbf{ of } man \end{aligned} \qquad (3.39)$$

It is plain that y is not strictly necessary since it represents information that is already present through the existence of x. In fact, the relations

$$x[y[w]] = w, \qquad y[x[m]] = m \qquad (3.40)$$

hold for all m, w who are married. Thus, the value $y[w]$ could be determined by a simple search of x; the array y, however, clearly improves the efficiency of the algorithm. The information represented by x and y is needed to determine stability of a proposed set of marriages. Since this set is constructed stepwise by marrying individuals and testing stability after each proposed marriage, x and y are needed even before all their components are defined. In order to keep track of defined components, we may introduce Boolean arrays

```
singlem : array [man] of boolean
singlew : array [woman] of boolean                    (3.41)
```

with the meaning that

$$\neg singlem[m] \quad \textit{implies that} \quad x[m] \quad \textit{is defined,}$$
$$\neg singlew[w] \quad \textit{implies that} \quad y[w] \quad \textit{is defined.}$$

An inspection of the proposed algorithm, however, quickly reveals that the marital status of a man is determined by the value m in a simple manner, namely that

$$\neg singlem[k] \equiv k < m \tag{3.42}$$

This suggests that the array *singlem* be omitted; accordingly, we will simplify the name *singlew* to *single*.

These conventions lead to the refinement shown in (3.43). The predicate *acceptable* can be refined into the conjunction of *single* and *stable*, where *stable* is still a function to be further elaborated.

```
procedure try(m: man);
    var r: rank;  w: woman;
begin for r := 1 to n do
    begin w := wmr[m,r];
        if single[w] ∧ stable then
        begin x[m] := w;  y[w] := m;  single[w] := false;      (3.43)
            if m < n then try(succ(m))
                else record stable set;
            single[w] := true
        end
    end
end
```

At this point, the strong similarity of this solution with Program 3.5 is still noticeable.

The crucial task is now the refinement of the algorithm to determine stability. Unfortunately, it is not possible to represent stability by such a simple expression as the safety of a queen's position in Program 3.5. The first detail which should be kept in mind is that stability follows by definition from comparisons of preferences or ranks. The ranks of men or women, however, are nowhere explicitly available in our collection of data established so far. Surely, the rank of woman w in the mind of man m can be computed, but only by a costly search of w in $wmr[m]$.

Since the computation of stability is a very frequent operation, it is advisable to make this information more directly accessible. To this end, we introduce the two matrices

$$\begin{aligned} &rmw : \textbf{array } [man, woman] \textbf{ of } rank; \\ &rwm : \textbf{array } [woman, man] \textbf{ of } rank \end{aligned} \tag{3.44}$$

such that $rmw[m, w]$ denotes woman w's rank in the preference list of man m, and $rwm[w, m]$ denotes the tank of man m in the list of w. It is plain that the values of these auxiliary arrays are constant and can initially be determined from the values of *wmr* and *mwr*.

The process of determining the predicate *stable* now proceeds strictly according to its original definition. Recall that we are trying the feasibility of marrying m and w, where $w = wmr[m, r]$, i.e., w has rank r in m's list of preferences. Being optimistic, we first presume that stability still prevails, and then we set out to find possible sources of trouble. Where could they be hidden? There are two symmetrical possibilities:

1. There might be a woman pw, preferred to w by m, who herself prefers m to her husband.
2. There might be a man pm, preferred to m by w, who himself prefers w to his wife.

Pursuing trouble source 1, we compare ranks $rwm[pw, m]$ and $rwm[pw, y[pw]]$ for all women preferred to w by m, i.e., for all $pw = wmr[m, i]$ such that $i < r$. We happen to know that all these candidate women are already married because, were anyone of them still single, m would have picked her beforehand. The described process can be formulated by a simple linear search; s denotes stability.

```
s := true;  i := 1;
while (i < r) ∧ s do
   begin pw := wmr[m,i];  i := i+1;                          (3.45)
      if ¬single[pw] then s := rwm[pw, m] < rwm[pw, y[pw]]
   end
```

Following trouble source 2, we must investigate all candidates pm who are preferred by w to their current assignation m, i.e., the investigators are all preferred men $pm = mwr[w, i]$ such that $i < rwm[w, m]$. In analogy to tracing trouble source 1, comparison between ranks $rmw[pm, w]$ and $rmw[pm, x[pm]]$ will be necessary. We must be careful, however, to omit comparisons involving $x[pm]$ where pm is still single. The necessary safeguard is a test $pm < m$, since we know that all men preceding m are already married.

Program 3.6 Stable Marriages.

```
program marriage (input,output);
{problem of the stable marriages}
const n = 8;
type man = 1 .. n; woman = 1 .. n; rank = 1 .. n;
var m: man; w: woman; r: rank;
    wmr: array [man, rank] of woman;
    mwr: array [woman, rank] of man;
    rmw: array [man, woman] of rank;
    rwm: array [woman, man] of rank;
    x:   array [man] of woman;
    y:   array [woman] of man;
    single: array [woman] of boolean;

procedure print;
    var m: man; rm, rw: integer;
begin rm := 0; rw := 0;
    for m := 1 to n do
    begin write (x[m]:4);
        rm := rm + rmw[m,x[m]]; rw := rw + rwm[x[m],m]
    end ;
    writeln (rm:8,rw:4);
end {print} ;

procedure try(m: man);
    var r: rank; w: woman;

    function stable: boolean;
        var pm: man; pw: woman;
            i, lim: rank;  s: boolean;
    begin s := true; i := 1;
        while (i<r) ∧ s do
        begin pw := wmr[m,i]; i := i+1;
            if ¬single[pw] then s := rwm[pw,m] < rwm[pw,y[pw]]
        end ;
        i := 1; lim := rwm[w,m];
        while (i<lim) ∧ s do
        begin pm := mwr[w,i]; i := i+1;
            if pm < m then s := rmw[pm,w] > rmw[pm,x[pm]]
        end ;
        stable := s
    end {stable} ;
```

```
begin {try}
    for r := 1 to n do
    begin w := wmr[m,r];
        if single[w] then
            if stable then
            begin x[m] := w; y[w] := m; single[w] := false;
                if m < n then try(succ(m)) else print;
                single[w] := true
            end
    end
end {try} ;
begin {main program}
    for m := 1 to n do
        for r := 1 to n do
        begin read(wmr[m,r]); rmw[m,wmr[m,r]] := r
        end ;
    for w := 1 to n do
        for r := 1 to n do
        begin read(mwr[w,r]); rwm[w,mwr[w,r]] := r
        end ;
    for w := 1 to n do single[w] := true;
    try (1)
end .
```

Program 3.6 (Continued)

The complete algorithm is shown in Program 3.6. Table 3.3 is a set of input data representing arrays *wmr* and *mwr*. Finally, Table 3.4 gives the nine computed stable solutions.

This algorithm is notably based on a straightforward backtracking scheme. Its efficiency primarily depends on the sophistication of the solution tree pruning scheme. A somewhat faster, but more complex and less transparent algorithm has been presented by McVitie and Wilson [3-1 and 3-2], who also have extended it to the case of sets (of men and women) of unequal size.

Algorithms of the kind of the last two examples, which generate all possible solutions to a problem (given certain constraints), are often used to select one or several of the solutions which are optimal in some sense. In the present example, one might, for instance, be interested in the solution which on the average best satisfies the men—or the women—or all persons.

Notice that Table 3.4 indicates the sums of the ranks of all women in the preference lists of their husbands, and the sums of the ranks of all men in

	x_1	x_2	x_3	x_4	x_5	x_6	x_7	x_8	rm	rw	c*
Solution 1	7	4	3	8	1	5	2	6	16	32	21
2	2	4	3	8	1	5	7	6	22	27	449
3	2	4	3	1	7	5	8	6	31	20	59
4	6	4	3	8	1	5	7	2	26	22	62
5	6	4	3	1	7	5	8	2	35	15	47
6	6	3	4	8	1	5	7	2	29	20	143
7	6	3	4	1	7	5	8	2	38	13	47
8	3	6	4	8	1	5	7	2	34	18	758
9	3	6	4	1	7	5	8	2	43	11	34

*c = number of evaluations of stability.
Solution 1 = male optimal solution.
Solution 9 = female optimal solution.

Table 3.4 Result of the Stable Marriage Problem.

the preference lists of their wives. These are the values

$$rm = \sum_{m=1}^{n} rmw[m, x[m]], \qquad rw = \sum_{m=1}^{n} rwm[x[m], m] \tag{3.46}$$

The solution with the least value *rm* is called the male-optimal stable solution; the one with the smallest *rw* is the female-optimal stable solution. It lies in the nature of the chosen search strategy that good solutions from the men's point of view are generated first and that the good solutions from the women's perspective appear toward the end. In this sense, the algorithm is biased toward the male population. This can quickly be changed by systematically interchanging the role of men and women, i.e., by merely interchanging *mwr* with *wmr* and interchanging *rmw* with *rwm*.

We refrain from extending this program further and leave the incorporation of a search for an optimal solution to the next and last example of a backtracking algorithm.

3.7. THE OPTIMAL SELECTION PROBLEM

The last example of a backtracking algorithm is a logical extension of the previous two examples represented by the general schema (3.35). First we were using the principle of backtracking to find a *single* solution to a given problem. This was exemplified by the knight's tour and the eight queens. Then we tackled the goal of finding *all* solutions to a given problem; the examples were those of the eight queens and the stable marriages. Now we wish to find *the optimal solution.*

To this end, it is necessary to generate all possible solutions, and in the course of generating them to retain the one which is optimal in some specific sense. Assuming that optimality is defined in terms of some positive valued function $f(s)$, the algorithm is derived from schema (3.35) by replacing the statement *print solution* by the statement

$$\textbf{if } f(solution) > f(optimum) \textbf{ then } optimum := solution \tag{3.47}$$

The variable *optimum* records the best solution so far encountered. Naturally, it has to be properly initialized; moreover, it is customary to record the value $f(optimum)$ by another variable in order to avoid its frequent recomputation.

An example of the general problem of finding an optimal solution to a given problem follows: We choose the important and frequently encountered problem of finding an *optimal selection* out of a given set of objects subject to constraints. Selections that constitute acceptable solutions are gradually built up by investigating individual objects from the base set. A procedure *try* describes the process of investigating the suitability of one individual object, and it is called recursively (to investigate the next object) until all objects have been considered.

We note that the consideration of each object (called candidates in previous examples) has two possible outcomes, namely, either the *inclusion* of the investigated object in the current selection or its *exclusion*. This makes the use of a **repeat** or **for** statement inappropriate; instead, the two cases may as well be explicitly written out. This is shown in (3.48) (assume that the objects are numbered $1, 2, \ldots, n$).

```
procedure try(i: integer);
begin
1: if inclusion is acceptable then
   begin include i-th object;
      if i < n then try(i+1) else check optimality;      (3.48)
      eliminate i-th object
   end;
2: if exclusion is acceptable then
         if i < n then try(i+1) else check optimality
end
```

From this pattern it is evident that there are 2^n possible sets; clearly, appropriate acceptability criteria must be employed to reduce the number of investigated candidates very drastically. In order to elucidate this process, let us choose a concrete example for a selection problem: Let each of the n objects $a_1, \ldots, a_n$ be characterized by its weight w and its value v. Let the optimal set be the one with the largest sum of the values of its components, and let the constraint be a limit on the sum of their weight. This is a problem well-known to all travellers who pack suitcases by selecting from n items in such a way that their total value is optimal and that their total weight does not exceed a specific allowance.

We are now in a position to decide upon the representation of the given facts in terms of data. The choices of (3.49) are easily derived from the foregoing developments.

```
type index  = 1 .. n;
     object = record w,v: integer end
var a: array [index] of object;                    (3.49)
    limw, totv, maxv: integer;
    s, opts: set of index
```

The variables *limw* and *totv* denote the weight limit and the total value of all n objects. These two values are actually constant during the entire selection process. s represents the current selection of objects in which each object is represented by its name (index). *opts* is the optimal selection so far encountered, and *maxv* is its value.

Which are now the criteria for acceptability of an object for the current selection? If we consider *inclusion*, then an object is selectable if it fits into the weight allowance. If it does not fit, we may stop trying to add further objects to the current selection. If, however, we consider *exclusion*, then the criterion for acceptability, i.e., for the continuation of building up the current selection, is that the total value which is still achievable after this exclusion is not less than the value of the optimum so far encountered. For, if it is less, continuation of the search, although it may produce some solutions, will not yield the optimal solution. Hence any further search on the current path is fruitless. From these two conditions we determine the relevant quantities to be computed for each step in the selection process:

1. The total weight tw of the selection s so far made.
2. The still achievable value av of the current selection s.

These two entities are appropriately represented as parameters of the procedure *try*.

The condition *inclusion is acceptable* in (3.48) can now be formulated as

$$tw + a[i].w \leq limw \tag{3.50}$$

and the subsequent check for optimality as

```
if av > maxv then
    begin {new optimum, record it}
        opts := s;  maxv := av                    (3.51)
    end
```

The last assignment is based on the reasoning that the achievable value is the achieved value, once all n objects have been dealt with.

The condition *exclusion is acceptable* in (3.48) is expressed by

$$av - a[i].v > maxv \tag{3.52}$$

Since it is used again thereafter, the value $av - a[i].v$ is given the name *av*1 in order to circumvent its re-evaluation.

The entire program now follows from (3.48) through (3.52) with the addition of appropriate initialization statements for the global variables. The ease of expressing inclusion and exclusion from the set s by use of the set operators is noteworthy. The results of execution of Program 3.7 with weight allowances ranging from 10 to 120 are listed in Table 3.5.

Program 3.7 Optimal Selection.

```
program selection (input,output);
{find optimal selection of objects under constraint}
const n = 10;
type index = 1 .. n;
    object = record v,w: integer end;
var i: index;
    a: array [index] of object;
    limw, totv, maxv: integer;
    w1, w2, w3: integer;
    s, opts: set of index;
    z: array [boolean] of char;

procedure try(i: index; tw,av: integer);
    var av1: integer;
begin {try inclusion of object i}
    if tw + a[i] .w ≤ limw then
    begin s := s + [i];
        if i < n then try(i+1, tw+a[i].w, av) else
            if av > maxv then
            begin maxv := av; opts := s
            end ;
        s := s - [i]
    end ;
```

```
        {try exclusion of object i}  av1 := av - a[i] .v;
        if av1 > maxv then
        begin if i < n then try(i+1, tw, av1) else
                  begin maxv := av1; opts := s
                  end
        end
    end {try} ;

    begin totv := 0;
        for i := 1 to n do
            with a[i] do
            begin read(w,v); totv := totv + v
            end ;
        read(w1,w2,w3);
        z[true] := '*'; z[false] := ' ';
        write(' WEIGHT    ');
        for i := 1 to n do write (a[i] .w: 4);
        writeln; write (' VALUE     ');
        for i := 1 to n do write (a[i] .v: 4);
        writeln;
        repeat limw := w1; maxv := 0; s := [ ]; opts := [ ];
            try(1,0,totv);
            write(limw);
            for i := 1 to n do write('     ', z[i in opts]);
            writeln; w1 := w1 + w2
        until w1 > w3
    end .
```

Program 3.7 (Continued)

Weight	10	11	12	13	14	15	16	17	18	19
Value	18	20	17	19	25	21	27	23	25	24
10	*									
20							*			
30					*		*			
40	*				*		*			
50	*	*		*			*			
60	*	*	*	*	*					
70	*	*			*		*		*	
80	*	*	*		*		*	*		
90	*	*			*		*		*	*
100	*	*		*	*		*	*	*	
110	*	*	*	*	*	*	*		*	
120	*	*			*	*	*	*	*	*

Table 3.5 Sample Output from Optimal Selection Program.

This backtracking scheme with a limitation factor curtailing the growth of the potential search tree, is also known as *branch and bound algorithm*.

EXERCISES

3.1. (Towers of Hanoi). Given are three rods and n disks of different sizes. The disks can be stacked up on the rods, thereby forming "towers." Let the n disks initially be placed on rod A in the order of decreasing size, as shown in Fig. 3.10 for $n = 3$. The task is to move the n disks from rod A to rod C

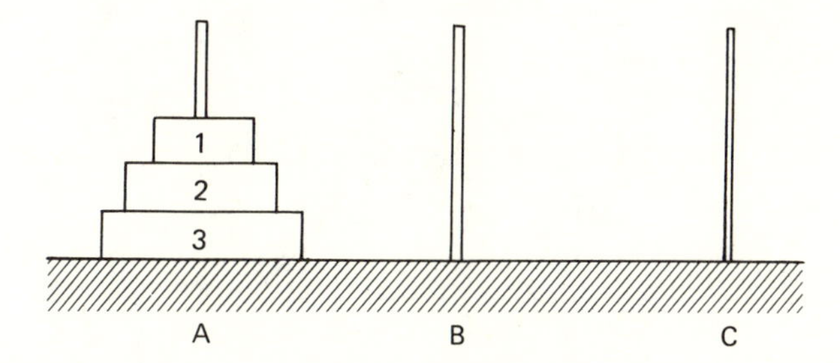

Fig. 3.10 Towers of Hanoi.

such that they are ordered in the original way. This has to be achieved under the constraints that

1. In each step exactly one disk is moved from one rod to another rod.
2. A disk may never be placed on top of a smaller disk.
3. Rod B may be used as an auxiliary store.

Find an algorithm which performs this task. Note that a tower may conveniently be considered as consisting of the single disk at the top and the tower consisting of the remaining disks. Describe the algorithm as a recursive program.

3.2. Write a procedure that generates all $n!$ permutations of n elements $a_1, \ldots, a_n$ *in situ*, i.e., without the aid of another array. Upon generating the next permutation, a parametric procedure Q is to be called which may, for instance, output the generated permutation.

Hint: Consider the task of generating all permutations of the elements $a_1, \ldots, a_m$ as consisting of the m subtasks of generating all permutations of $a_1, \ldots, a_{m-1}$ followed by a_m, where in the ith subtask the two elements a_i and a_m had initially been interchanged.

3.3. Deduce the recursion scheme of Fig. 3.11 which is a superposition of the four curves W_1, W_2, W_3, W_4. The structure is similar to that of the Sierpinski curves (3.21) and (3.22). From the recursion pattern, derive a recursive program that plots these curves.

Fig. 3.11 *W*-curves of order 1 through 4.

3.4. Only 12 of the 92 solutions computed by the Eight Queens Program 3.5 are essentially different. The other ones can be derived by reflections about axes or the center point. Devise a program that determines the 12 principal solutions. Note that, for example, the search in column 1 may be restricted to positions 1-4.

3.5. Change the Stable Marriage Program so that it determines the optimal solution (male or female). It therefore becomes a "branch and bound" program of the type represented by Program 3.7.

3.6. A certain railway company serves n stations $S_1, \ldots, S_n$. It intends to improve its customer information service by computerized information terminals. A customer types in his departure station S_A and his destination S_D, and he is supposed to be (immediately) given the schedule of the train connections with minimum total time of the journey.

Devise a program to compute the desired information. Assume that the timetable (which is your data bank) is provided in a suitable data structure containing departure (= arrival) times of all available trains. Naturally, not all stations are connected by direct lines (see also Exercise 1.8).

3.7. The Ackermann Function A is defined for all non-negative integer arguments m and n as follows:

$$A(0, n) = n + 1$$
$$A(m, 0) = A(m - 1, 1) \qquad (m > 0)$$
$$A(m, n) = A(m - 1, A(m, n - 1)) \qquad (m, n > 0)$$

Design a program that computes $A(m, n)$ without the use of recursion.

As a guideline, use Program 2.11, the non-recursive version of Quicksort. Devise a set of rules for the transformation of recursive into iterative programs in general.

REFERENCES

3-1. McVitie, D. G. and Wilson, L. B., "The Stable Marriage Problem," *Comm. ACM*, **14**, No. 7 (1971), 486–92.

3-2. ______ ______, "Stable Marriage Assignment for Unequal Sets," *BIT*, **10**, (1970), 295–309.

3-3. "Space Filling Curves, or How to Waste Time on a Plotter," *Software-Practice and Experience*, **1**, No. 4 (1971), 403–40.

3-4. Wirth, N., "Program Development by Stepwise Refinement," *Comm. ACM*, **14**, No. 4 (1971), 221–27.

4 DYNAMIC INFORMATION STRUCTURES

4.1. RECURSIVE DATA TYPES

In Chap. 2 the array, record, and set structures were introduced as fundamental data structures. They are called fundamental because they constitute the building blocks out of which more complex structures are formed and because in practice they do occur most frequently. The purpose of defining a data type, and of thereafter specifying that certain variables be of that type, is that the range of values assumed by these variables, and thereby their storage pattern, is fixed once and for all. Hence, variables declared in this way are said to be *static*. However, there are many problems which involve far more complicated information structures. The characteristic of these problems is that their structures change during the computation. They are therefore called *dynamic* structures. Naturally, the components of such structures are—at some level of detail—static, i.e., of one of the fundamental data types. This chapter is devoted to the construction, analysis, and management of dynamic information structures.

It is noteworthy that there exist some close analogies between the methods used for structuring algorithms and those for structuring data. As with all analogies, there remain some differences (otherwise they would be identities), but a comparison of structuring methods for programs and data is nevertheless illuminating.

The elementary, unstructured statement is the assignment. Its corresponding member in the family of data structures is the scalar, unstructured type. These two are the atomic building blocks for composite statements and data types. The simplest structures, obtained through enumeration or sequencing, are the compound statement and the record structure. They both consist of a finite (usually small) number of explicitly enumerated components, which may themselves all be different from each other. If all components are identical, they need not be written out individually: we use the **for** statement and the **array** structure to indicate replication by a known, finite factor. A choice among two or more variants is expressed by the conditional or the case statement and by the variant record structure, respectively. And finally, a repetition by an initially unknown (and potentially infinite) factor is expressed by the **while** or **repeat** statements. The corresponding data structure is the sequence (file), the simplest kind which allows the construction of types of infinite cardinality.

The question arises whether or not there exists a data structure that corresponds in a similar way to the procedure statement. Naturally, the most interesting and novel property of procedures in this respect is *recursion*. Values of such a recursive data type would contain one or more components belonging to the same type as itself, in analogy to a procedure containing one or more calls to itself. Like procedures, such data type definitions might be directly or indirectly recursive.

A simple example of an object that would most appropriately be represented as a recursively defined type is the arithmetic expression found in programming languages. Recursion is used to reflect the possibility of nesting, i.e., of using parenthesized subexpressions as operands in expressions. Hence, let an expression here be defined informally as follows:

> An *expression* consists of a term, followed by an operator, followed by a term. (The two terms constitute the operands of the operator.) A *term* is either a variable—represented by an identifier—or an expression enclosed in parentheses.

A data type whose values represent such expressions can easily be described by using the tools already available with the addition of recursion:

```
type expression = record op: operator;
                          opd1, opd2: term
                  end;
type term = record                                   (4.1)
                if t then (id: alfa)
                     else (subex: expression)
            end
```

Hence, every variable of type term consists of two components, namely, the tagfield *t* and, if *t* is true, the field *id*, or of the field *subex* otherwise.

Consider now, for example, the following four expressions:

$$
\begin{aligned}
&1.\ x + y \\
&2.\ x - (y * z) \\
&3.\ (x + y) * (z - w) \\
&4.\ (x/(y + z)) * w
\end{aligned}
\tag{4.2}
$$

These expressions may be visualized by the patterns in Fig. 4.1, which exhibit their nested, recursive structure, and they determine the layout or mapping of these expressions onto a store.

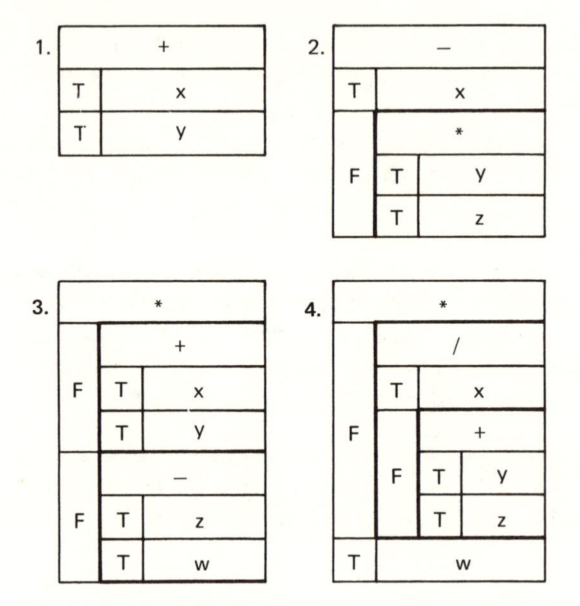

Fig. 4.1 Storage pattern for recursive record structures.

A second example of a recursive information structure is the family pedigree: Let a *pedigree* be defined by (the name of) a person and the two pedigrees of his parents. This definition leads inevitably to an infinite structure. Real pedigrees are bounded because at some level of ancestry information is missing. This can be taken into account by again using a variant structure as shown in (4.3).

```
type ped = record
             if known then
               (name: alfa;                    (4.3)
                father, mother: ped)
           end
```

(Note that every variable of type *ped* has at least one component, namely, the tagfield called *known*. If its value is true, then there are three more fields; otherwise, there are none.) The particular value denoted by the (recursive) record constructor

$$x = (T, Ted, (T, Fred, (T, Adam, (F), (F)), (F)),\\ (T, Mary, (F), (T, Eva, (F), (F)))$$

is depicted in Fig. 4.2 in a way that may suggest a possible storage pattern. (Since only a single record type is involved, we have omitted the type identifier *ped* preceding each constructor).

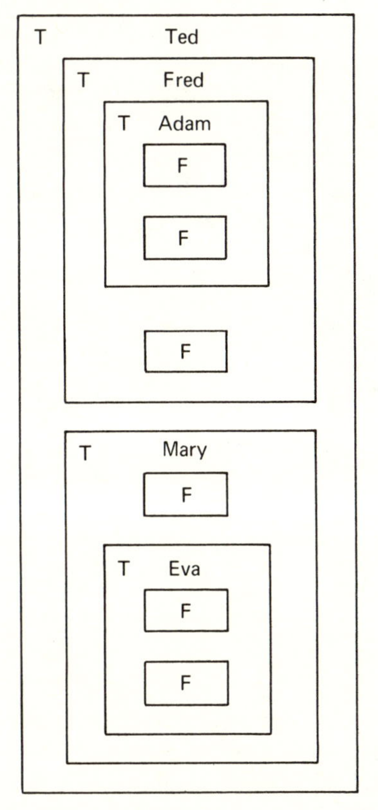

Fig. 4.2 Pedigree structure.

The important role of the variant facility becomes clear; it is the only means by which a recursive data structure can be bounded, and it is therefore an inevitable companion of every recursive definition. The analogy between program and data structuring concepts is particularly pronounced in this case. A conditional statement must necessarily be part of every recursive procedure in order that execution of the procedure can terminate. Termination of execution evidently corresponds to finite cardinality.

4.2. POINTERS OR REFERENCES

The characteristic property of recursive structures which clearly distinguishes them from the fundamental structures (arrays, records, sets) is their ability to vary in size. Hence, it is impossible to assign a fixed amount of storage to a recursively defined structure, and as a consequence a compiler cannot associate specific addresses to the components of such variables. The technique most commonly used to master this problem involves a *dynamic allocation* of storage, i.e., allocation of store to individual components at the time when they come into existence during program execution, instead of at translation time. The compiler then allocates a fixed amount of storage to hold the *address* of the dynamically allocated component instead of the component itself. For instance, the pedigree illustrated in Fig. 4.2 would be represented by individual—quite possibly non-contiguous—records, one for each person. These persons are then linked by their addresses assigned to the respective "father" and "mother" fields. Graphically, this situation is best expressed by the use of arrows or pointers (see Fig. 4.3).

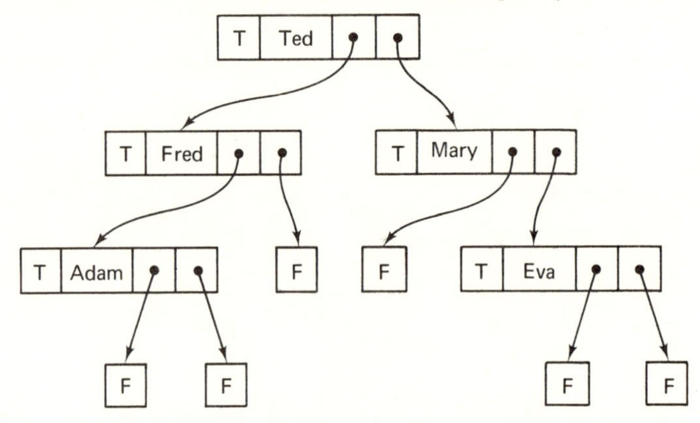

Fig. 4.3 Structure linked by pointers.

It must be emphasized that the use of pointers to implement recursive structures is merely a technique. The programmer need not be aware of their existence. Storage may be allocated automatically the first time a new component is referenced. However, if the technique of using references or pointers is made explicit, more general data structures can be constructed than those definable by purely recursive data definition. In particular, it is then possible to define "infinite" or circular structures and to dictate that certain structures are *shared*. It has therefore become common in advanced programming languages to make possible the explicit manipulation of references to data in addition to the data themselves. This implies that a clear notational distinction must exist between data and references to data and that

consequently data types must be introduced whose values are pointers (references) to other data. The notation we use for this purpose is the following;

$$\textbf{type } T_p = \uparrow T \tag{4.4}$$

The type declaration (4.4) expresses that values of type T_p are pointers to data of type T. Thus, the arrow in (4.4) is verbalized as "pointer to." It is fundamentally important that the type of elements pointed to is evident from the declaration of T_p. We say that T_p is *bound* to T. This binding distinguishes pointers in higher-level languages from addresses in assembly codes, and it is a most important facility to increase security in programming through redundancy of the underlying notation.

Values of pointer types are generated whenever a data item is dynamically allocated. We will adhere to the convention that such an occasion be explicitly mentioned at all times. This is in contrast to the situation in which the first time that an item is mentioned it is automatically (assumed to be) allocated. For this purpose, we introduce the intrinsic procedure *new*. Given a pointer variable p of type T_p, the statement

$$new(p) \tag{4.5}$$

effectively allocates a variable of type T, generates a pointer of type T_p referencing this new variable, and assigns this pointer to the variable p (see Fig. 4.4). The pointer value itself can now be referred to as p (i.e., as the value of the pointer variable p). In contrast, the variable which is referenced by p is denoted by $p\uparrow$. This is the dynamically allocated variable of type T.

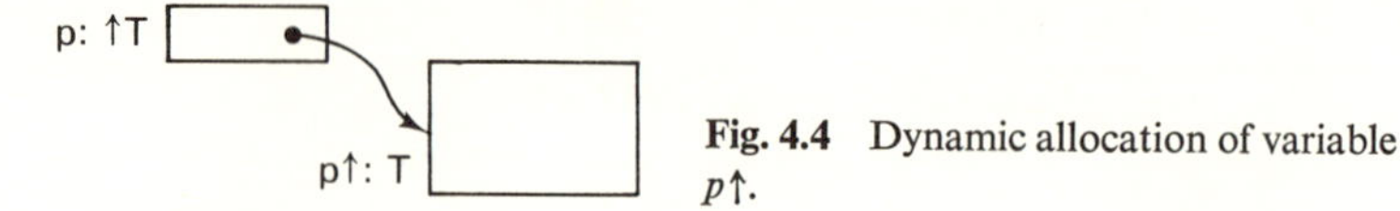

Fig. 4.4 Dynamic allocation of variable $p\uparrow$.

It was mentioned above that a variant component is essential in every recursive type to ensure finite cardinality. The example of the family predigree is of a pattern that exhibits a most frequently occurring constellation [see (4.3)], namely, the case in which the tagfield is two-valued (Boolean) and in which its value being *false* implies the absence of any further components. This is expressed by the *declaration schema* (4.6).

$$\textbf{type } T = \textbf{record if } p \textbf{ then } S(T) \textbf{ end} \tag{4.6}$$

$S(T)$ denotes a sequence of field definitions which includes one or more fields of type T, thereby ensuring recursivity. All structures of a type patterned after (4.6) will exhibit a tree (or list) structure similar to that shown in Fig. 4.3. Its peculiar property is that it contains pointers to data components with a tagfield only, i.e., without further relevant information. The implementation technique using pointers suggests an easy way of saving storage space by

letting the tagfield information be included in the value of the pointer itself. The common solution is to *extend* the range of values of a type T_p by a single value that is pointing to no element at all. We denote this value by the special symbol **nil**, and we understand that **nil** is automatically an element of all pointer types declared. This extension of the range of pointer values explains why finite structures may be generated *without* the explicit presence of variants (conditions) in their (recursive) declaration.

The new formulations of the data types declared in (4.1) and (4.3)—based on explicit pointers—are given in (4.7) and (4.8), respectively. Note that in the latter case (which originally corresponded to the schema (4.6)) the variant record component has vanished, since *p*↑.*known* = *false* is now expressed as *p* = **nil**. The renaming of the type *ped* to *person* reflects the difference in the viewpoint brought about by the introduction of explicit pointer values. Instead of first considering the given structure in its entirety and then investigating its substructure and its components, attention is focused on the components in the first place, and their interrelationship (represented by pointers) is not evident from any fixed declaration.

```
type expression = record op: operator;
                     opd1, opd2: ↑term
                  end;
type term       = record                                   (4.7)
                     if t then (id: alfa)
                          else (sub: ↑expression)
                  end

type person     = record name: alfa;
                     father, mother: ↑person               (4.8)
                  end
```

The data structure representing the pedigree shown in Figs. 4.2 and 4.3 is again shown in Fig. 4.5 in which pointers to unknown persons are denoted by **nil**. The ensuing improvement in storage economy is obvious.

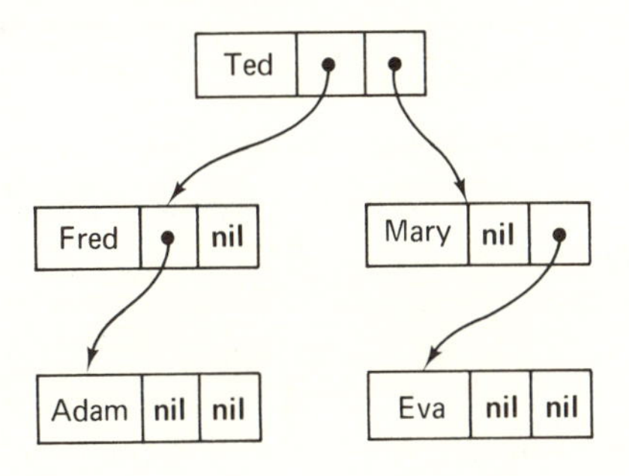

Fig. 4.5 Structure with **nil** pointers.

Again referring to Fig. 4.5, assume that Fred and Mary are siblings, i.e., have the same father and mother. This situation is easily expressed by replacing the two **nil** values in the respective fields of the two records. An implementation that hides the concept of pointers or uses a different technique of storage handling would force the programmer to represent the records of Adam and Eva twice each. Although in accessing their data for inspection it does not matter whether the two fathers (and the two mothers) are duplicated or represented by a single record, the difference is *essential when selective updating* is permitted. Treating pointers as explicit data items instead of as hidden implementation aids allows the programmer to express clearly where *storage sharing* is intended.

A further consequence of the explicitness of pointers is that it is possible to define and manipulate cyclic data structures. This additional flexibility yields, of course, not only increased power but also requires increased care by the programmer because the manipulation of cyclic data structures may easily lead to non-terminating processes.

This phenomenon of power and flexibility being intimately coupled with the danger of misuse is well-known in programming, and it particularly recalls the **goto** statement. Indeed, if the analogy between program structures and data structures is to be extended, the purely recursive data structure could well be placed at the level corresponding with the procedure, whereas the introduction of pointers is comparable to the use of **goto** statements. For, as the **goto** statement allows the construction of any kind of program pattern (including loops), so do pointers allow for the composition of any kind of data structure (including cycles). The parallel development of corresponding program and data structures is shown in concise form in Table 4.1.

In Chap. 3 it was shown that iteration is a special case of recursion and

Construction Pattern	Program Statement	Data Type
Atomic element	Assignment	Scalar type
Enumeration	Compound statement	Record type
Repetition by a known factor	For statement	Array type
Choice	Conditional statement	Variant record, type union
Repetition by an unknown factor	While or repeat statement	Sequence or file type
Recursion	Procedure statement	Recursive data type
General "graph"	Go to statement	Structure linked by pointers

Table 4.1 Correspondences of Program and Data Structures.

that a call of a recursive procedure P defined according to schema (4.9)

```
procedure P;
begin
    if B then begin P0; P end
end                                   (4.9)
```

where P_0 is a statement not involving P, is equivalent to and replaceable by the iterative statement

while B **do** P_0

The analogies outlined in Table 4.1 reveal that a similar relationship holds between recursive data types and the sequence. In fact, a recursive type defined according to the schema

```
type T = record
             if B then (t0: T0; t: T)          (4.10)
         end
```

where T_0 is a type not involving T, is equivalent and replaceable by the sequential data type

file of T_0

This shows that recursion can be replaced by iteration in program *and* data definitions if (and only if) the procedure or type name occurs recursively only once at the end (or the beginning) of its definition.

The remainder of this chapter is devoted to the generation and manipulation of data structures whose components are linked by explicit pointers. Structures with specific simple patterns are emphasized in particular; recipes for handling more complex structures may be derived from those for manipulating basic formations. These are the linear list or chained sequence—the most simple case—and trees. Our preoccupation with these "building blocks" of data structuring does not imply that more involved structures do not occur in practice. In fact, the following story which appeared in a Zürich newspaper in July 1922 is a proof that irregularity may even occur in cases which usually serve as examples for regular structures, such as (family) trees. The story tells of a man who describes the misery of his life in the following words:

> I married a widow who had a grown-up daughter. My father, who visited us quite often, fell in love with my step-daughter and married her. Hence, my father became my son-in-law, and my step-daughter became my mother. Some months later, my wife gave birth to a son, who became the brother-in-law of my father as well as my uncle. The wife of my father, that is my step-daughter, also had a son. Thereby, I got a brother and at the same time a grandson. My wife is my grandmother, since she is my mother's mother. Hence, I am my wife's husband and at the same time her step-grandson; in other words, I am my own grandfather.

4.3. LINEAR LISTS

4.3.1. Basic Operations

The simplest way to interrelate or link a set of elements is to line them up in a single *list* or *queue*. For, in this case, only a single link is needed for each element to refer to its successor.

Assume that a type T is defined as shown in (4.11). Every variable of this type consists of three components, namely, an identifying key, the pointer to its successor, and possibly further associated information omitted in (4.11).

```
type T = record key: integer;
                next: ↑T;
                ......
         end                                    (4.11)
```

A list of T's, with a pointer to its first component being assigned to a variable p, is illustrated in Fig. 4.6. Probably the simplest operation to be performed with a list as shown in Fig. 4.6 is the *insertion of an element at its head.* First, an element of type T is allocated, its reference (pointer) being assigned to an auxiliary pointer variable, say q. Thereafter, a simple reassignment of pointers completes the operation, which is programmed in (4.12).

```
new(q); q↑.next := p; p := q                    (4.12)
```

Note that the order of these three statements is essential.

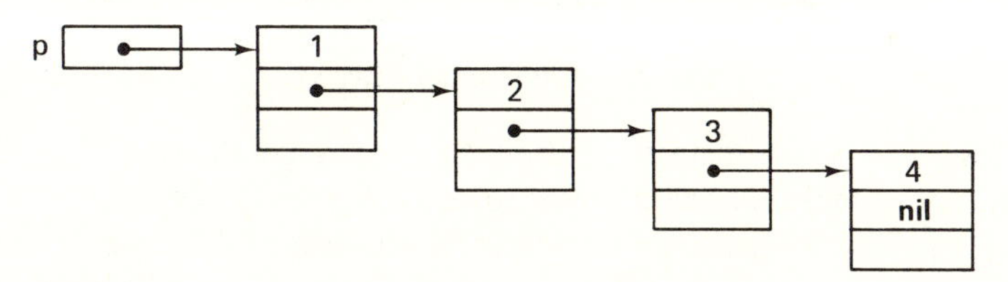

Fig. 4.6 Example of a list.

The operation of inserting an element at the head of a list immediately suggests how such a list can be generated: starting with the empty list, a heading element is added repeatedly. The process of *list generation* is expressed in (4.13); here the number of elements to be linked is n.

```
p := nil; {start with empty list}
while n > 0 do
    begin new(q); q↑.next := p; p := q;         (4.13)
        q↑.key := n; n := n-1
    end
```

This is the simplest way of forming a list. However, the resulting order of elements is the inverse of the order of their "arrival." In some applications this is undesirable; consequently, new elements must be appended at the *end* of the list. Although the end can easily be determined by a scan of the list, this naive approach involves an effort that may as well be saved by using a second pointer, say q, always designating the last element. This method is, for example, applied in Program 4.4 which generates cross-references to a given text. Its disadvantage is that the first element inserted has to be treated differently from all later ones.

The explicit availability of pointers makes certain operations very simple which are otherwise cumbersome; among the elementary list operations are those of inserting and deleting elements (selective updating of a list), and, of course, the traversal of a list. We first investigate *list insertion*.

Assume that an element designated by a pointer (variable) q is to be inserted in a list *after* the element designated by the pointer p. The necessary pointer assignments are expressed in (4.14), and their effect is visualized by Fig. 4.7.

$$q\uparrow.next := p\uparrow.next;\ p\uparrow.next := q \tag{4.14}$$

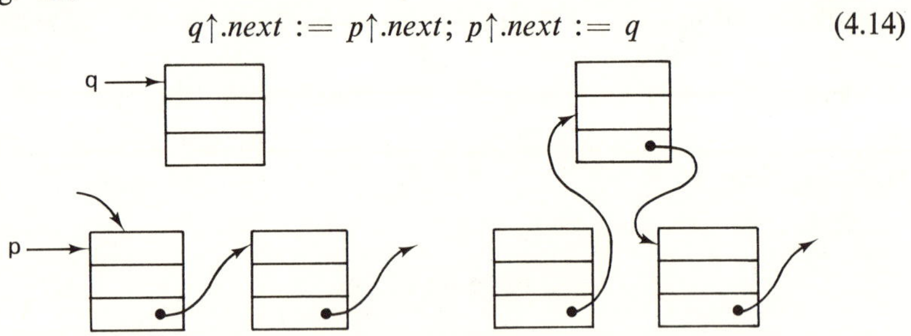

Fig. 4.7 List insertion after $p\uparrow$.

If insertion *before* instead of after the designated element $p\uparrow$ is desired, the one-directional link chain seems to cause a problem because it does not provide any kind of path to an element's predecessors. However, a simple "trick" solves our dilemma: it is expressed in (4.15) and illustrated in Fig. 4.8. Assume that the key of the new element is $k = 8$.

$$\begin{aligned} &new(q);\ q\uparrow := p\uparrow; \\ &p\uparrow.key := k;\ p\uparrow.next := q \end{aligned} \tag{4.15}$$

The "trick" evidently consists of actually inserting a new component *after* $p\uparrow$, but then interchanging the values of the new element and of $p\uparrow$.

Next, we consider the process of *list deletion*. Deleting the successor of a $p\uparrow$ is straightforward. In (4.16) it is shown in combination with the re-insertion of the deleted element at the head of another list (designated by q). r is an auxiliary variable of type $\uparrow T$.

$$\begin{aligned} &r := p\uparrow.next;\ p\uparrow.next := r\uparrow.next; \\ &r\uparrow.next := q;\ q := r \end{aligned} \tag{4.16}$$

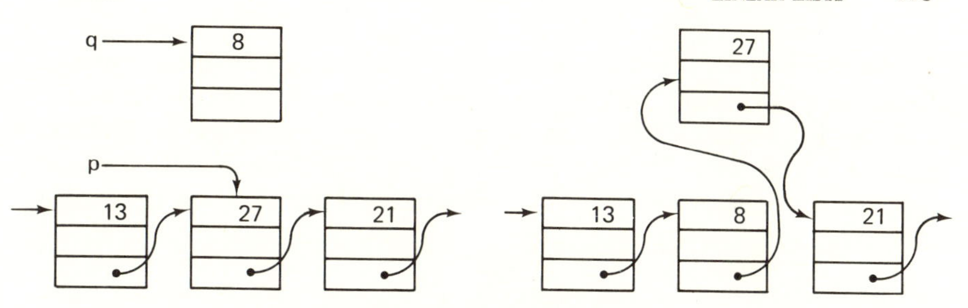

Fig. 4.8 List insertion before $p\uparrow$.

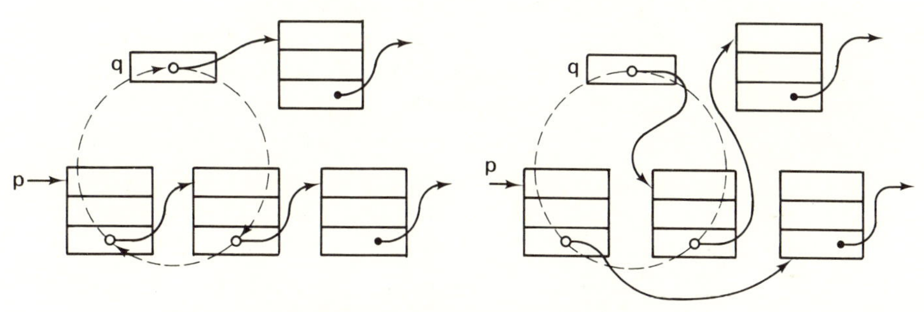

Fig. 4.9 List deletion and re-insertion.

Figure 4.9 illustrates process (4.16) and shows that it consists of a cyclic exchange of three pointers.

More difficult is the removal of a designated element itself (instead of its successor), since we encounter the same problem as with insertion in front of a $p\uparrow$: backtracking to the denoted element's predecessor is impossible. But deleting the successor after moving its value forward is a relatively obvious and simple solution. It can be applied whenever $p\uparrow$ has a successor, i.e., is not the last element on the list.

We now turn to the fundamental operation of *list traversal.* Let us assume that an operation $P(x)$ has to be performed for every element of the list whose first element is $p\uparrow$. This task is expressible as follows:

```
while list designated by p is not empty do
begin perform operation P;
      proceed to the successor
end
```

In detail, this operation is described by statement (4.17).

```
while p ≠ nil do
   begin P(p↑); p := p↑.next                    (4.17)
   end
```

It follows from the definitions of the **while** statement and of the linking structure that P is applied to all elements of the list and to no other ones.

A very frequent operation performed is *list searching* for an element with a given key x. As with file structures, the search is purely sequential. The search terminates either if an element is found or if the end of the list is reached. Again, we assume that the head of the list is designated by a pointer p. A first attempt to formulate this simple search results in the following:

$$\textbf{while } (p \neq \textbf{nil}) \wedge (p\uparrow.key \neq x) \textbf{ do } p := p\uparrow.next \tag{4.18}$$

However, it must be noticed that $P = \textbf{nil}$ implies that $p\uparrow$ does not exist. Hence, evaluation of the termination condition may imply access to a non-existing variable (in contrast to a variable with undefined value) and may cause failure of program execution. This can be remedied either by using an explicit break of the repetition expressed by a **goto** statement (4.19) or by introducing an auxiliary Boolean variable to record whether or not a desired key was found (4.20).

```
while p ≠ nil do
    if p↑.key = x then goto Found                (4.19)
                  else p := p↑.next
```

The use of the **goto** statement requires the presence of a destination label at some place; note that its incompatibility with the **while** statement is evidenced by the fact that the while clause becomes misleading: the controlled statement is *not* necessarily executed as long as $p \neq \textbf{nil}$.

```
b := true;
while (p ≠ nil) ∧ b do
   if p↑.key = x then b := false                 (4.20)
                 else p := p↑.next
{(p=nil) ∨ ¬b}
```

4.3.2. Ordered Lists and Re-organizing Lists

Algorithm (4.20) strongly recalls the search routines for scanning an array or a file. In fact, a file is nothing but a linear list in which the technique of linkage to the successor is left unspecified or implicit. Since the primitive file operators do not allow insertion of new elements (except at the end) or deletion (except removal of *all* elements), the choice of representation is left wide open to the implementor, and he may well use sequential allocation, leaving successive components in contiguous storage areas. Linear lists with explicit pointers provide *more flexibility*, and therefore they should be used whenever this additional flexibility is needed.

To exemplify, we will now consider a problem that will re-occur throughout this chapter in order to illustrate alternative solutions and techniques. It is the problem of reading a text, collecting all its words, and counting the frequency of their occurrence. It is called the construction of a *concordance*.

An obvious solution is to construct a *list* of words found in the text. The list is scanned for each word. If the word is found, its frequency count is incremented; otherwise the word is added to the list. We shall simply call this process *search*, although it may apparently also include an *insertion*.

In order to be able to concentrate our attention on the essential part of list handling, we assume that the words have already been extracted from the text under investigation, have been encoded as integers, and are available in the form of an input file.

The formulation of the procedure called *search* follows in a straightforward manner from (4.20). The variable *root* refers to the head of the list in which new words are inserted according to (4.12). The complete algorithm is listed as Program 4.1; it includes a routine for tabulating the constructed concordance list. The tabulation process is an example in which an action is executed once for each element of the list, as shown in schematic form in (4.17).

The linear scan algorithm of Program 4.1 recalls the search procedure for arrays and files, and in particular the simple technique used to simplify the loop termination condition: the use of a *sentinel*. A sentinel may as well

Program 4.1 Straight List Insertion.

```
program list (input,output);
{straight list insertion}
   type ref = ↑word;
      word = record key: integer;
                    count: integer;
                    next: ref
             end ;
   var k: integer; root: ref;

   procedure search (x: integer; var root: ref);
      var w: ref; b: boolean;
   begin w := root; b := true;
      while (w≠nil) ∧ b do
         if w↑.key = x then b := false else w := w↑.next;
      if b then
      begin {new entry} w := root; new (root);
         with root↑ do
         begin key := x; count := 1; next := w
         end
      end else
      w↑.count := w↑.count + 1
   end {search} ;
```

```
    procedure printlist (w: ref);
    begin while w ≠ nil do
          begin writeln (w↑.key, w↑.count);
             w := w↑.next
          end
    end {printlist} ;
begin root := nil; read(k);
    while k ≠ 0 do
       begin search (k, root); read(k)
       end ;
    printlist(root)
end .
```

Program 4.1 (Continued)

be used in list search; it is represented by a dummy element at the end of the list. The new procedure is (4.21), which replaces the search procedure of Program 4.1, provided that a global variable *sentinel* is added and that the initialization of *root* is replaced by the statements

new(*sentinel*); *root* := *sentinel*;

which generate the element to be used as sentinel.

```
procedure search(x: integer; var root: ref);
   var w: ref;
begin w := root; sentinel↑.key := x;
   while w↑.key ≠ x do w := w↑.next;
   if w ≠ sentinel then w↑.count := w↑.count + 1 else
      begin {new entry} w := root; new(root);                   (4.21)
         with root↑ do
         begin key := x; count := 1; next := w
         end
      end
end {search}
```

Obviously, the power and flexibility of the linked list are ill used in this example, and the linear scan of the entire list can only be accepted in cases in which the number of elements is limited. An easy improvement, however, is readily at hand: the *ordered list search*. If the list is ordered (say by increasing keys), then the search may be terminated at the latest upon encountering the first key which is larger than the new one. Ordering of the list is achieved by inserting new elements at the appropriate place instead of at the head. In effect, ordering is practically obtained free of charge. This is because of the ease by which insertion in a linked list is achieved, i.e., by making full use of its flexibility. It is a possibility not provided by the array and file structures.

(Note, however, that even in ordered lists no equivalent to the binary search of arrays is available.)

Ordered list search is a typical example of the situation described in (4.15) in which an element must be inserted *ahead* of a given item, namely, in front of the first one whose key is too large. The technique shown here, however, differs from the one used in (4.15). Instead of copying values, two pointers are carried along in the list traversal; *w*2 lags one step behind *w*1, and it thus identifies the proper insertion place when *w*1 has found too large a key. The general insertion step is shown in Fig. 4.10. Before proceeding we

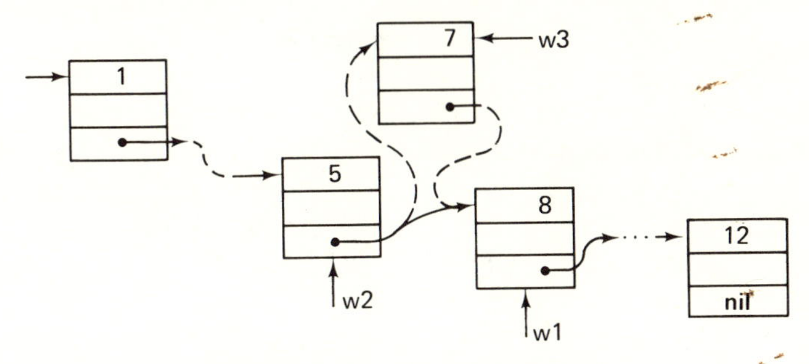

Fig. 4.10 Ordered list insertion.

must consider two circumstances:

1. The pointer to the new element (*w*3) is to be assigned to *w*2↑.*next*, except when the list is still empty. For reasons of simplicity and effectiveness, we prefer not to make this distinction by using a conditional statement. The only way to avoid this is to introduce a dummy element at the list head.
2. The scan with two pointers descending down the list one step apart requires that the list contain at least one element (in addition to the dummy). This implies that insertion of the first element be treated differently from the rest.

A proposal that follows these guidelines and hints is expressed in (4.23). It uses an auxiliary procedure *insert*, to be declared local to *search*. It generates and initializes the new element *w* and is shown in (4.22).

```
procedure insert(w: ref);
   var w3: ref;
begin new(w3);
   with w3↑ do
      begin key := x; count := 1; next := w
      end ;                                              (4.22)
   w2↑.next := w3
end {insert}
```

The initializing statement "*root* := **nil**" in Program 4.1 is accordingly replaced by

$$new(root);\ root\uparrow.next := \textbf{nil}$$

Referring to Fig. 4.10, we determine the condition under which the scan continues to proceed to the next element; it consists of two factors, namely,

$$(w1\uparrow.key < x) \land (w1\uparrow.next \neq \textbf{nil})$$

The resulting search procedure is shown in (4.23).

```
procedure search(x: integer; var root: ref);
   var w1,w2: ref;
begin w2 := root; w1 := w2↑.next;
   if w1 = nil then insert (nil) else
   begin
      while (w1↑.key < x) ∧ (w1↑.next ≠ nil) do
         begin w2 := w1; w1 := w2↑.next                    (4.23)
         end ;
      if w1↑.key = x then w1↑.count := w1↑.count + 1 else
        insert(w1)
   end
end {search} ;
```

Unfortunately, this proposal contains a logical flaw. In spite of our care, a "bug" has crept in! The reader is urged to try to identify the oversight before proceeding. For those who choose to skip this detective's assignment, it may suffice to say that (4.23) will always push the element first inserted to the tail of the list. The flaw is corrected by taking into account that if the scan terminates because of the second factor, the new element must be inserted *after* $w1\uparrow$ instead of *before* it. Hence, the statement "*insert*($w1$)" is replaced by

```
begin if w1↑.next = nil then
            begin w2 := w1; w1 := nil
            end;                                            (4.24)
         insert(w1)
end
```

Maliciously, the trustful reader has been fooled once more, for (4.24) is still incorrect. To recognize the mistake, assume that the new key lies between the last and the second last keys. This will result in both factors of the continuation condition being false when the scan reaches the end of the list, and consequently the insertion being made behind the tail element. If the same key occurs again later on, it will be inserted correctly and thus appear twice in the tabulation. The remedy lies in replacing the condition

$$w1\uparrow.next = \textbf{nil}$$

in (4.24) by

$$w1\uparrow.key < x$$

In order to speed up the search, the continuation condition of the **while** statement can once again be simplified by using a sentinel. This requires the initial presence of a *dummy header* as well as a sentinel at the tail. Hence, the list must be initialized by the following statements

new(*root*); *new*(*sentinel*); *root*↑.*next* := *sentinel*;

and the search procedure becomes noticeably simpler as evidenced by (4.25).

```
procedure search(x: integer; var root: ref);
   var w1,w2,w3: ref;
begin w2 := root; w1 := w2↑.next; sentinel↑.key := x;
   while w1↑.key < x do
      begin w2 := w1; w1 := w2↑.next
      end ;
   if (w1↑.key = x) ∧ (w1 ≠ sentinel) then
      w1↑.count := w1↑.count + 1 else                              (4.25)
   begin new(w3); {insert w3 between w1 and w2}
      with w3↑ do
         begin key := x; count := 1; next := w1
         end ;
      w2↑.next := w3
   end
end {search}
```

It is now high time to ask what gain can be expected from ordered list search. Remembering that the additional complexity incurred is small, one should not expect an overwhelming improvement.

Assume that all words in the text occur with equal frequency. In this case the gain through lexicographical ordering is indeed also nil, once all words are listed, for the position of a word does not matter if only the *total* of all access steps is significant and if all words have the same frequency of occurrence. However, a gain is obtained whenever a new word is to be inserted. Instead of first scanning the entire list, on the average only half the list is to be scanned. Hence, ordered list insertion pays off only if a concordance is to be generated with many distinct words compared to their frequency of occurrence. The preceding examples are therefore suitable primarily as programming exercises rather than for practical applications.

The arrangement of data in a linked list is recommended when the number of elements is relatively small (say < 100), varies, and, moreover, when no information is given about their frequencies of access. A typical example is the symbol table in compilers of programming languages. Each declaration causes the addition of a new symbol, and upon exit from its scope of validity, it is deleted from the list. The use of simple linked lists is appropriate for applications with relatively short programs. Even in this case a considerable improvement in access method can be achieved by a very simple technique which is mentioned here again primarily because it constitutes a pretty example for demonstrating the flexibilities of the linked list structure.

A characteristic property of programs is that occurrences of the same identifier are very often clustered, that is, one occurrence is often followed by one or more re-occurrences of the same word. This information is an invitation to re-organize the list after each access by moving the word that was found to the top of the list, thereby minimizing the length of the search path the next time it is sought. This method of access is called *list search with re-ordering*, or—somewhat pompously—*self-organizing list search*. In presenting the corresponding algorithm in the form of a procedure which may be substituted in Program 4.1, we take advantage of our experience made so far and introduce a sentinel right from the start. In fact, a sentinel not only speeds up the search, but in this case it also simplifies the program. The list is, however, not initially empty, but contains the sentinel element already. The initial statements are

$$new\ (sentinel);\ root := sentinel;$$

Note that the main difference between the new algorithm and the straight list search (4.21) is the action of re-ordering when an element has been found. It is then detached or deleted from its old position and re-inserted at the top. This deletion again requires the use of two chasing pointers, such that the predecessor $w2\uparrow$ of an identified element $w1\uparrow$ is still locatable. This, in turn, calls for the special treatment of the first element (i.e., the empty list). To conceive the re-linking process, we refer to Fig. 4.11. It shows the two pointers

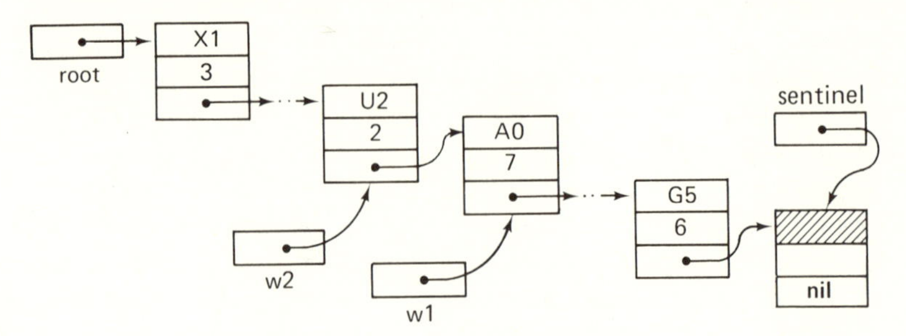

Fig. 4.11 List before re-ordering.

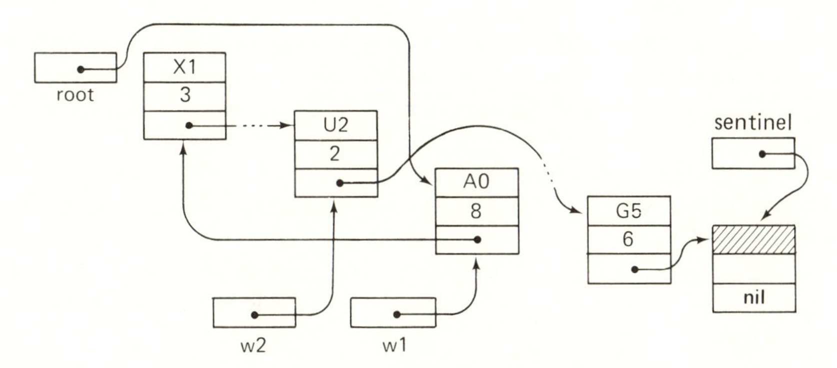

Fig. 4.12 List after re-ordering.

when $w1\uparrow$ was identified as the desired element. The configuration after correct re-ordering is represented in Fig. 4.12, and the complete new search procedure is shown in (4.26).

```
procedure search(x: integer; var root: ref);
   var w1,w2: ref;
begin w1 := root; sentinel↑.key := x;
   if w1 = sentinel then
   begin {first element} new(root);
      with root↑ do
         begin key := x; count := 1; next := sentinel
         end
   end else
   if w1↑.key = x then w1↑.count := w1↑.count + 1 else
   begin {search}
      repeat w2 := w1; w1 := w2↑.next
      until w1↑.key = x;
      if w1 = sentinel then                                      (4.26)
      begin {insert}
         w2 := root; new(root);
         with root↑ do
            begin key := x; count := 1; next := w2
            end
      end else
      begin {found, now reorder}
         w1↑.count := w1↑.count + 1;
         w2↑.next := w1↑.next; w1↑.next := root; root := w1
      end
   end
end {search}
```

The improvement in this search method strongly depends on the degree of clustering in the input data. For a given factor of clustering, the improvement will be more pronounced for large lists. To provide an idea of what order of magnitude an improvement can be expected, an empirical measurement was made by applying the above concordance program to a short text and to a relatively long text and then comparing the methods of linear list ordering (4.21) and of list reorganization (4.26). The measured data are condensed into Table 4.2. Unfortunately, the improvement is greatest when a different data organization is needed anyway. We will return to this example in Sect. 4.4.

	Test 1	Test 2
Number of distinct keys	53	582
Number of occurrences of keys	315	14341
Time for search with ordering	6207	3200622
Time for search with re-ordering	4529	681584
Improvement factor	1.37	4.70

Table 4.2 Comparison of List Search Methods.

4.3.3. An Application: Topological Sorting

An appropriate example of the use of a flexible, dynamic data structure is the process of *topological sorting*. This is a sorting process of items over which a *partial ordering* is defined, i.e., where an ordering is given over *some* pairs of items but not among all of them. This is a common situation. Following are examples of partial orderings:

1. In a dictionary or glossary, words are defined in terms of other words. If a word v is defined in terms of a word w, we denote this by $v \prec w$. Topological sorting of the words in a dictionary means arranging them in an order such that there will be no forward references.
2. A task (e.g., an engineering project) is broken up into subtasks. Completion of certain subtasks must usually precede the execution of other subtasks. If a subtask v must precede a subtask w, we write $v \prec w$. Topological sorting means their arrangement in an order such that upon initiation of each subtask all its prerequisite subtasks have been completed.
3. In a university curriculum, certain courses must be taken before others since they rely on the material presented in their prerequisites. If a course v is a prerequisite for course w, we write $v \prec w$. Topological sorting means arranging the courses in such an order that no course lists a later course as prerequisite.

4. In a program, some procedures may contain calls of other procedures. If a procedure v is called by a procedure w, we write $v \prec w$. Topological sorting implies the arrangement of procedure declarations in such a way that there are no forward references.

In general, a partial ordering of a set S is a relation between the elements of S. It is denoted by the symbol $\prec$, verbalized by "precedes," and satisfies the following three properties (axioms) for any distinct elements x, y, and z of S:

$$\begin{aligned} &(1)\ \text{if } x \prec y \text{ and } y \prec z, \text{ then } x \prec z \text{ (transitivity)} \\ &(2)\ \text{if } x \prec y, \text{ then not } y \prec x \text{ (asymmetry)} \\ &(3)\ \text{not } x \prec x \text{ (irreflexivity)} \end{aligned} \tag{4.27}$$

For evident reasons, we will assume that the sets S to be topologically sorted by an algorithm are finite. Hence, a partial ordering can be illustrated by drawing a diagram or graph in which the vertices denote the elements of S and the directed edges represent ordering relationships. An example is shown in Fig. 4.13.

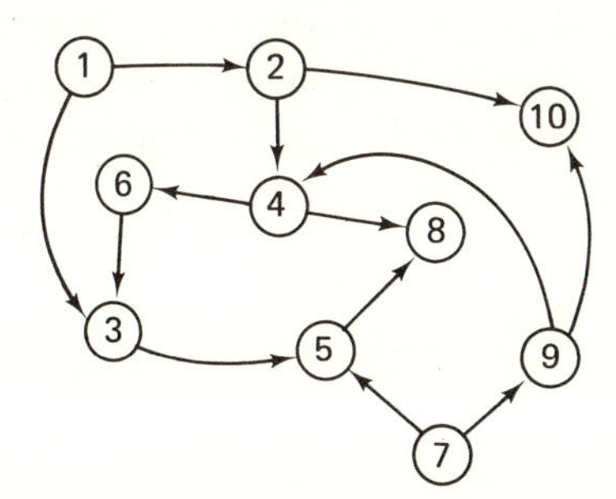

Fig. 4.13 Partially ordered set.

The problem of topological sorting is to embed the partial order in a linear order. Graphically, this implies the arrangement of the vertices of the graph in a row, such that all arrows point to the right, as shown in Fig. 4.14. Properties (1) and (2) of partial orderings ensure that the graph contains no loops. This is exactly the prerequisite condition under which such an embedding in a linear order is possible.

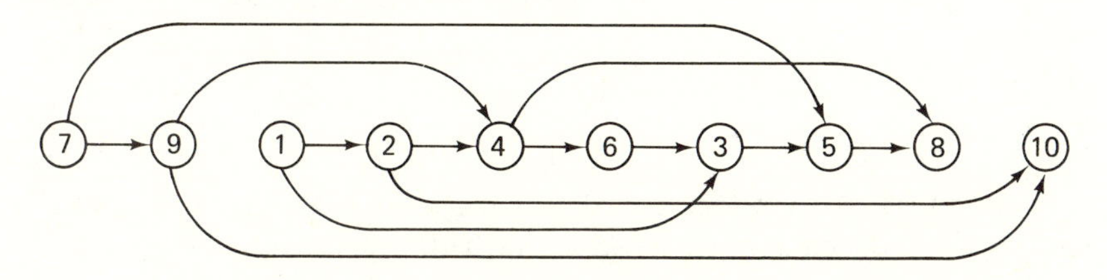

Fig. 4.14 Linear arrangement of partially ordered set of Fig. 4.13.

How do we proceed to find one of the possible linear orderings? The recipe is quite simple. We start by choosing any item that is not preceded by another item (there must be at least one; otherwise there would have to exist a loop). This object is placed at the head of the resulting list and removed from the set S. The remaining set is still partially ordered, and so the same algorithm can be applied again until the set is empty.

In order to describe this algorithm more rigorously, we must settle on a data structure and representation of S and its ordering. The choice of this representation is determined by the operations to be performed, particularly the operation of selecting elements with zero predecessors. Every item should therefore be represented by three characteristics: its identification key, its set of successors, and a count of its predecessors. Since the number n of elements in S is not given *a priori*, the set is conveniently organized as a linked list. Consequently, an additional entry in the description of each item contains the link to the next item in the list. We will assume that the keys are integers (but not necessarily the consecutive integers from 1 to n). Analogously, the set of each item's successors is conveniently represented as a linked list. Each element of the successor list is described by an identification and a link to the next item on this list. If we call the descriptors of the main list, in which each item of S occurs exactly once, *leaders*, and the descriptors of elements on the successor chains *trailers*, we obtain the following declarations of data types:

```
type lref   = ↑leader;
     tref   = ↑trailer;
     leader = record key, count: integer;
                 trail: tref;
                 next: lref                          (4.28)
              end;
     trailer = record id: lref;
                 next: tref
              end
```

Assume that the set S and its ordering relations are initially represented as a sequence of pairs of keys in the input file. The input data for the example in Fig. 4.13 are shown in (4.29) in which the symbols $\prec$ are added for the sake of clarity.

$$\begin{array}{ccccccc} 1 \prec 2 & 2 \prec 4 & 4 \prec 6 & 2 \prec 10 & 4 \prec 8 & 6 \prec 3 & 1 \prec 3 \\ 3 \prec 5 & 5 \prec 8 & 7 \prec 5 & 7 \prec 9 & 9 \prec 4 & 9 \prec 10 & \end{array} \qquad (4.29)$$

The first part of the topological sort program must read the input file and transform the data into a list structure. This is performed by successively reading a pair of keys x and y ($x \prec y$). Let us denote the pointers to their representations on the linked list of leaders by p and q. These records must be located by a list search and, if not yet present, be inserted in the list. This task is performed by a function procedure called L. Subsequently, a new entry is added in the list of trailers of x, along with an identification of y; the count of predecessors of y is incremented by 1. This algorithm is called *input phase* (4.30). Figure 4.15 illustrates the data structure generated during processing the input data (4.29) by (4.30) .This piece of program refers to a function $L(w)$ yielding the reference to the list component with key w (see also Program 4.2). We assume that the sequence of input key pairs is terminated by an additional zero.

```
{input phase} read(x);
new(head); tail := head; z := 0;
while x ≠ 0 do
begin read(y); p := L(x); q := L(y);
    new(t); t↑.id := q; t↑.next := p↑.trail;          (4.30)
    p↑.trail := t; q↑.count := q↑.count + 1;
    read(x)
end
```

After the data structure of Fig. 4.15 has been constructed in this input phase, the actual process of topological sorting can be taken up as described above. But since it consists of repeatedly selecting an element with a zero count of predecessors, it seems sensible to first gather all such elements in a linked chain. Since we note that the original chain of leaders will afterwards no longer be needed, the same field called *next* may be re-used to link the zero predecessor leaders. This operation of replacing one chain by another chain occurs frequently in list processing. It is expressed in detail in (4.31), and for reasons of convenience it builds up the new chain in the *reverse direction.*

```
{search for leaders with 0 predecessors}
   p := head; head := nil;
   while p ≠ tail do
      begin q := p; p := q↑.next;
         if q↑.count = 0 then                          (4.31)
         begin {insert q↑ in new chain}
            q↑.next := head; head := q
         end
      end
```

Referring to Fig. 4.15, we see that the *next* chain of leaders is replaced by the one of Fig. 4.16 in which the pointers not depicted are left unchanged.

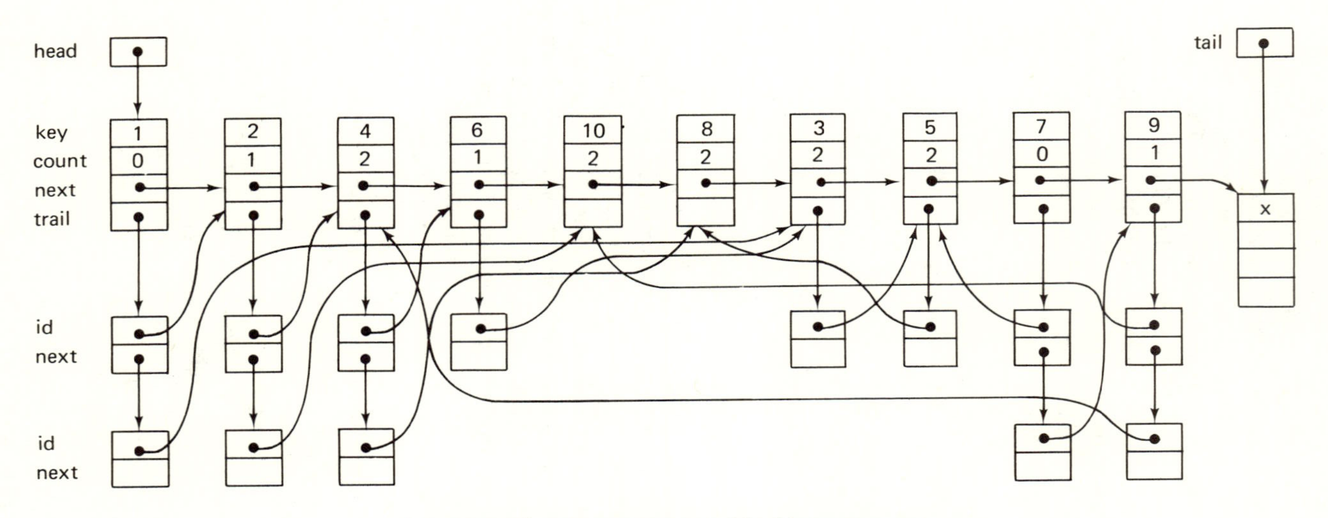

Fig. 4.15 List structure generated by Topsort program.

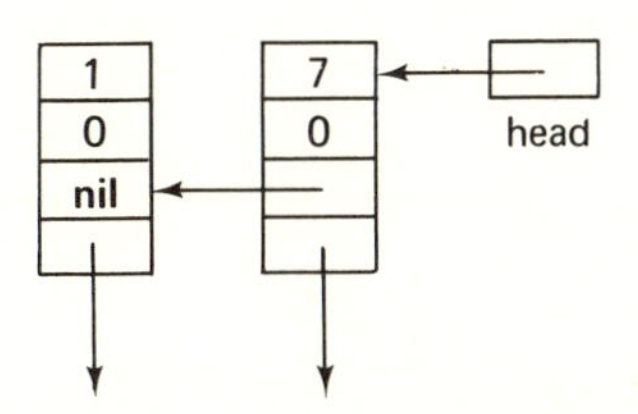

Fig. 4.16 List of leaders with zero counts.

After all this preparatory setting up of a convenient representation of the partially ordered set S, we can finally proceed to the actual task of topological sorting, i.e., of generating the output sequence. In a first rough version it can be described as follows:

```
q := head;
while q ≠ nil do
begin {output this element, then delete it}
   writeln(q↑.key); z := z-1;
   t := q↑.trail; q := q↑.next;                              (4.32)
   "decrement the predecessor count of all its successors
    on trailer list t; if any count becomes 0, insert this
    element in the leader list q"
end
```

The statement in (4.32) that is to be still further refined constitutes one more scan of a list [see schema (4.17)]. In each step, the auxiliary variable p designates the leader element whose count has to be decremented and tested.

```
while t ≠ nil do
begin p := t↑.id; p↑.count := p↑.count-1;
      if p↑.count = 0 then
      begin {insert p↑ in leader list}
           p↑.next := q; q := p                              (4.33)
      end;
      t := t↑.next
end
```

This completes the program for topological sorting. Note that a counter z was introduced to count the leaders generated in the input phase. This count is decremented each time a leader element is output in the output phase. It should therefore return to 0 at the end of the program. Its failure to return to 0 is an indication that there are elements left in the structure when none is without predecessor. In this case the set S is evidently not partially ordered.

The output phase programmed above is an example of a process that maintains a list which pulsates, i.e., in which elements are inserted and removed in an unpredictable order. It is therefore an example of a process which utilizes the full flexibility afforded by the explicitly linked list.

Program 4.2 Topological Sorting.

```
program topsort(input,output);
type lref = ↑leader;
    tref = ↑trailer;
    leader = record key: integer;
                    count: integer;
                    trail:  tref;
                    next:  lref;
                end ;
    trailer = record id: lref;
                    next: tref
                end ;
var head, tail, p,q: lref;
    t: tref; z: integer;
    x,y: integer;
function L(w: integer): lref;
    {reference to leader with key w}
    var h: lref;
begin h := head; tail↑.key := w;  {sentinel}
    while h↑.key ≠ w do h := h↑.next;
    if h = tail then
        begin {no element with key w in the list}
            new(tail); z := z+1;
            h↑.count := 0; h↑.trail := nil; h↑.next := tail
        end ;
    L := h
end {L} ;
begin {initialize list of leaders with a dummy}
    new(head); tail := head; z := 0;
{input phase}  read(x);
    while x ≠ 0 do
    begin read(y); writeln(x,y);
        p := L(x); q := L(y);
        new (t); t↑.id := q; t↑.next :≐ p↑.trail;
        p↑.trail := t; q↑.count := q↑.count + 1;
        read(x)
    end ;
{search for leaders with count = 0}
    p := head; head := nil;
    while p ≠ tail do
    begin q := p; p := p↑.next;
        if q↑.count = 0 then
            begin q↑.next := head; head := q
            end
    end ;
```

```
{output phase}  q := head;
    while q ≠ nil do
    begin writeln(q↑.key); z := z-1;
       t := q↑.trail; q := q↑.next;
       while t ≠ nil do
       begin p := t↑.id; p↑.count := p↑.count - 1;
          if p↑.count = 0 then
          begin {insert p↑ in q-list}
             p↑.next := q; q := p
          end ;
          t := t↑.next
       end
    end ;
    if z ≠ 0 then writeln ('THIS SET IS NOT PARTIALLY ORDERED')
end .
```

Program 4.2 (Continued)

4.4. TREE STRUCTURES

4.4.1. Basic Concepts and Definitions

We have seen that sequences and lists may conveniently be defined in the following way: A sequence (list) with base type T is either

1. The empty sequence (list).
2. The concatenation (chain) of a T and a sequence with base type T.

Hereby recursion is used as an aid in defining a structuring principle, namely, sequencing or iteration. Sequences and iterations are so common that they are usually considered as fundamental patterns of structure and behavior. But it should be kept in mind that they *can* be defined in terms of recursion, whereas the reverse is not true, for recursion may be effectively and elegantly used to define much more sophisticated structures. Trees are a well-known example. Let a tree structure be defined as follows: A *tree structure* with base type T is either

1. The empty structure.
2. A node of type T with a finite number of associated disjoint tree structures of base type T, called *subtrees*.

From the similarity of the recursive definitions of sequences and tree structures it is evident that the sequence (list) is a tree structure in which each node has at most one "subtree." The sequence (list) is therefore also called a *degenerate tree*.

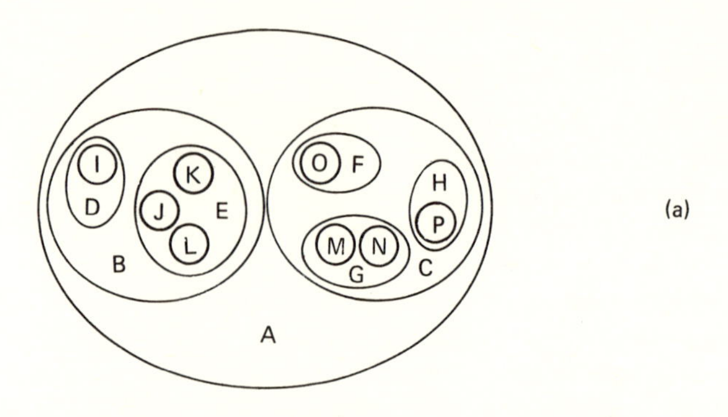

(A (B (D (I), E (J, K, L)), C (F (O), G (M, N), H (P)))) (b)

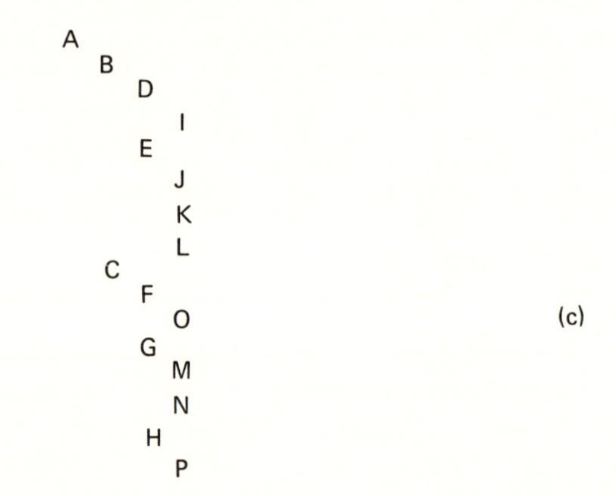

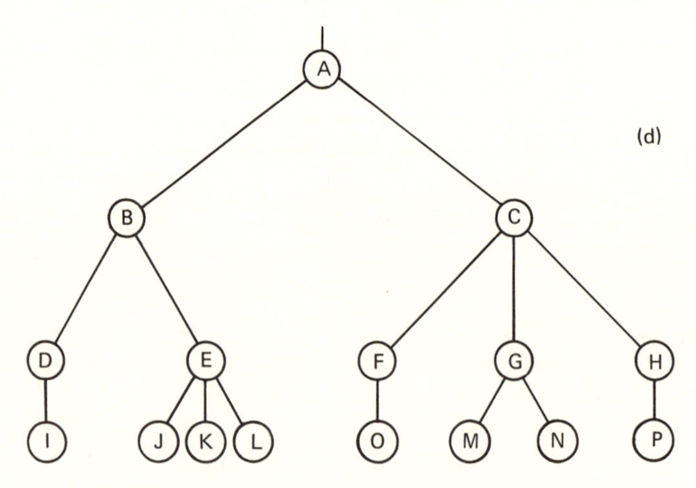

Fig. 4.17 Representation of a tree structure: (a) Nested sets; (b) nested parentheses; (c) indentation; (d) graph.

There are several ways to represent a tree structure. For example, let the base type T range over the letters; such a tree structure is shown in various ways in Fig. 4.17. These representations all show the same structure and are therefore equivalent. It is the graph structure that explicitly illustrates the branching relationships which, for obvious reasons, led to the generally used name "tree." Strangely enough, it is customary to depict trees upside down, or—if one prefers to express this fact differently—to show the roots of trees. The latter formulation, however, is misleading, since the top node (A) is commonly called *the root*. Although we recognize that trees in nature are somewhat more complicated creations than our abstractions, henceforth we will call our tree structures simply *trees*.

An *ordered tree* is a tree in which the branches of each node are ordered. Hence the two ordered trees in Fig. 4.18 are distinct, different objects. A node y which is directly below node x is called a (direct) *descendant* of x; if x is at *level* i, then y is said to be a level $i + 1$. Inversely, node x is said to be the (direct) *ancestor* of y. The root of a tree is defined to be at level 1. The maximum level of any element of a tree is said to be its *depth* or *height*.

Fig. 4.18 Two distinct binary trees.

If an element has no descendants, it is called a *terminal* element or a *leaf*; and an element which is not terminal is an *interior node*. The number of (direct) descendants of an interior node is called its *degree*. The maximum degree over all nodes is the degree of the tree. The number of branches or edges which have to be traversed in order to proceed from the root to a node x is called the *path length* of x. The root has path length 1, its direct descendants have path length 2, etc. In general, a node at level i has path length i. The path length of a tree is defined as the sum of the path lengths of all its components. It is also called its *internal path* length. The internal path length of the tree shown in Fig. 4.17, for instance, is 52. Evidently, the average path length P_I is

$$P_I = \frac{1}{n} \sum_i n_i \cdot i \tag{4.34}$$

where n_i is the number of nodes at level i. In order to define what is called the external path length, we extend the tree by a special node wherever a null subtree was present in the original tree. In doing so, we assume that all nodes are supposed to have the same degree, namely, the degree of the tree. Extending the tree in this way therefore amounts to filling up empty branches, whereby the special nodes, of course, have no further descendants. The tree of

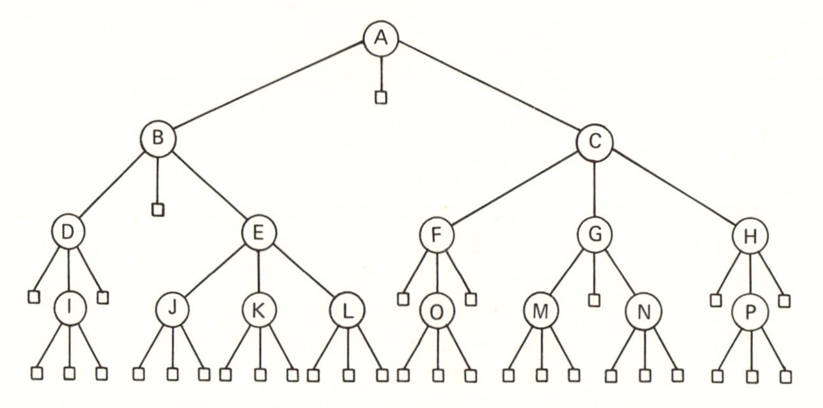

Fig. 4.19 Ternary tree extended with special nodes.

Fig. 4.17 extended with special nodes is shown in Fig. 4.19 in which the special nodes are represented by square boxes.

The *external path length* is now defined as the sum of the path lengths over all special nodes. If the number of special nodes at level i is m_i, then the average external path length P_E is

$$P_E = \frac{1}{m} \sum_i m_i \cdot i \tag{4.35}$$

In the tree shown in Fig. 4.19 the external path length is 153.

The number of special nodes m to be added in a tree of degree d directly depends on the number n of original nodes. Note that every node has exactly one edge pointing to it. Thus, there are $m + n$ edges in the extended tree. On the other hand, d edges are emanating from each original node, none from the special nodes. Therefore, there exist $dn + 1$ edges, the 1 resulting from the edge pointing to the root. The two results yield the following equation between the number m of special nodes and n of original nodes: $dn + 1 = m + n$, or

$$m = (d - 1)n + 1 \tag{4.36}$$

The maximum number of nodes in a tree of a given height h is reached if all nodes have d subtrees, except those at level h, all of which have none. For a tree of degree d, level 1 then contains 1 node (namely, the root), level 2 contains its d descendants, level 3 contains the d^2 descendants of the d nodes at level 2, etc. This yields

$$N_d(h) = 1 + d + d^2 + \cdots + d^{h-1} = \sum_{i=0}^{h-1} d^i \tag{4.37}$$

as the maximum number of nodes for a tree with height h and degree d. For $d = 2$, we obtain

$$N_2(h) = \sum_{i=0}^{h-1} 2^i = 2^h - 1 \tag{4.38}$$

Of particular importance are the ordered trees of degree 2. They are called *binary trees*. We define an ordered binary tree as *a finite set of elements (nodes) which either is empty or consists of a root (node) with two disjoint binary trees* called the *left* and the *right subtree* of the root. In the following sections we shall exclusively deal with binary trees, and we therefore shall use the word "tree" to mean "ordered binary tree." Trees with degree greater than 2 are called *multiway trees* and are discussed in Sect. 5 of this chapter.

Familiar examples of *binary* trees are the family tree (pedigree) with a person's father and mother as his descendants (!), the history of a tennis tournament, with each game being a node denoted by its winner and the two previous games of the combatants as its descendants, or an arithmetic expression with dyadic operators, with each operator denoting a branch node with its operands as subtrees (see Fig. 4.20).

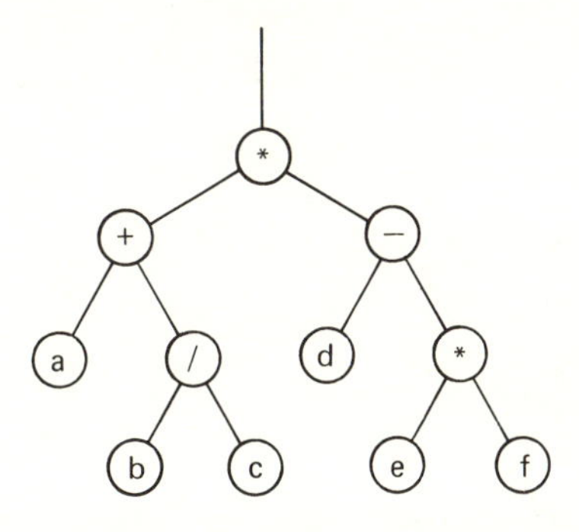

Fig. 4.20 Tree representation of the expression $(a + b/c)*(d - e*f)$.

We now turn to the problem of representation of trees. It is plain that the illustration of such recursive structures in terms of branching structures immediately suggests the use of our pointer facility. There is evidently no use in declaring variables with a fixed tree structure; instead, we define the *nodes* as variables with a fixed structure, i.e., of a fixed type, in which the degree of the tree determines the number of pointer components referring to the node's subtrees. Evidently, the reference to the empty tree is denoted by **nil**. Hence, the tree of Fig. 4.20 consists of components of a type defined as follows

```
type node = record op: char;
                   left,right: ↑node          (4.39)
            end
```

and may then be constructed as shown in Fig. 4.21.

There clearly exist ways of representing the abstract idea of a tree structure in terms of other available data types, such as arrays. This is common in all languages that do not provide the facility for allocating components dynamically and for referencing them via pointers. In this case, the tree in

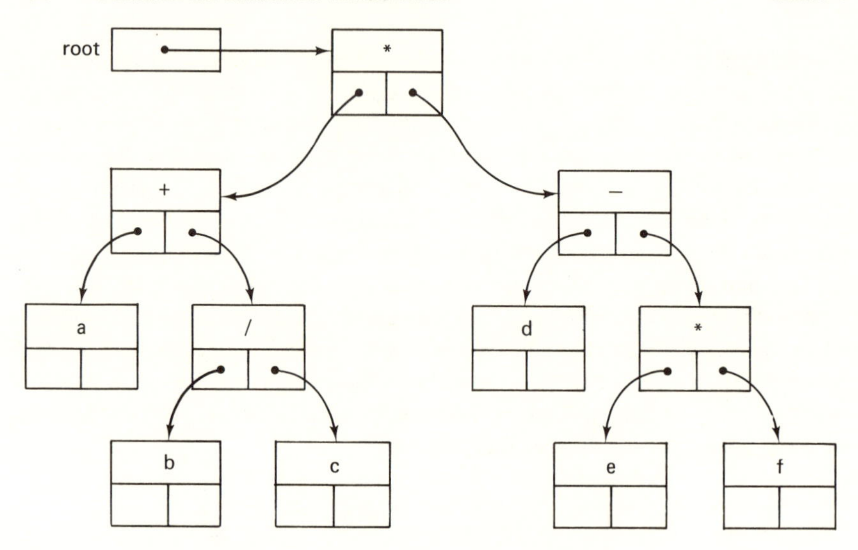

Fig. 4.21 Tree represented as data structure.

Fig. 4.20 might be represented by an array variable declared as

$$t\text{: } \mathbf{array}[1\;..\;11]\ \mathbf{of}\ \mathbf{record}\ op\text{: } char;\ left,right\text{: } integer\ \mathbf{end} \tag{4.40}$$

and with component values as shown in Table 4.3.

1	*	2	3
2	+	6	4
3	−	9	5
4	/	7	8
5	*	10	11
6	*a*	0	0
7	*b*	0	0
8	*c*	0	0
9	*d*	0	0
10	*e*	0	0
11	*f*	0	0

Table 4.3 Tree represented by an array.

Although the underlying, abstract structure of the data represented by the array *t* is a tree, we shall not call this a tree but rather an array according to the explicit declaration. We shall not further discuss other possibilities of representing trees in systems that lack a dynamic allocation facility, for we assume that programming systems and languages including this feature are or will become commonly available.

Before investigating how trees might be used advantageously and how to perform operations on trees, we give an example of how a tree may be constructed by a program. Assume that a tree is to be generated containing nodes of the type defined in (4.39), with the values of the nodes being n numbers read from an input file. In order to make the problem more challenging, let the task be the construction of a tree with n nodes and minimal height.

In order to obtain a minimal height for a given number of nodes, one has to allocate the maximum possible number of nodes of all levels except the lowest one. This can clearly be achieved by distributing incoming nodes equally to the left and right at each node. This implies that we structure the tree for given n as shown in Fig. 4.22, for $n = 1, \ldots, 7$.

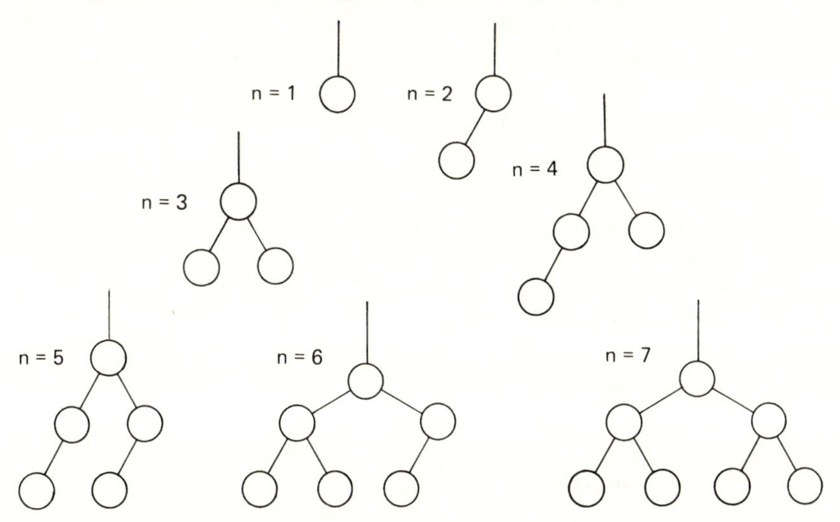

Fig. 4.22 Perfectly balanced trees.

The rule of equal distribution under a known number n of nodes is best formulated in recursive terms:

1. Use one node for the root.
2. Generate the left subtree with $nl = n$ **div** 2 nodes in this way.
3. Generate the right subtree with $nr = n - nl - 1$ nodes in this way.

The rule is expressed as a recursive procedure as part of Program 4.3 which reads the input file and constructs the perfectly balanced tree. We note the following definition:

> A tree is *perfectly balanced* if for each node the numbers of nodes in its left and right subtrees differ by at most 1.

```
program buildtree(input,output);
type ref = ↑node;
   node = record key: integer;
                 left, right: ref
          end ;
var n: integer; root: ref;

function tree(n: integer): ref;
   var newnode: ref;
      x, nl, nr: integer;
begin {construct perfectly balanced tree with n nodes}
   if n = 0 then tree := nil else
   begin nl := n div 2; nr := n-nl-1;
      read(x); new(newnode);
      with newnode↑ do
         begin key := x; left := tree(nl); right := tree(nr)
         end ;
      tree := newnode
   end
end {tree} ;

procedure printtree(t: ref; h: integer);
   var i: integer;
begin {print tree t with indentation h}
   if t ≠ nil then
   with t↑ do
   begin printtree(left, h+1);
      for i := 1 to h do write('     ');
      writeln(key);
      printtree(right, h+1)
   end
end {printtree} ;

begin {first integer is number of nodes}
   read(n);
   root := tree(n);
   printtree(root,0)
end .
```

Program 4.3 Construct Perfectly Balanced Tree.

Assume, for example, the following input data for a tree with 21 nodes.

21 8 9 11 15 19 20 21 7 3 2 1 5 6 4 13 14
10 12 17 16 18

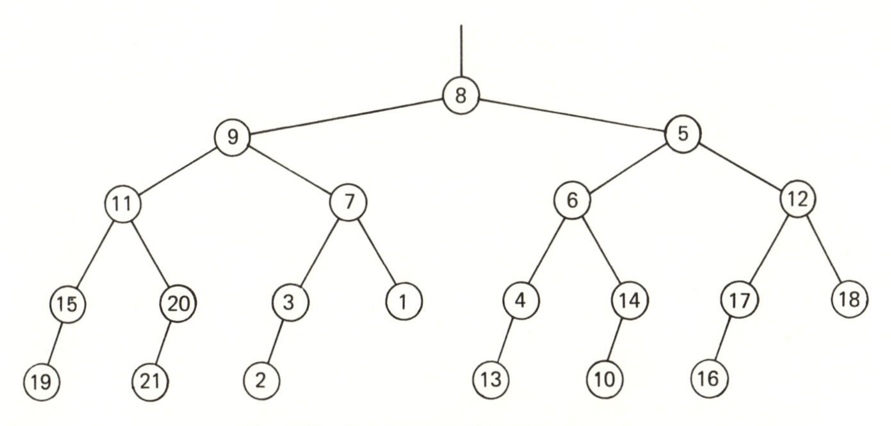

Fig. 4.23 Tree generated by Program 4.3.

Program 4.3 then constructs the perfectly balanced tree shown in Fig. 4.23.

Note the simplicity and transparence of this program that is obtained through the use of recursive procedures. It is obvious that recursive algorithms are particularly suitable when a program is to manipulate information whose structure is itself defined recursively. This is again manifested in the procedure which prints the resulting tree: the empty tree results in no printing, the subtree at level L in first printing its own left subtree, then the node, properly indented by preceding it with L blanks, and finally in printing its right subtree.

The advantage of the recursive algorithm becomes particularly plain by comparing it with a non-recursive formulation. The reader is explicity urged to use his ingenuity in writing a non-recursive equivalent of the above tree generator before looking at (4.41). This program is listed without further comments and may serve as a challenge for the reader to discover how and why it works.

```
program buildtree(input,output);
type ref = ↑node;
   node = record key: integer;
                 left, right: ref
          end ;
var i,n,nl,nr,x: integer;                        (4.41)
   root,p,q,r,dmy: ref;
   s: array [1 .. 30] of  {stack}
        record n: integer; rf: ref
        end ;
```

```
begin {first integer is number of nodes}
    read(n); new(root); new(dmy);  {dummy}
    i := 1; s[1].n := n; s[1].rf := root;
    repeat n := s[i] .n; r := s[i] .rf; i := i-1; {pop}
        if n = 0 then r↑.right := nil else
        begin p := dmy;
            repeat nl := n div 2; nr := n-nl-1;
                read(x); new(q); q↑.key := x;
                i := i+1; s[i].n := nr; s[i].rf := q; {push}
                n := nl; p↑.left := q; p := q
            until n = 0;
            q↑.left := nil; r↑.right := dmy↑.left
        end
    until i = 0;
    printtree (root↑.right,0)
end .
```

4.4.2. Basic Operations on Binary Trees

There are many tasks that may have to be perfomed on a tree structure; a common one is that of executing a given operation P on each element of the tree. P is then understood to be a parameter of the more general task of visiting all nodes or, as it is usually called, of *tree traversal.*

If we consider the task as a single sequential process, then the individual nodes are visited in some specific order and may be considered as being laid out in a linear arrangement. In fact, the description of many algorithms is considerably facilitated if we can talk about processing the *next* element in the tree based on an underlying order.

There are three principal orderings that emerge naturally from the structure of trees. Like the tree structure itself, they are conveniently expressed in recursive terms. Referring to the binary tree in Fig. 4.24 in which R denotes the root and A and B denote the left and right subtrees, the three orderings are

1. *Preorder* : R,A,B (visit root *before* the subtrees)
2. *Inorder* : A,R,B
3. *Postorder*: A,B,R (visit root *after* the subtrees)

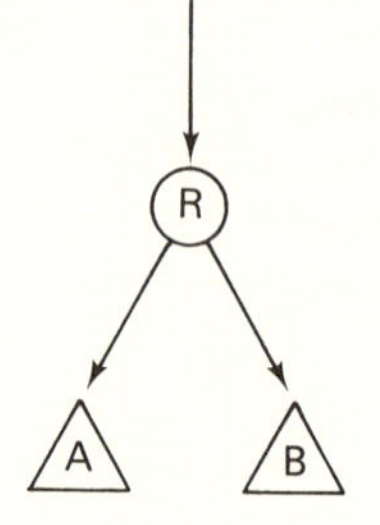

Fig. 4.24 Binary tree.

Traversing the tree of Fig. 4.20 and recording the characters seen at the nodes in the sequence of encounter, we obtain the following orderings:

1. *Preorder* : $* + a / b\ c - d * e\ f$
2. *Inorder* : $a + b / c * d - e * f$
3. *Postorder*: $a\ b\ c / + d\ e\ f * - *$

We recognize the three forms of expressions: preorder traversal of the expression tree yields *prefix* notation; postorder traversal generates *postfix* notation; and inorder traversal yields conventional *infix* notation, although without the parentheses necessary to express operator precedences.

Let us now formulate the three methods of traversal by three concrete programs with the explicit parameter t denoting the tree to be operated upon and with the implicit parameter p denoting the operation to be performed on each node. Assume the following definitions:

```
type ref  = ↑node
     node = record ...
               left,right: ref                    (4.42)
            end
```

The three methods are now readily formulated as *recursive* procedures; they demonstrate again the fact that operations on recursively defined data structures are most conveniently defined as recursive algorithms.

```
procedure preorder(t: ref);
begin if t ≠ nil then
      begin P(t);
            preorder(t↑.left);                    (4.43)
            preorder(t↑.right)
      end
end

procedure inorder(t: ref);
begin if t ≠ nil then
      begin inorder(t↑.left);
            P(t);                                 (4.44)
            inorder(t↑.right)
      end
end

procedure postorder(t: ref);
begin if t ≠ nil then
      begin postorder(t↑.left);
            postorder(t↑.right);                  (4.45)
            P(t)
      end
end
```

Note that the pointer t is passed as a value parameter. This expresses the fact that the relevant entity is the *reference* to the considered subtree and *not* the variable whose value is the pointer, and which could be changed in case t were passed as a variable parameter.

An example of a tree traversing routine is that of printing a tree, with appropriate indentation indicating each node's level (see Program 4.3).

Binary trees are frequently used to represent a set of data whose elements are to be retrievable through a unique key. If a tree is organized in such a way that for each node t_i, all keys in the left subtree of t_i are less than the key of t_i, and those in the right subtree are greater than the key of t_i, then this tree is called a *search tree*. In a search tree it is possible to locate an arbitrary key by starting at the root and proceeding along a search path switching to a node's left or right subtree by a decision based on inspection of that node's key only. As we have seen, n elements may be organized in a binary tree of a height as little as $\log n$. Therefore, a search among n items may be performed with as few as $\log n$ comparisons if the tree is perfectly balanced. Obviously, the tree is a much more suitable form for organizing such a set of data than the linear list used in the previous section.

As this search follows a single path from the root to the desired node, it can readily be programmed by iteration (4.46)

```
function loc(x: integer; t: ref): ref;
   var found: boolean;
begin found := false;
   while (t ≠ nil) ∧ ¬found do
   begin                                                    (4.46)
      if t↑.key = x then found := true else
      if t↑.key > x then t := t↑.left else t := t↑.right
   end;
   loc := t
end
```

The function $loc(x, t)$ has the value **nil** if no key with value x is found in the tree with root t. As in the case of the search through a list, the complexity of the termination condition gives rise to a search for a better solution. It consists in the use of a *sentinel* at the end of the list. This technique is equally applicable in the case of a tree. The use of pointers makes it possible for all branches of the tree to terminate with the same, identical sentinel. The resulting structure is no longer a tree, but rather a tree with all leaves tied down by strings to a single anchor point (Fig. 4.25). The sentinel may be considered as a common representative of all external nodes by which the original tree was extended (see Fig. 4.19). The resulting, simplified search routine is shown in (4.47).

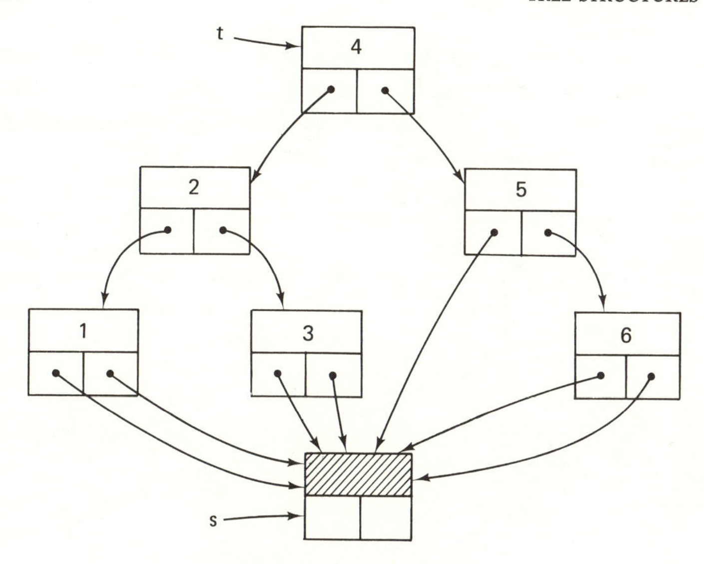

Fig. 4.25 Search tree with sentinel.

```
function loc(x: integer; t: ref): ref;
begin s↑.key := x; {sentinel}
   while t↑.key ≠ x do                                        (4.47)
      if x < t↑.key then t := t↑.left else t := t↑.right;
   loc := t
end
```

Note that in this case $loc(x, t)$ obtains the value s, i.e., the pointer to the sentinel, if no key with value x is found in the tree with root t. s simply assumes the role of the **nil** pointer.

4.4.3. Tree Search and Insertion

The full power of the dynamic allocation technique with access through pointers is hardly displayed by those examples in which a given set of data is built, and thereafter kept unchanged. More suitable examples are those applications in which the structure of the tree itself varies, i.e., grows and/or shrinks during the execution of the program. This is also the case in which other data representations, such as the array, fail and in which the tree with elements linked by pointers emerges as *the* appropriate solution.

We shall first consider only the case of a steadily growing but never shrinking tree. An appropriate example is the concordance problem which was already investigated in connection with linked lists. It is now to be revisited. In this problem a sequence of words is given, and the number of occurrences of each word has to be determined. This means that—starting out with an empty tree—each word is searched in the tree. If it is found, its occurrence count is incremented; otherwise it is inserted as a new word (with a count initialized to 1). We call the underlying task *tree search with insertion*. The following data type definitions are assumed:

```
type ref = ↑word;
     word = record
               key: integer;
               count: integer;                    (4.48)
               left, right: ref
            end
```

Assuming moreover a source file f of keys and a variable denoting the root of the search tree, we may formulate the program as

```
reset(f);
while ¬eof(f) do                                   (4.49)
   begin read(f, x); search(x,root) end
```

Finding the search path is again straightforward. However, if it leads to a "dead end" (i.e., to an empty subtree designated by a pointer value **nil**), then the given word must be inserted in the tree at the place of the empty subtree. Consider, for example, the binary tree shown in Fig. 4.26 and the insertion of the word "Paul." The result is shown in dotted lines in the same picture.

The entire operation is shown in Program 4.4. The search process is formulated as a recursive procedure. Note that its parameter p is a variable

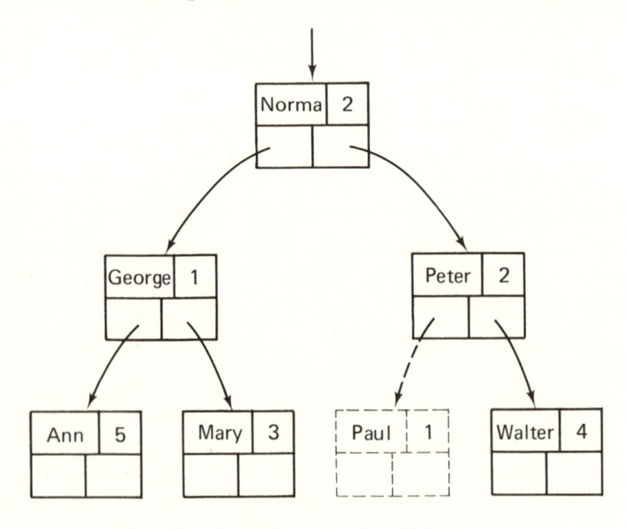

Fig. 4.26 Insertion in ordered binary tree.

parameter and *not* a value parameter. This is essential because in the case of insertion a new pointer value must be assigned to the *variable* which previously held the value **nil**. Using the input sequence of 21 numbers that had been applied to Program 4.3 to construct the tree of Fig. 4.23, Program 4.4 yields the binary search tree shown in Fig. 4.27.

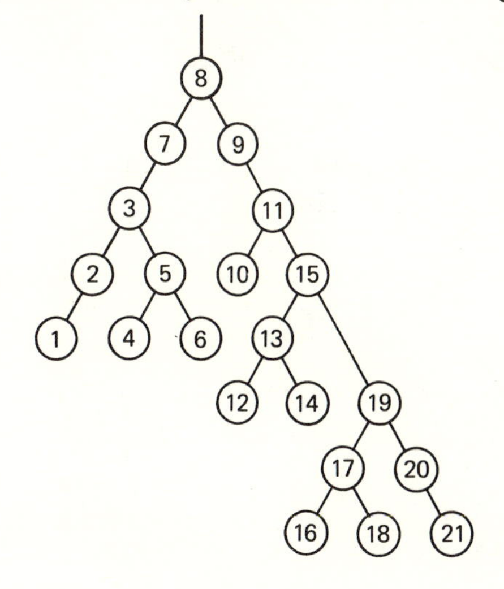

Fig. 4.27 Search tree generated by Program 4.4.

Program 4.4 Tree Search and Insertion.

```
program treesearch(input,output);
{binary tree search and insertion}
type ref = ↑word;
    word = record key: integer;
                  count: integer;
                  left, right: ref;
           end ;
var root: ref; k: integer;

procedure printtree(w: ref; l: integer);
    var i: integer;
begin if w ≠ nil then
        with w↑ do
        begin printtree(left, l+1);
              for i := 1 to l do write('      ');
              writeln(key);
              printtree(right, l+1)
        end
end ;
```

```
procedure search(x: integer; var p: ref);
begin
   if p = nil then
   begin {word is not in tree; insert it}
      new (p);
      with p↑ do
         begin key := x; count := 1; left := nil; right := nil
         end
   end else
   if x < p↑.key then search(x, p↑.left) else
   if x > p↑.key then search(x, p↑.right) else
      p↑.count := p↑.count + 1
end {search} ;

begin root := nil;
   while ¬eof(input) do
      begin read(k); search(k, root)
      end ;
   printtree(root,0)
end .
```

Program 4.4 (Continued)

The use of a sentinel again simplifies the task somewhat, as is shown in (4.50). Clearly, at the start of the program the variable *root* must be initialized by the pointer to the sentinel instead of the value **nil**, and before each search the searched value x must be assigned to the key field of the sentinel.

```
procedure search(x: integer; var p: ref);
begin
   if x < p↑.key then search(x,p↑.left) else
   if x > p↑.key then search(x,p↑.right) else
   if p ≠ s then p↑.count := p↑.count + 1 else
     begin {insert} new(p);                                  (4.50)
        with p↑ do
        begin key := x; left := s; right := s; count := 1
        end
     end
end
```

Once again, and for the last time, we will develop an alternative version of this program, refraining from the use of recursion. Avoiding recursion is not as trivial now as without insertion, for if an insertion has to be performed, the traversed path must be remembered at least one step backward. This function of remembering has been automatically achieved in Program 4.4 through the use of a variable parameter.

In order to link the inserted component correctly, we must know the reference to its ancestor and know whether it is to be inserted as the ancestor's left or right subtree. For this purpose, two variables *p*2 and *d* (for *d*irection) are introduced.

```
procedure search(x: integer; root: ref);
    var p1,p2: ref; d: integer;
begin p2 := root; p1 := p2↑.right; d := 1;
    while (p1≠nil) ∧ (d≠0) do
    begin p2 := p1;
        if  x < p1↑.key then
            begin p1 := p1↑.left; d := −1 end else
        if  x > p1↑.key then
            begin p1 := p1↑.right; d := 1 end else
        d := 0
    end ;                                                    (4.51)
    if d = 0 then p1↑.count := p1↑.count + 1 else
        begin {insert} new(p1);
            with p1↑ do
            begin key := x; left := nil; right := nil; count := 1
            end ;
            if d < 0 then p2↑.left := p1 else p2↑.right := p1
        end
end
```

As in the case of list search and insertion, two pointers *p*1 and *p*2 are used which traverse the search path such that *p*2 always designates the ancestor of *p*1↑. In order to start the search process with this condition being satisfied, an auxiliary and dummy element is introduced, designated by the pointer called *root*. The origin of the actual search tree is designated by the pointer *root*↑.*right*. The program must therefore start with the statements

new(*root*); *root*↑.*right* := **nil**

in place of the original assignment

root := **nil**

Although the purpose of this algorithm is searching, it can be used for sorting as well. In fact, it resembles the sorting by insertion method quite strongly, and because of the use of a tree structure instead of an array, the need for relocation of the components above the insertion point vanishes. Tree sorting can be programmed to be almost as efficient as the best array sorting methods known. But a few precautions must be taken. Of course, the case of encountering a matching key must be treated differently now. If the case $x = p\uparrow.key$ is handled identically to the case $x > p\uparrow.key$, then the algorithm represents a stable sorting method, i.e., items with identical keys turn up in the same sequence when scanning the tree in normal order as when they were inserted.

In general, there are better ways to sort, but in applications in which searching and sorting are both needed, the tree search and insertion algorithm is strongly recommended. It is, in fact, very often applied in compilers and in data banks to organize the objects to be stored and retrieved. An appropriate example is the construction of a *cross-reference index* for a given text. Let us pursue this problem in detail.

Our task is to construct a program that (while reading a text f and printing it after supplying consecutive line numbers) collects all words of this text, thereby retaining the numbers of the lines in which each word occurred. When this scan is terminated, a table is to be generated containing all collected words in alphabetical order with lists of their occurrences.

Obviously, the search tree (also called a *lexicographic* tree) is a most suitable candidate for representing the words encountered in the text. Each node now not only contains a word as key value, but it is also the head of a list of line numbers. We shall call each recording of an occurrence an *item*. Hence, we encounter both trees and linear lists in this example. The program consists of two main parts (see Program 4.5), namely, the scanning

Program 4.5 Cross Reference Generator.

```
program crossref (f,output);
{cross reference generator using binary tree}
const c1  =  10;     {length of words}
      c2  =   8;     {numbers per line}
      c3  =   6;     {digits per number}
      c4  =  9999;   {max line number}
type alfa  =  packed array [1 .. c1] of char;
      wordref  =  ↑word;
      itemref  =  ↑item;
      word  =  record key: alfa;
                     first, last: itemref;
                     left, right: wordref
               end ;
      item  =  packed record
                     lno: 0 .. c4;
                     next: itemref
               end ;
var root: wordref;
      k,k1: integer;
      n: integer;           {current line number}
      id: alfa;
      f: text;
      a: array [1 .. c1] of char;
```

```
procedure search (var w1: wordref);
   var w: wordref; x: itemref;
begin w := w1;
   if w = nil then
   begin new(w); new(x);
      with w↑ do
      begin key := id; left := nil; right := nil;
         first := x; last := x
      end ;
      x↑.lno := n; x↑.next := nil; w1 := w
   end else
   if id < w↑.key then search(w↑.left) else
   if id > w↑.key then search(w↑.right) else
   begin new(x); x↑.lno := n; x↑.next := nil;
      w↑.last↑.next := x; w↑.last := x
   end
end {search} ;
procedure printtree(w: wordref);
   procedure printword(w: word);
      var l: integer; x: itemref;
   begin write ('   ', w.key);
      x := w.first; l := 0;
      repeat if l = c2 then
               begin writeln;
                  l := 0; write ('   ':c1+1)
               end ;
            l := l+1; write (x↑.lno:c3); x := x↑.next
      until x = nil;
      writeln
   end {printword} ;
begin if w ≠ nil then
   begin printtree(w↑.left);
         printword(w↑); printtree(w↑.right)
   end
end {printtree} ;

begin root := nil; n := 0; k1 := c1;
   page (output); reset(f);
   while ¬eof (f) do
   begin if n = c4 then n := 0;
      n := n+1; write (n:c3);          {next line}
      write ('   ');
```

Program 4.5 (Continued)

```
      while ¬eoln (f) do
      begin {scan non-empty line}
         if f↑ in ['A' .. 'Z'] then
         begin k := 0;
            repeat if k < c1 then
               begin k := k+1; a[k] := f↑;
               end ;
               write (f↑); get (f)
            until ¬(f↑ in ['A' .. 'Z','0' .. '9']);
            if k ≥ k1 then k1 := k else
            repeat a[k1] := ' '; k1 := k1−1
            until k1 = k;
            pack (a,1,id); search(root)
         end else
         begin {check for quote or comment}
            if f↑ = '''' then
               repeat write(f↑); get(f)
               until f↑ = '''' else
            if f↑ = '{' then
               repeat write(f↑); get(f)
               until f↑ = '}' ;
            write (f↑); get (f)
         end
      end ;
      writeln; get(f)
   end ;
   page(output); printtree(root);
end .
```

Program 4.5 (Continued)

phase and the table printing phase. The latter is a straightforward application of a tree traversal routine in which visiting each node implies the printing of the key value (word) and the scanning of its associated list of line numbers (items). Following are further clarifications regarding the *Cross-Reference Generator* of Program 4.5:

1. A word is considered as any sequence of letters and digits starting with a letter.
2. Only the first $c1$ characters are retained as key. Thus, two words not differing in their $c1$ first characters are considered identical.
3. The $c1$ characters are packed into an array *id* (type *alfa*). If $c1$ is sufficiently small, many computers will be able to compare such packed arrays by a single instruction.
4. The variable $k1$ is used as an index that maintains the following invariant

condition about the character buffer a:

$$a[i] = ' ' \qquad \text{for } i = k1 + 1 \ldots c1$$

Words consisting of fewer than $c1$ characters are extended by an appropriate number of blanks.

5. It is desirable that the line numbers be printed in ascending order in the cross-reference index. Therefore, the item lists must be generated in the same order as they are scanned upon printing. This requirement suggests the use of two pointers in each word node, one referring to the first, and one referring to the last item on the list.
6. The scanner is constructed so that words within quotes and within comments are omitted from the index, assuming that quotations and comments do not extend over line ends.

Table 4.4 shows the results of processing a short program text.

Table 4.4 Sample Output of Program 4.5.

```
 1 PROGRAM PERMUTE (OUTPUT);
 2    CONST N = 4;
 3    VAR I: INTEGER;
 4        A: ARRAY [1 .. N] OF INTEGER;
 5
 6    PROCEDURE PRINT;
 7       VAR I: INTEGER;
 8    BEGIN FOR I := 1 TO N DO WRITE (A[I]:3);
 9       WRITELN
10    END {PRINT} ;
11
12    PROCEDURE PERM (K: INTEGER);
13       VAR I,X: INTEGER;
14    BEGIN
15       IF K = 1 THEN PRINT ELSE
16       BEGIN PERM (K-1);
17          FOR I := 1 TO K-1 DO
18          BEGIN X := A[I]; A[I] := A[K]; A[K] := X;
19             PERM (K-1);
20                X := A[I]; A[I] := A[K]; A[K] := X;
21          END
22       END
23    END {PERM} ;
24
25 BEGIN
26    FOR I := 1 TO N DO A[I] := I;
27    PERM (N)
28 END .
```

ARRAY	4							
A	4	8	18	18	18	18	20	20
	20	20	26					
BEGIN	8	14	16	18	25			
CONST	2							
DO	8	17	26					
ELSE	15							
END	10	21	22	23	28			
FOR	8	17	26					
IF	15							
INTEGER	3	4	7	12	13			
I	3	7	8	8	13	17	18	18
	20	20	26	26	26			
K	12	15	16	17	18	18	19	20
	20							
N	2	4	8	26	27			
OF	4							
OUTPUT	1							
PERMUTE	1							
PERM	12	16	19	27				
PRINT	6	15						
PROCEDURE	6	12						
PROGRAM	1							
THEN	15							
TO	8	17	26					
VAR	3	7	13					
WRITELN	9							
WRITE	8							
X	13	18	18	20	20			

Table 4.4 (Continued)

4.4.4. Tree Deletion

We now turn to the inverse problem of insertion, namely, deletion. Our task is to define an algorithm for deleting, i.e., removing the node with key x in a tree with ordered keys. Unfortunately, removal of an element is not generally as simple as insertion. It is straightforward if the element to be deleted is a terminal node or one with a single descendant. The difficulty lies in removing an element with *two* descendants, for we cannot point in two directions with a single pointer. In this situation, the deleted element is to be replaced by either the rightmost element of its left subtree or by the left-most node of its right subtree, both of which have at most one descendant. The details are shown in the recursive procedure called *delete* (4.52). This procedure distinguishes among three cases:

1. There is no component with a key equal to x.

2. The component with key x has at most one descendant.
3. The component with key x has two descendants.

```
procedure delete (x: integer; var p: ref);
    var q: ref;
    procedure del (var r: ref);
    begin if r↑.right ≠ nil then del (r↑.right) else
            begin q↑.key := r↑.key; q↑.count := r↑.count;
                q := r; r := r↑.left
            end
    end ;
begin {delete}                                                        (4.52)
    if p = nil then writeln (' WORD IS NOT IN TREE') else
    if x < p↑.key then delete(x, p↑.left) else
    if x > p↑.key then delete(x, p↑.right) else
    begin {delete p↑}  q := p;
        if q↑.right = nil then p := q↑.left else
        if q↑.left = nil then p := q↑.right else del (q↑.left);
        {dispose(q)}
    end
end {delete}
```

The auxiliary, recursive procedure *del* is activated in case 3 only. It "descends" along the rightmost branch of the left subtree of the element $q\uparrow$ to be deleted, and then it replaces the relevant information (key and count) in $q\uparrow$ by the corresponding values of the rightmost component $r\uparrow$ of that left subtree, whereafter $r\uparrow$ may be disposed of. The unspecified procedure *dispose*(q) may be considered the inverse or opposite of *new*(q). The latter allocates storage for a new component, but the former may be used to indicate to a computer system that storage occupied by $q\uparrow$ is again disposable and reusable (sort of recycling of storage).

In order to illustrate the functioning of procedure (4.52), we refer to Fig. 4.28. Consider the tree (a); then delete successively the nodes with keys 13, 15, 5, 10. The resulting trees are shown in Fig. 4.28 (b-e).

4.4.5. Analysis of Tree Search and Insertion

It is a natural—and healthy—reaction to be suspicious of the algorithm of tree search and insertion. At least one should retain some skepticism until having been given a few more details about its behavior. What worries many programmers at first is the peculiar fact that generally we do not know how the tree will grow; we have no idea about the shape that it will assume. We can only guess that it will most probably not be the perfectly balanced

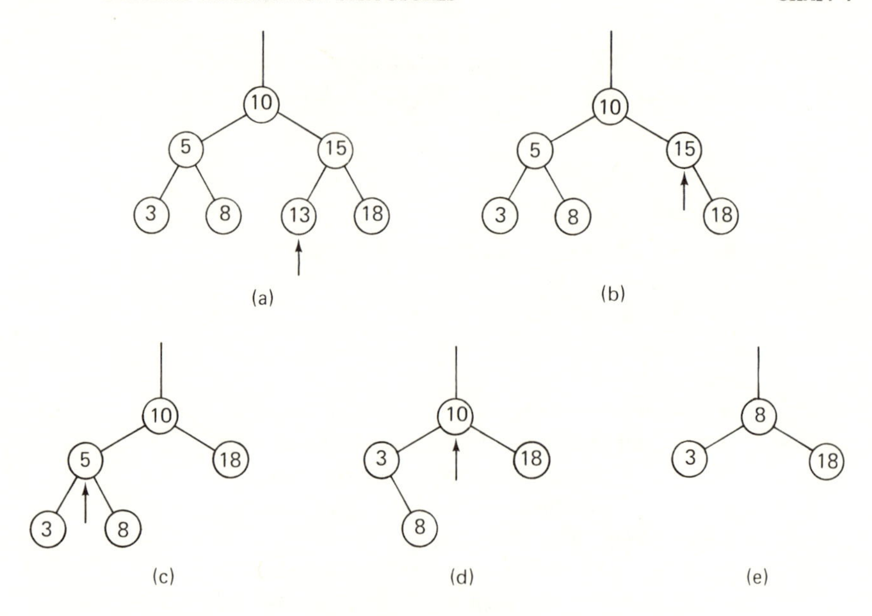

Fig. 4.28 Tree deletion.

tree. Since the average number of comparisons needed to locate a key in a perfectly balanced tree with n nodes is approximately $h = \log n$, the number of comparisons in a tree generated by this algorithm will be greater than h. But how much greater?

First of all, it is easy to find the worst case. Assume that all keys arrive in already strictly ascending (or descending) order. Then each key is appended immediately to the right (left) of its predecessor, and the resulting tree becomes completely degenerate, i.e., it turns out to be a linear list. The average search effort is then $n/2$ comparisons. This worst case evidently leads to a very poor performance of the search algorithm, and it seems to fully justify our skepticism. The remaining question is, of course, how likely this case will be. More precisely, we should like to know the length a_n of the search path averaged over all n keys and averaged over all $n!$ trees which are generated from the $n!$ permutations of the original n distinct keys. This problem of algorithmic analysis turns out to be fairly straightforward, and it is presented here as a typical example of analyzing an algorithm as well as for the practical importance of its result.

Given are n distinct keys with values $1, 2, \ldots, n$. Assume that they arrive in a random order. The probability of the first key—which notably becomes the root node—having the value i is $1/n$. Its left subtree will eventually contain $i - 1$ nodes, and its right subtree $n - i$ nodes (see Fig. 4.29). Let the average path length in the left subtree be denoted by a_{i-1}, and the one in the right subtree is a_{n-i}, again assuming that all possible permutations of the remaining $n - 1$ keys are equally likely. The average path length in a tree

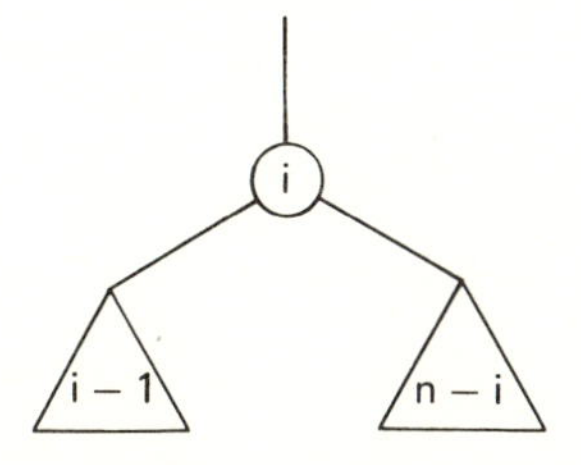

Fig. 4.29 Weight distribution of branches.

with n nodes is the sum of the products of each node's level and its probability of access. If all nodes are assumed to be wanted with equal likelihood, then

$$a_n = \frac{1}{n} \sum_{i=1}^{n} p_i \tag{4.53}$$

where p_i is the path length of node i.

In the tree in Fig. 4.29 we divide the nodes into three classes:

1. The $i - 1$ nodes in the left subtree have an average path length $a_{i-1} + 1$.
2. The root has a path length of 1.
3. The $n - i$ nodes in the right subtree have an average path length $a_{n-i} + 1$.

Hence, (4.53) can be expressed as a sum of three terms

$$a_n^{(i)} = (a_{i-1} + 1)\frac{i-1}{n} + 1 \cdot \frac{1}{n} + (a_{n-i} + 1)\frac{n-i}{n} \tag{4.54}$$

The desired quantity a_n is now derived as the average of $a_n^{(i)}$ over all $i = 1 \ldots n$, i.e., over all trees with the key $1, 2, \ldots, n$ at the root.

$$\begin{aligned} a_n &= \frac{1}{n} \sum_{i=1}^{n} \left[(a_{i-1} + 1)\frac{i-1}{n} + \frac{1}{n} + (a_{n-1} + 1)\frac{n-i}{n} \right] \\ &= 1 + \frac{1}{n^2} \sum_{i=1}^{n} [(i-1)a_{i-1} + (n-i)a_{n-i}] \\ &= 1 + \frac{2}{n^2} \sum_{i=1}^{n} (i-1)a_{i-1} = 1 + \frac{2}{n^2} \sum_{i=1}^{n-1} i \cdot a_i \end{aligned} \tag{4.55}$$

Equation (4.55) is a recurrence relation for a_n of the form $a_n = f_1(a_1, a_2, \ldots, a_{n-1})$. From this we can derive a simpler recurrence relation of the form $a_n = f_2(a_{n-1})$ as follows:

From (4.55) we derive directly

(1) $$a_n = 1 + \frac{2}{n^2} \sum_{i=1}^{n-1} i \cdot a_i = 1 + \frac{2}{n^2}(n-1)a_{n-1} + \frac{2}{n^2} \sum_{i=1}^{n-2} i \cdot a_i$$

(2) $$a_{n-1} = 1 + \frac{2}{(n-1)^2} \sum_{i=1}^{n-2} i \cdot a_i$$

Multiplying (2) by $((n-1)/2)^2$, we obtain

(3) $$\frac{2}{n^2} \sum_{i=1}^{n-2} i \cdot a_i = \frac{(n-1)^2}{n^2}(a_{n-1} - 1)$$

and substituting (3) in (1), we find

$$a_n = \frac{1}{n^2}((n^2 - 1)a_{n-1} + 2n - 1) \tag{4.56}$$

It turns out that a_n can be expressed in non-recursive, closed form in terms of the harmonic function

$$\begin{aligned} H_n &= 1 + \frac{1}{2} + \frac{1}{3} + \cdots + \frac{1}{n} \\ a_n &= 2\frac{n+1}{n}H_n - 3 \end{aligned} \tag{4.57}$$

[The skeptical reader should verify that (4.57) satisfies the recurrence relation (4.56).]

From Euler's formula (using Euler's constant $\gamma \cong 0.577$)

$$H_n = \gamma + \ln(n) + \frac{1}{12n^2} + \cdots$$

we derive, for large n, the relationship

$$a_n \cong 2[\ln(n) + \gamma] - 3 = 2\ln(n) - c$$

Since the average path length in the perfectly balanced tree is approximately

$$a'_n = \log(n) - 1 \tag{4.58}$$

we obtain, neglecting the constant terms which become insignificant for large n,

$$\lim_{n\to\infty} \frac{a_n}{a'_n} = \frac{2\ln n}{\log n} = 2 \cdot \ln 2 \cong 1.386 \tag{4.59}$$

What does the result (4.59) of this analysis teach us? It tells us that by taking the pains of always constructing a perfectly balanced tree instead of the "random" tree obtained from Program 4.4, we could—always provided that all keys are looked up with equal probability—expect an average improvement in the search path length of at most 39%. Emphasis is to be put on the word "average," for the improvement may of course be very much greater in the unhappy case in which the generated tree had completely degenerated into a list, which, however, is very unlikely to occur (if all permutations of the n keys to be inserted are equally probable). In this connection it is noteworthy that the expected average path length of the "random" tree grows also strictly logarithmically with the number of its nodes, even though in the worst case the path length grows linearly.

The figure of 39% imposes a limit on the amount of additional effort that may be spent profitably on any kind of re-organization of the tree's structure upon insertion of keys. Naturally, the ratio r between the frequencies of access (retrieval) of nodes (information) and of insertion significantly influences the payoff limits of any such undertaking. The higher this ratio, the higher is the payoff of a re-organization procedure. The 39% figure is low enough that in most applications improvements of the straight tree insertion algorithm do not pay off unless the number of nodes *and* the access vs. insertion ratio are large (or if one is afraid of the worst case).

4.4.6. Balanced Trees

From the preceding discussion it is clear that an insertion procedure that always restores the trees' structure to perfect balance has hardly any chance of being profitable because the restoration of perfect balance after a random insertion is a fairly intricate operation. Possible improvements lie in the formulation of less strict definitions of "balance." Such "imperfect" balance criteria should lead to simpler tree re-organization procedures at the cost of only a slight deterioration of average search performance.

One such definition of balance has been postulated by Adelson-Velskii and Landis [4-1]. The balance criterion is the following:

> A tree is *balanced* if and only if for every node the heights of its two subtrees differ by at most 1.

Trees satisfying this condition are often called AVL-trees (after their inventors). We shall simply call them *balanced trees* because this balance criterion appears a most suitable one. (Note that all perfectly balanced trees are also AVL-balanced.)

The definition is not only simple, but it also leads to a manageable rebalancing procedure and an average search path length practically identical to that of the perfectly balanced tree.

The following operations can be performed on balanced trees with O (log n) units of time, even in the worst case:

1. Locate a node with a given key.
2. Insert a node with a given key.
3. Delete the node with a given key.

These statements are direct consequences of a theorem proved by Adelson-Velskii and Landis, which guarantees that a balanced tree will never be more than 45% higher than its perfectly balanced counterpart, no matter how many nodes there are. If we denote the height of a balanced tree with n nodes by $h_b(n)$, then

$$\log (n + 1) \leq h_b(n) \leq 1.4404 \cdot \log (n + 2) - 0.328 \qquad (4.60)$$

The optimum is of course reached if the tree is perfectly balanced for $n = 2^k - 1$. But which is the structure of the worst AVL-balanced tree?

In order to find the maximum height h of all balanced trees with n nodes, let us consider a fixed h and try to construct the balanced tree with the minimum number of nodes. This strategy is recommended because, as in the case of the minimal h, the value can be attained only for certain specific values of n. Let this tree of height h be denoted by T_h. Clearly, T_0 is the empty tree, and T_1 is the tree with a single node. In order to construct the tree T_h for $h > 1$, we will provide the root with two subtrees which again have a minimal number of nodes. Hence, the subtrees are also T's. Evidently, one subtree *must* have height $h - 1$, and the other is then allowed to have a height of one less, i.e., of $h - 2$. Figure 4.30 shows the trees with height 2,

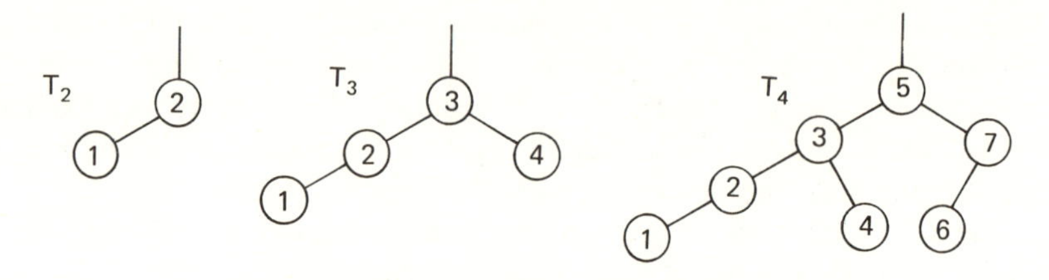

Fig. 4.30 Fibonacci-trees of height 2, 3, and 4.

3, and 4. Since their composition principle very strongly resembles that of Fibonacci numbers, they are called *Fibonacci-trees*. They are defined as follows:

1. The empty tree is the Fibonacci-tree of height 0.
2. A single node is the Fibonacci-tree of height 1.
3. If T_{h-1} and T_{h-2} are Fibonacci-trees of heights $h - 1$ and $h - 2$, then $T_h = \langle T_{h-1}, x, T_{h-2} \rangle$ is a Fibonacci-tree of height h.
4. No other trees are Fibonacci-trees.

The number of nodes of T_h is defined by the following simple recurrence relation:

$$N_0 = 0, \qquad N_1 = 1$$
$$N_h = N_{h-1} + 1 + N_{h-2} \tag{4.61}$$

The N_i are those numbers of nodes for which the worst case (upper limit of h) of (4.60) can be attained.

4.4.7. Balanced Tree Insertion

Let us now consider what may happen when a new node is inserted in a balanced tree. Given a root r with the left and right subtrees L and R, three cases must be distinguished. Assume that the new node is inserted in L causing its height to increase by 1:

1. $h_L = h_R$: L and R become of unequal height, but the balance criterion is not violated.

2. $h_L < h_R$: L and R obtain equal height, i.e., the balance has even been improved.
3. $h_L > h_R$: the balance criterion is violated, and the tree must be restructured.

Consider the tree in Fig. 4.31. Nodes with keys 9 and 11 may be inserted without rebalancing; the tree with root 10 will become one-sided (case 1); the one with root 8 will improve its balance (case 2). Insertion of nodes 1, 3, 5, or 7, however, requires subsequent rebalancing.

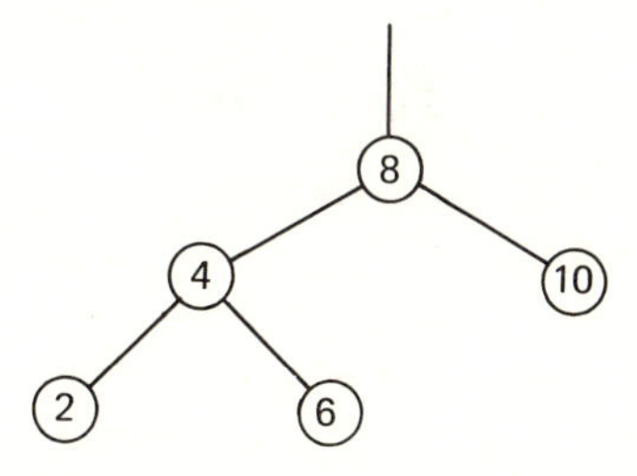

Fig. 4.31 Balanced tree.

Some careful scrutiny of the situation reveals that there are only two essentially different constellations needing individual treatment. The remaining ones can be derived by symmetry considerations from those two. Case 1 is characterized by inserting keys 1 or 3 in the tree of Fig. 4.31, Case 2 by inserting nodes 5 or 7.

The two cases are generalized in Fig. 4.32 in which rectangular boxes denote subtrees, and the height added by the insertion is indicated by crosses. Simple transformations of the two structures restore the desired balance. Their result is shown in Fig. 4.33; note that the only movements allowed are those occurring in the vertical direction, whereas the relative horizontal positions of the shown nodes and subtrees must remain unchanged.

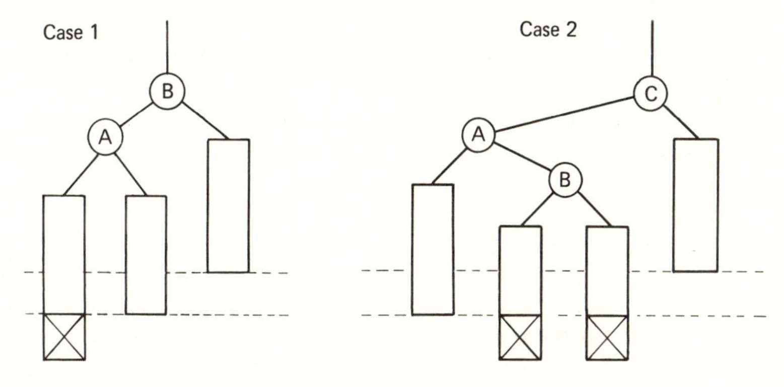

Fig. 4.32 Imbalance resulting from insertion.

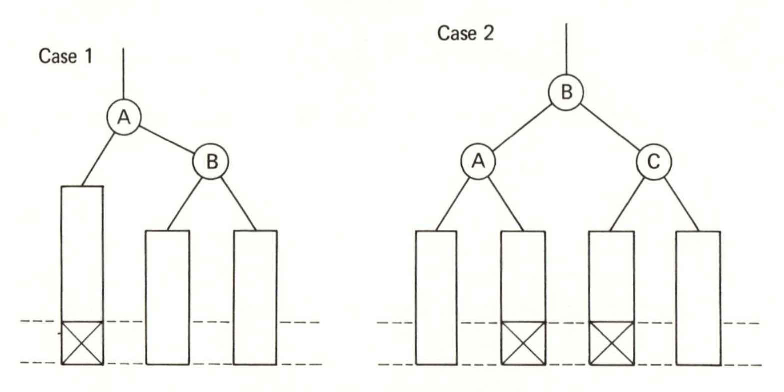

Fig. 4.33 Restoring the balance.

An algorithm for insertion and rebalancing critically depends on the way information about the tree's balance is stored. An extreme solution lies in keeping balance information entirely implicit in the tree structure itself. In this case, however, a node's balance factor must be rediscovered each time it is affected by an insertion, resulting in an excessively high overhead. The other extreme is to attribute an explicitly stored balance factor to every node. The definition (4.48) of the node type is then extended into

```
type node = record key: integer;
                   count: integer;
                   left,right: ref;                (4.62)
                   bal: -1 .. +1
            end
```

We shall subsequently interpret a node's balance factor as the height of its right subtree minus the height of its left subtree, and we shall base the resulting algorithm on the node type (4.62).

The process of node insertion consists essentially of the following three consecutive parts:

1. Follow the search path until it is verified that the key is not already in the tree.
2. Insert the new node and determine the resulting balance factor.
3. Retreat along the search path and check the balance factor at each node.

Although this method involves some redundant checking (once balance is established, it need not be checked on that node's ancestors), we shall first adhere to this evidently correct schema because it can be implemented through a pure extension of the already established search and insertion procedure of Program 4.4. This procedure describes the search operation needed at each single node, and because of its recursive formulation it can easily accommodate an additional operation "on the way back along the search path." At each step, information must be passed as to whether or not the height of the subtree (in which the insertion had been performed) had increased. We therefore extend the procedure's parameter list by the Boolean h with the meaning "*the subtree height has increased.*" Clearly, h must denote a variable parameter since it is used to transmit a result.

Assume now that the process is returning to a node $p\uparrow$ from the left branch (see Fig. 4.32), with the indication that it has increased its height. We now must distinguish between the three situations involving the subtree heights prior to insertion:

1. $h_L < h_R$, $p\uparrow.bal = +1$, the previous imbalance at p has been equilibrated.
2. $h_L = h_R$, $p\uparrow.bal = 0$, the weight is now slanted to the left.
3. $h_L > h_R$, $p\uparrow.bal = -1$, rebalancing is necessary.

In the third case, inspection of the balance factor of the root of the left subtree (say, $p1\uparrow.bal$) determines whether case 1 or case 2 of Fig. 4.32 is present. If that node has also a higher left than right subtree, then we have to deal with case 1, otherwise with case 2. (Convince yourself that a left subtree with a balance factor equal to 0 at its root cannot occur in this case.) The rebalancing operations necessary are entirely expressed as sequences of pointer re-assignments. In fact, pointers are cyclically exchanged, resulting in either a single or a double rotation of the two or three nodes involved. In addition to pointer rotation, the respective node balance factors also have to be adjusted. The details are shown in the search, insertion, and rebalancing procedure (4.63).

The working principle is shown by Fig. 4.34. Consider the binary tree

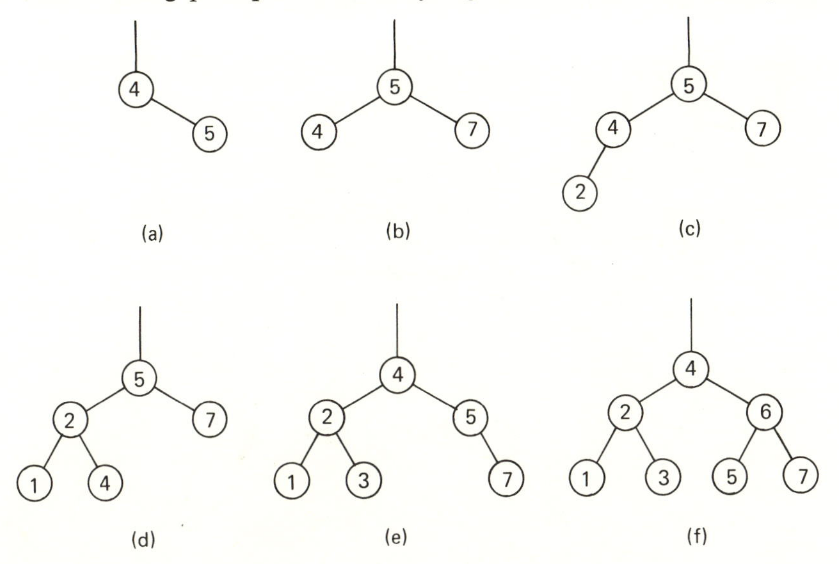

Fig. 4.34 Insertions in balanced tree.

(a) which consists of two nodes only. Insertion of key 7 first results in an unbalanced tree (i.e., a linear list). Its balancing involves a *RR* single rotation, resulting in the perfectly balanced tree (b). Further insertion of nodes 2 and 1 result in an imbalance of the subtree with root 4. This subtree is balanced by an *LL* single rotation (d). The subsequent insertion of key 3 immediately offsets the balance criterion at the root node 5. Balance is thereafter re-established by the more complicated *LR* double rotation; the outcome is tree (e). The only candidate for loosing balance after a next insertion is node 5. Indeed, insertion of node 6 must invoke the fourth case of rebalancing outlined in (4.63), the *RL* double rotation. The final tree is shown in Fig. 4.34(f).

```
procedure search(x: integer; var p: ref; var h: boolean);
   var p1,p2: ref;      {h = false}
begin
   if p = nil then
   begin {word is not in tree; insert it}
      new(p); h := true;
      with p↑ do
      begin key := x; count := 1;
         left := nil; right := nil; bal := 0
      end
   end else
   if x < p↑.key then
   begin search(x, p↑.left, h);
      if h then     {left branch has grown higher}
      case p↑.bal of                                              (4.63)
    1: begin p↑.bal := 0; h := false
       end ;
    0: p↑.bal := -1;
   -1: begin {rebalance} p1 := p↑.left;
          if p1↑.bal = -1 then
          begin {single LL rotation}
             p↑.left := p1↑.right; p1↑.right := p;
             p↑.bal := 0; p := p1
          end else
          begin {double LR rotation}  p2 := p1↑.right;
             p1↑.right := p2↑.left; p2↑.left := p1;
             p↑.left := p2↑.right; p2↑.right := p;
             if p2↑.bal = -1 then p↑.bal := +1 else p↑.bal := 0;
             if p2↑.bal = +1 then p1↑.bal := -1 else p1↑.bal := 0;
             p := p2
          end ;
```

```
            p↑.bal := 0; h := false
         end
      end
   end else
   if x > p↑.key then
   begin search(x, p↑.right, h);
      if h then      {right branch has grown higher}
      case p↑.bal of
   -1:  begin p↑.bal := 0; h := false
         end ;
    0:  p↑.bal := +1;
    1:  begin {rebalance} p1 := p↑.right;
            if p1↑.bal = +1 then
            begin {single RR rotation}
               p↑.right := p1↑.left; p1↑.left := p;
               p↑.bal := 0; p := p1
            end else
            begin {double RL rotation}  p2 := p1↑.left;
               p1↑.left := p2↑.right; p2↑.right := p1;
               p↑.right := p2↑.left; p2↑.left := p;
               if p2↑.bal = +1 then p↑.bal := -1 else p↑.bal := 0;
               if p2↑.bal = -1 then p1↑.bal := +1 else p1↑.bal := 0;
               p := p2
            end ;
            p↑.bal := 0; h := false
         end
      end
   end
   else
   begin p↑.count := p↑.count + 1; h := false
   end
end {search}
```

Two particularly interesting questions concerning the performance of the balanced tree insertion algorithm are the following:

1. If all $n!$ permutations of n keys occur with equal probability, what is the expected height of the constructed balanced tree?
2. What is the probability that an insertion requires rebalancing?

Mathematical analysis of this complicated algorithm is still an open problem. Empirical tests support the conjecture that the expected height of the balanced tree generated by (4.63) is $h = \log(n) + c$, where c is a small constant ($c \cong 0.25$). This means that in practice the AVL-balanced tree behaves as well as the perfectly balanced tree, although it is much simpler to maintain. Empirical evidence also suggests that, on the average, rebalancing is necessary once for approximately every two insertions. Here single and double rotations are equally probable. The example of Fig. 4.34 has evidently been carefully chosen to demonstrate as many rotations as possible in a minimum number of insertions!

The complexity of the balancing operations suggests that balanced trees should be used only if information retrievals are considerably more frequent than insertions. This is particularly true because the nodes of such search trees are usually implemented as densely packed records in order to economize storage. The speed of access and of updating the balance factors—each requiring two bits only—is therefore often a decisive factor to the efficiency of the rebalancing operation. Empirical evaluations show that balanced trees lose much of their appeal if tight record packing is mandatory. It is indeed difficult to beat the straightforward, simple tree insertion algorithm!

4.4.8. Balanced Tree Deletion

Our experience with tree deletion suggests that in the case of balanced trees deletion will also be more complicated than insertion. This is indeed true, although the rebalancing operation remains essentially the same as for insertion. In particular, rebalancing consists of either a single or a double rotation of nodes.

The basis for balanced tree deletion is algorithm (4.52). The easy cases are terminal nodes and nodes with only a single descendant. If the node to be deleted has two subtrees, we will again replace it by the rightmost node of its left subtree. As in the case of insertion (4.63), a Boolean variable parameter h is added with the meaning "the height of the subtree has been reduced." Rebalancing has to be considered only when h is true. h is assigned the value true upon finding and deleting a node or if rebalancing itself reduces the height of a subtree. In (4.64) we introduce the two (symmetric) balancing operations in the form of procedures since they have to be invoked from more than one place in the deletion algorithm. Note that *balance1* is applied when the left, *balance2* after the right branch had been reduced in height.

The operation of the procedure is illustrated in Fig. 4.35. Given the balanced tree (a), successive deletion of the nodes with keys 4, 8, 6, 5, 2, 1, and 7 results in the trees (b) . . . (h).

The deletion of key 4 is simple in itself since it represents a terminal node. However, it results in an unbalanced node 3. Its rebalancing operation involves an *LL* single rotation. Rebalancing becomes again necessary after the deletion of node 6. This time the right subtree of the root (7) is rebalanced

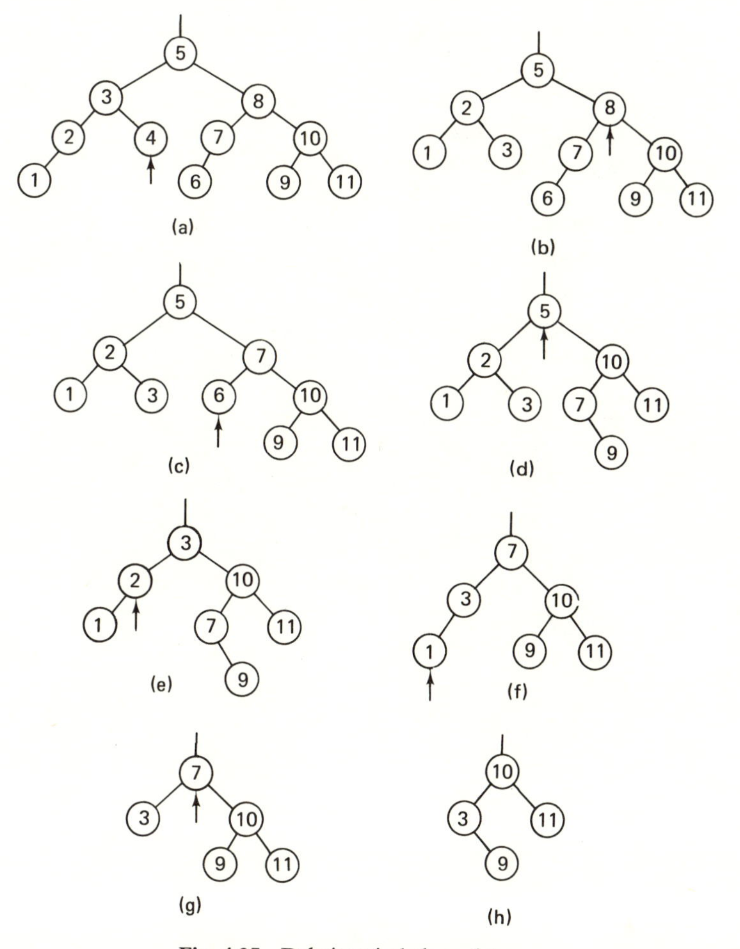

Fig. 4.35 Deletions in balanced tree.

by an *RR* single rotation. Deletion of node 2, although in itself straightforward since it has only a single descendant, calls for a complicated *RL* double rotation. The fourth case, an *LR* double rotation, is finally invoked after the removal of node 7, which at first was replaced by the rightmost element of its left subtree, i.e., by the node with key 3.

```
procedure delete(x: integer; var p: ref; var h: boolean);
   var q: ref;      {h = false}
   procedure balance1(var p: ref; var h: boolean);
      var p1,p2: ref; b1,b2: -1 .. +1;
   begin {h = true, left branch has become less high}
      case p↑.bal of                                              (4.64)
   -1:  p↑.bal := 0;
    0:  begin p↑.bal := +1; h := false
        end ;
```

```
    1:  begin {rebalance} p1 := p↑.right; b1 := p1↑.bal;
          if b1 ≥ 0 then
          begin {single RR rotation}
            p↑.right := p1↑.left; p1↑.left := p;
            if b1 = 0 then
            begin p↑.bal := +1; p1↑.bal := -1; h := false
            end else
            begin p↑.bal := 0; p1↑.bal := 0
            end ;
            p := p1
          end else
          begin {double RL rotation}
            p2 := p1↑.left; b2 := p2↑.bal;
            p1↑.left := p2↑.right; p2↑.right := p1;
            p↑.right := p2↑.left; p2↑.left := p;
            if b2 = +1 then p↑.bal := -1 else p↑.bal := 0;
            if b2 = -1 then p1↑.bal := +1 else p1↑.bal := 0;
            p := p2; p2↑.bal := 0
          end
        end
    end                                                        (4.64)
end {balance 1} ;


procedure balance2(var p: ref; var h: boolean);
    var p1,p2: ref; b1,b2: -1 . . +1;
begin {h = true, right branch has become less high}
    case p↑.bal of
    1:  p↑.bal := 0;
    0:  begin p↑.bal := -1; h := false
        end ;
   -1:  begin {rebalance} p1 := p↑.left; b1 := p1↑.bal;
          if b1 ≤ 0 then
          begin {single LL rotation}
            p↑.left := p1↑.right; p1↑.right := p;
            if b1 = 0 then
            begin p↑.bal := -1; p1↑.bal := +1; h := false
            end else
            begin p↑.bal := 0; p1↑.bal := 0
            end ;
            p := p1
          end else
```

```
            begin {double LR rotation}
              p2 := p1↑.right; b2 := p2↑.bal;
              p1↑.right := p2↑.left; p2↑.left := p1;
              p↑.left := p2↑.right; p2↑.right := p;
              if b2 = -1 then p↑.bal := +1 else p↑.bal := 0;
              if b2 = +1 then p1↑.bal := -1 else p1↑.bal := 0;
              p := p2; p2↑.bal := 0
            end
          end
        end
      end {balance2} ;

      procedure del(var r: ref; var h: boolean);
      begin {h = false}
        if r↑.right ≠ nil then
        begin del(r↑.right,h); if h then balance2(r,h)
        end else
        begin q↑.key := r↑.key; q↑.count := r↑.count;
          r := r↑.left; h := true
        end
      end ;

    begin {delete}
      if p = nil then
        begin writeln ('KEY IS NOT IN TREE'); h := false                (4.64)
        end else
      if x < p↑.key then
        begin delete(x,p↑.left,h); if h then balance1(p,h)
        end else
      if x > p↑.key then
        begin delete(x,p↑.right,h); if h then balance2(p,h)
        end else
      begin {delete p↑} q := p;
        if q↑.right = nil then
          begin p := q↑.left; h := true
          end else
        if q↑.left = nil then
          begin p := q↑.right; h := true
          end else
        begin del(q↑.left,h);
          if h then balance1(p,h)
        end ;
        {dispose(q)}
      end
    end {delete}
```

Evidently, deletion of an element in a balanced tree can also be performed with—in the worst case—O ($\log n$) operations. An essential difference between the behavior of the insertion and deletion procedures must not be overlooked, however. Whereas insertion of a single key may result in at most one rotation (of two or three nodes), deletion may require a rotation at *every* node along the search path. Consider, for instance, deletion of the rightmost node of a Fibonacci-tree. In this case the deletion of any single node leads to a reduction of the height of the tree; in addition, deletion of its rightmost node requires the maximum number of rotations. This therefore represents the worst choice of node in the worst case of a balanced tree, a rather unlucky combination of chances! How probable are rotations, then, in general? The surprising result of empirical tests is that whereas one rotation is invoked for approximately every two insertions, one is required for every five deletions only. Deletion in balanced trees is therefore about as easy—or as complicated—as insertion.

4.4.9. Optimal Search Trees

So far our consideration of organizing search trees has been based on the assumption that the frequency of access is equal for all nodes, that is, that all keys are equally probable to occur as a search argument. This is probably the best assumption if one has no idea of access distribution. However, there *are* cases (they are the exception rather than the rule) in which information on the probabilities of access to individual keys is available. These cases usually have the characteristic that the keys always remain the same, i.e., the search tree is subjected neither to insertions nor to deletions, but retains a constant structure. A typical example is the scanner of a compiler which determines for each word (identifier) whether or not it is a keyword (reserved word). Statistical measurements over hundreds of compiled programs may in this case yield accurate information on the relative frequencies of occurrence, and thereby of access, of individual keys.

Assume that in a search tree the probability with which node i is accessed is p_i.

$$Pr\{x = k_i\} = p_i, \qquad \sum_{i=1}^{n} p_i = 1 \tag{4.65}$$

We now wish to organize the search tree in a way that the total number of search steps—counted over sufficiently many trials—becomes minimal. For this purpose the definition of path length (4.34) is modified by attributing a certain weight to each node. Nodes which are frequently accessed become heavy nodes; those which are rarely visited become light nodes. The (internal) *weighted path length* is then the sum of all paths from the root to each node weighted by that node's probability of access.

$$P_I = \sum_{i=1}^{n} p_i h_i \tag{4.66}$$

h_i is the level of node i (or its distance from the root $+1$). The goal is now to *minimize the weighted path length* for a given probability distribution.

As an example, consider the set of keys 1, 2, 3, with probabilities of access $p_1 = 1/7, p_2 = 2/7$, and $p_3 = 4/7$. These three keys can be arranged in five different ways as search trees (see Fig. 4.36).

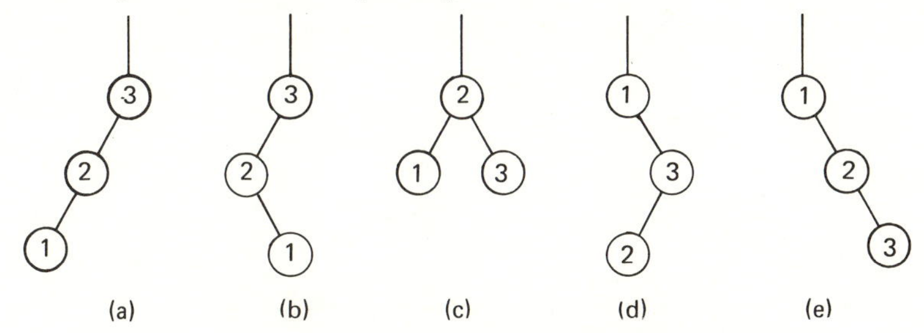

Fig. 4.36 Search trees with three nodes.

The weighted path lengths are computed according to (4.66) as

$$P_I^{(a)} = \frac{1}{7}(1\cdot 3 + 2\cdot 2 + 4\cdot 1) = \frac{11}{7}$$

$$P_I^{(b)} = \frac{1}{7}(1\cdot 2 + 2\cdot 3 + 4\cdot 1) = \frac{12}{7}$$

$$P_I^{(c)} = \frac{1}{7}(1\cdot 2 + 2\cdot 1 + 4\cdot 2) = \frac{12}{7}$$

$$P_I^{(d)} = \frac{1}{7}(1\cdot 1 + 2\cdot 3 + 4\cdot 2) = \frac{15}{7}$$

$$P_I^{(e)} = \frac{1}{7}(1\cdot 1 + 2\cdot 2 + 4\cdot 3) = \frac{17}{7}$$

Hence, in this example not the perfectly balanced but the degenerate tree (a) turns out to be the optimal arrangement.

The example of the compiler scanner immediately suggests that this problem should be viewed under a slightly more general condition: Words occurring in the source text are not always keywords; as a matter of fact, their being keywords is rather the exception. Finding that a given word k is *not* a key in the search tree can be considered as an access to a hypothetical "special node" inserted between the next lower and next higher key (see Fig. 4.19) with an associated external path length. If the probability q_i of a search argument x lying between the two keys k_i and k_{i+1} is also known, this information may considerably change the structure of the optimal search tree. Hence, we generalize the problem by also considering unsuccessful searches.

The overall average weighted path length is now

$$P = \sum_{i=1}^{n} p_i h_i + \sum_{j=0}^{m} q_j h'_j \tag{4.67}$$

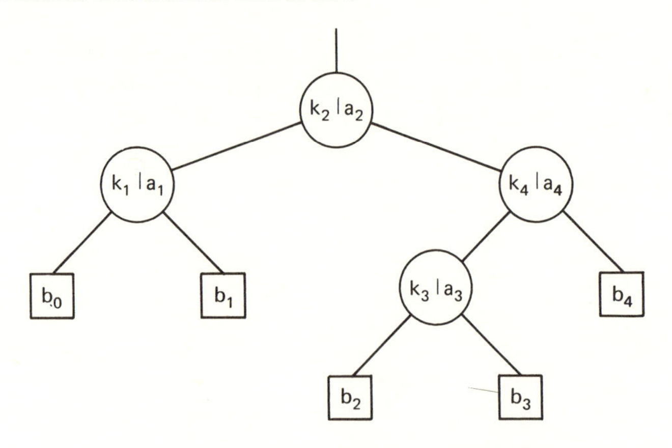

Fig. 4.37 Search tree with associated access frequencies.

where

$$\sum_{i=1}^{n} p_i + \sum_{j=0}^{m} q_j = 1$$

and where h_i is the level of the (internal) node i and h'_j is the level of the external node j. The average weighted path length may be called the "cost" of the search tree since it represents a measure for the expected amount of effort to be spent for searching. The search tree whose structure yields the minimal cost among all trees with a given set of keys k_i and probabilities p_i and q_j is called the *optimal tree*.

For finding the optimal tree, there is no need to require that the p's and q's sum up to 1. In fact, these probabilities are commonly determined by experiments in which the accesses to nodes are counted. Instead of using the probabilities p_i and q_j, we will subsequently use such frequency counts and denote them by

a_i = number of times the search argument x equals k_i
b_j = number of times the search argument x lies between k_j and k_{j+1}

By convention, b_0 is the number of times that x is less than k_1, and b_n is the frequency of x being greater than k_n (see Fig. 4.37). We will subsequently use P to denote the cumulated weighted path length instead of the average path length:

$$P = \sum_{i=1}^{n} a_i h_i + \sum_{j=0}^{n} b_j h'_j \tag{4.68}$$

Thus, apart from avoiding the computation of the probabilities from measured frequency counts, we gain the further advantage of being able to use integers only in our search for the optimal tree.

Considering the fact that the number of possible configurations of n nodes grows exponentially with n, the task of finding the optimum seems rather hopeless for large n. Optimal trees, however, have one significant property that helps to find them: all their subtrees are optimal too. For instance, if the tree in Fig. 4.37 is optimal for given a's and b's, then the subtree with keys k_3 and k_4 is also optimal as shown. This property suggests an algorithm that systematically finds larger and larger trees, starting with individual nodes as smallest possible subtrees. The tree thus grows "from the leaves to the root," which is, since we are used to drawing trees upside-down, the "bottom-up" direction [4-6].

The equation that is the key to this algorithm is (4.69). Let P be the weighted path length of a tree, and let P_L and P_R be those of the left and right subtrees of its root. Clearly, P is the sum of P_L and P_R and the number of times a search travels on the single leg to the root, which is simply the total number W of search trials.

$$P = P_L + W + P_R \tag{4.69}$$

$$W = \sum_{i=1}^{n} a_i + \sum_{j=0}^{n} b_j \tag{4.70}$$

We call W the *weight* of the tree. Its *average* path length is then P/W.

These considerations show the need for a denotation of the weights and the path lengths of any subtree consisting of a number of adjacent keys. Let w_{ij} denote the weight and let p_{ij} denote the path length of the optimal subtree T_{ij} consisting of nodes with keys $k_{i+1}, k_{i+2}, \ldots, k_j$. These quantities are defined by the recurrence relations (4.71) and (4.72)

$$\begin{aligned} w_{ii} &= b_i && (0 \leq i \leq n) \\ w_{ij} &= w_{i,j-1} + a_j + b_j && (0 \leq i < j \leq n) \end{aligned} \tag{4.71}$$

$$\begin{aligned} p_{ii} &= w_{ii} && (0 \leq i \leq n) \\ p_{ij} &= w_{ij} + \min_{i<k\leq j} (p_{i,k-1} + p_{kj}) && (0 \leq i < j \leq n) \end{aligned} \tag{4.72}$$

The last equation follows immediately from (4.69) and the definition of optimality.

Since there are approximately $(1/2)n^2$ values p_{ij}, and since (4.72) calls for a choice among $0 < j - i \leq n$ cases, the minimization operation will involve approximately $(1/6)n^3$ operations. Knuth pointed out that a factor n can be saved by the following consideration, which alone saves this algorithm for practical purposes.

Let r_{ij} be a value of k which achieves the minimum in (4.72). It is possible to limit the search for r_{ij} to a much smaller interval, i.e., to reduce the number of the $j - i$ evaluation steps. The key is the observation that if we have found the root r_{ij} of the optimal subtree T_{ij}, then neither extending the tree by adding a node to the right nor removing its leftmost node ever can cause the root to move to the left. This is expressed by the relation

$$r_{i,j-1} \leq r_{ij} \leq r_{i+1,j} \tag{4.73}$$

which limits the search for possible solutions for r_{ij} to the range $r_{i,j-1} \dots r_{i+1,j}$, and it results in a total number of elementary steps to $O(n^2)$. We are now ready to construct the optimization algorithm in detail. We recall the following definitions which are based on optimal trees T_{ij} consisting of nodes with keys $k_{i+1} \dots k_j$.

1. a_i: the frequency of a search for k_i.
2. b_j: the frequency of a search argument x between k_j and k_{j+1}.
3. w_{ij}: the weight of T_{ij}.
4. p_{ij}: the weighted path length of T_{ij}.
5. r_{ij}: the index of the root of T_{ij}.

Given

type *index* $= 0 \,..\, n$

we declare the following arrays:

```
  a: array[1 .. n] of integer;
  b: array[index] of integer;
p,w: array[index, index] of integer;                (4.74)
  r: array[index, index] of index
```

Assume that the weight w_{ij} has been computed from a and b in a straightforward way [see (4.71)]. Now consider w as the argument of the procedure to be developed and consider r as its result, for r describes the structure completely. p may be considered an intermediate result. Starting out by considering the smallest possible subtrees, namely, those consisting of no nodes at all, we proceed to larger and larger trees. Let us denote the width $j - i$ of the subtree T_{ij} by h. Then we can trivially determine the values p_{ii} for all trees with $h = 0$ according to (4.72).

```
for i := 0 to n do p[i,i] := w[i,i]                (4.75)
```

In the case $h = 1$ we deal with trees consisting of a single node, which plainly is also the root (Fig. 4.38).

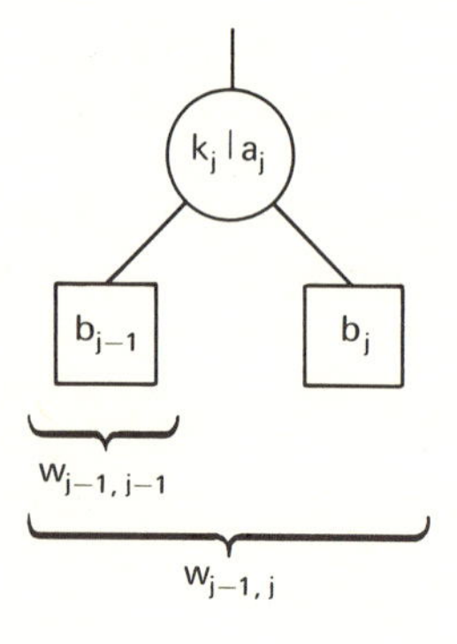

Fig. 4.38 Optimal tree with one node.

```
for i := 0 to n-1 do
    begin j := i+1; p[i,j] := p[i,i]+p[j,j]; r[i,j] := j        (4.76)
    end
```

Note that i denotes the left index limit and j the right index limit in the considered tree T_{ij}. For the cases $h > 1$ we use a repetitive statement with h ranging from 2 to n, the case $h = n$ spanning the entire tree $T_{0,n}$. In each case the minimal path length p_{ij} and the associated root index r_{ij} are determined by a simple repetitive statement with an index k ranging over the interval given by (4.73).

```
for h := 2 to n do
for i := 0 to n-h do
begin j := i+h;
      "find m and min = minimum(p[i,m-1]+p[m,j]) for all    (4.77)
       m such that r[i,j-1] ≤ m ≤ r[i+1,j]";
       p[i,j] := min + w[i,j];   r[i,j] := m
end
```

The details of the refinement of the statement within quotes can be found in Program 4.6. The average path length of $T_{0,n}$ is now given by the quotient $p_{0,n}/w_{0,n}$ and its root is the node with index $r_{0,n}$.

It is evident from algorithm (4.77) that the effort to determine the optimal structure is of the order $O(n^2)$; also, the amount of required storage is $O(n^2)$. This is unacceptable if n is very large. Algorithms with greater efficiency are therefore highly desirable. One of them is the algorithm developed by Hu and Tucker [4-5] which requires only $O(n)$ storage and $O(n \cdot \log n)$ computations. However, it considers only the case in which the key frequencies are zero ($a_i = 0$), i.e., where only the unsuccessful search trials are registered. Another algorithm, also requiring $O(n)$ storage elements and $O(n \cdot \log n)$ computations was described by Walker and Gotlieb [4-11]. Instead of trying to find the optimum, this algorithm merely promises to yield a nearly optimal tree. It can therefore be based on *heuristic* principles. The basic idea is the following.

Consider the nodes (true and special nodes) being distributed on a linear scale, weighted by their frequencies (or probabilities) of access. Then find the node which is closest to the "center of gravity." This node is called the *centroid*, and its index is

$$\frac{1}{w}\left(\sum_{i=1}^{n} i \cdot a_i + \sum_{j=0}^{n} j \cdot b_j\right) \tag{4.78}$$

rounded to the nearest integer. If all nodes have equal weight, then the root of the desired optimal tree evidently coincides with the centroid, and—so the reasoning goes—it will in most cases be in the close neighborhood of the centroid. A limited search is then used to find the local optimum, whereafter this procedure is applied to the resulting two subtrees. The likelihood of the root lying very close to the centroid grows with the size n of the tree. As soon as the subtrees have reached a "manageable" size, their optimum can be determined by the above exact algorithm.

4.4.10. Displaying a Tree Structure

We now turn to the associated programming problem of how to generate an output which *displays* the structure of the tree in a reasonably clear, graphic form, given only the means of the ordinary printer. That is, we should like to draw a picture of the tree, printing the keys as nodes and connecting them with appropriate horizontal and vertical bar characters.

On a line printer whose data we represent as a textfile, i.e., as a sequence of characters, we can proceed only in strict sequence from left to right and from top to bottom. Hence, it seems to be a reasonable idea to first build a representation of the *tree* that closely reflects its topological structure. The second step then is to *map* this picture in orderly fashion onto the printed page and to compute the precise coordinates of the nodes and arcs.

For the first task we can readily draw on our experience with tree generating algorithms and unhesitatingly we adopt a recursive solution to the recursively defined problem. We formulate a function procedure called *tree* similar to the one used in Program 4.3. The parameters i and j are the limiting indices of the nodes belonging to the tree. Its root is then defined as the node with index r_{ij}. Before proceeding, however, we need to define the type of the variables that are to represent the nodes. They must contain the two pointers to their subtrees and the key of the node. For purposes to be discussed in the second step, two additional fields, called *pos* and *link*, are also incorporated. The chosen definitions are shown in (4.79) and the resulting function procedure is listed in Program 4.6.

```
type ref  = ↑node;
     node = record key: alfa;
                   pos: lineposition;                    (4.79)
                   left,right,link: ref
            end
```

Note that this procedure counts the number of generated nodes by the global counter variable k. The kth node is assigned the kth key, and as the keys are alphabetically ordered, k multiplied by a constant scale factor yields the horizontal coordinate of each key, a value which is immediately stored along with the other information. Note also that we have departed from the convention of using integers as keys, and we assume them to be of a type *alfa*, standing for an array of characters of a given (maximum) length upon which alphabetical ordering is defined.

To visualize what we have obtained so far, refer to Fig. 4.39. Given the set of n keys and the computed matrix r_{ij}, the statements

$$k := 0; \quad root := tree(0,n)$$

will generate the preliminary linked tree structure with the node's horizontal positions recorded and their vertical position determined implicitly by their level in the tree.

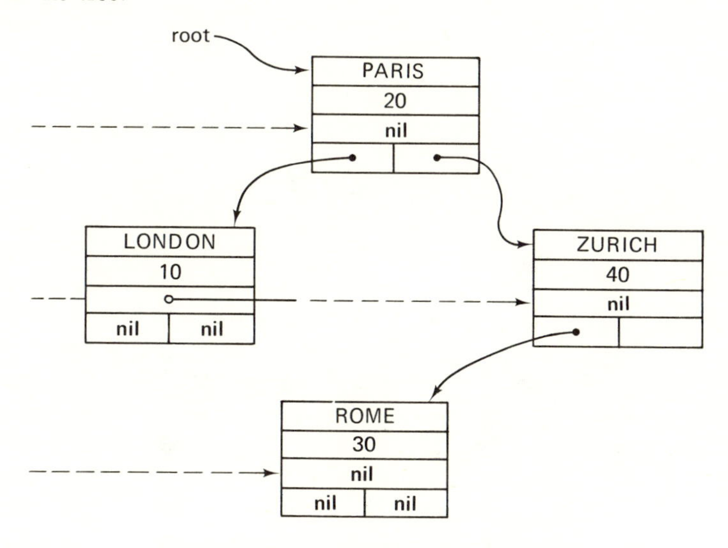

Fig. 4.39 Tree resulting from Program 4.6.

We may now proceed to the second step: mapping the tree onto paper. In this case, we must proceed strictly from the root level down, in each step processing one row of nodes. But how do we access the nodes lying on a row? For this purpose, namely, linking together nodes on the same row, we have previously introduced the record field called *link*. The chains to be established are shown as dotted links in Fig. 4.39. In each processing step we assume the presence of a chain linking the nodes to be printed—we call this chain *current* chain—and during handling each node we identify its descendants (if any), linking them into a second chain—which we call *next* chain. When proceeding one level down, the next chain becomes the current chain, and the next chain is marked empty.

The details of the algorithm may be taken from Program 4.6. The following remarks may help to clarify some points:

1. The chains of nodes on a row are generated from left to right, resulting in the leftmost node being last. Since the nodes are to be visited in the same sequence, the list must be inverted. This inversion is performed when the *next* list becomes the *current* list.
2. A printed line listing the keys—called master line—also contains the

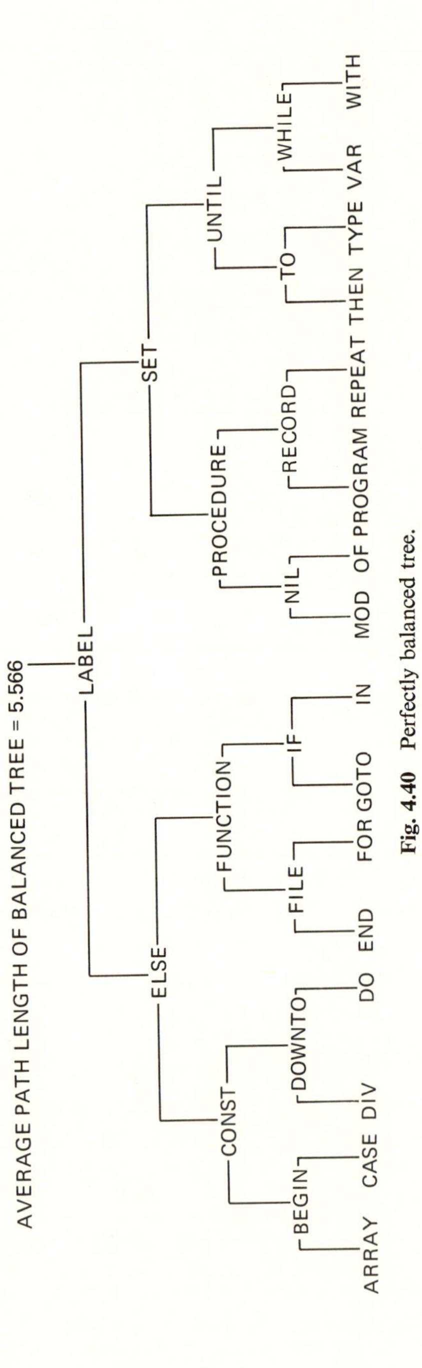

Fig. 4.40 Perfectly balanced tree.

horizontal arcs (see Fig. 4.40). The variables $u1$, $u2$, $u3$, $u4$ denote the beginning and end positions of the left and right horizontal arcs of a node.

3. The construction of each master line is preceded by three lines for marking the vertical parts of the arcs.

Let us now describe the structure of Program 4.6. Its two main components are the procedures to find the optimal search tree, given a weight distribution w, and to display a tree, given the indices r. The entire program

4	7	ARRAY
14	27	BEGIN
19	0	CASE
15	2	CONST
8	5	DIV
0	0	DOWNTO
0	20	DO
0	8	ELSE
0	28	END
1	0	FILE
0	12	FOR
0	2	FUNCTION
0	0	GOTO
9	13	IF
23	2	IN
208	0	LABEL
22	0	MOD
17	10	NIL
24	7	OF
17	2	PROCEDURE
0	1	PROGRAM
53	1	RECORD
6	8	REPEAT
16	0	SET
10	13	THEN
0	12	TO
6	2	TYPE
1	8	UNTIL
39	5	VAR
0	8	WHILE
0	0	WITH
37		
549	203	

Table 4.5 Keys and Frequencies of Occurrence

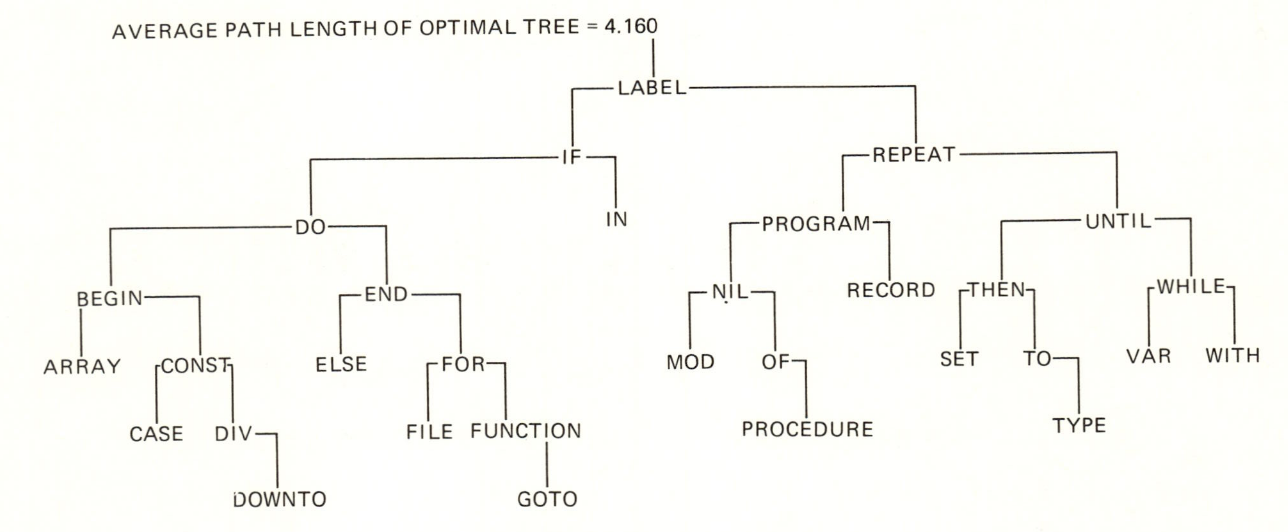

Fig. 4.41 Optimal search tree.

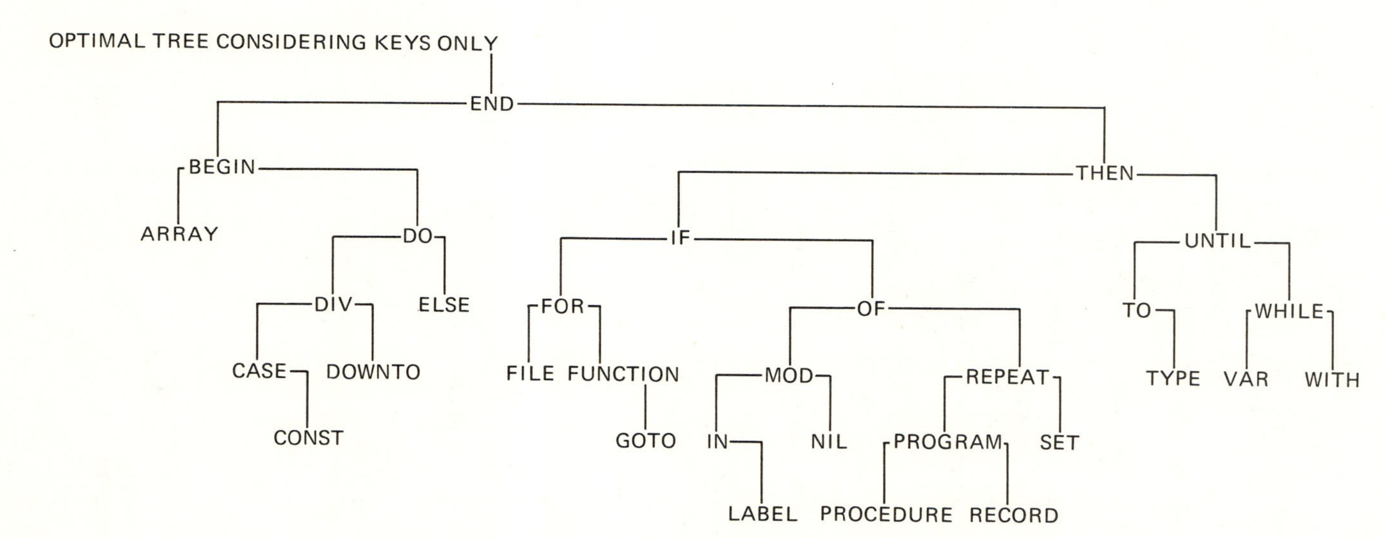

Fig. 4.42 Optimal tree considering keys only.

is adapted to processing program texts, PASCAL programs in particular. In the first part, such a program is read and its identifiers and keywords are recognized, yielding the counts a_i and b_j for finding a keyword k_i and identifiers between k_j and k_{j+1}. After printing the frequency statistics, the program proceeds to compute the path length of the perfectly balanced tree, in passing also determining the roots of its subtrees. Thereafter, the average weighted path length is printed and the tree is displayed.

In the third part, procedure *opttree* is activated in order to compute the optimal search tree; thereafter, it is also displayed. And, finally, the same procedures are used to compute and display the optimal tree under consideration of the key frequencies only.

Table 4.5 and Figs. 4.40 through 4.42 show the results generated by Program 4.6 when applied to its own program text. The differences in the three figures demonstrate that the balanced tree cannot even be considered as nearly optimal and that the frequencies of the non-keys crucially influence the choice of the optimal structure.

Program 4.6 Find Optimal Search Tree.

```
program optimaltree(input,output);
const n = 31; {no. of keys}
   kln = 10; {max keylength}
type index = 0 .. n;
   alfa = packed array [1 .. kln] of char;
var ch: char;
       k1, k2: integer;
       id: alfa;      {identifier or key}
       buf: array [1 .. kln] of char;      {character buffer}
       key: array [1 .. n] of alfa;
       i,j,k: integer;
       a: array [1 .. n] of integer;
       b: array [index] of integer;
       p,w: array [index,index] of integer;
       r: array [index,index] of index;
       suma, sumb: integer;
function baltree(i,j: index): integer;
   var k: integer;
begin k := (i+j+1) div 2; r[i,j] := k;
   if i ≥ j then baltree := b[k] else
       baltree := baltree(i,k-1) + baltree(k,j) + w[i,j]
end {baltree} ;
procedure opttree;
   var x, min: integer;
       i,j,k,h,m: index;
```

```
begin {argument: w, result: p,r}
   for i := 0 to n do p[i,i] := w[i,i];  {width of tree h = 0}
   for i := 0 to n-1 do            {width of tree h = 1}
   begin j := i+1;
      p[i,j] := p[i,i] + p[j,j];  r[i,j] := j
   end ;
   for h := 2 to n do                  { h = width of considered tree }
   for i := 0 to n-h do                { i = left index of considered tree }
   begin j := i+h;                     { j = right index of considered tree }
      m := r[i,j-1];  min := p[i,m-1] + p[m,j];
      for k := m+1 to r[i+1,j] do
      begin x := p[i,k-1] + p[k,j];
         if x < min then
            begin m := k;  min := x
            end
      end ;
      p[i,j] := min + w[i,j];  r[i,j] := m
   end
end {opttree} ;

procedure printtree;
   const lw = 120;     {line width of printer}
   type ref = ↑node;
      lineposition = 0 . . lw;
      node = record key: alfa;
                    pos: lineposition;
                    left, right, link: ref
             end ;
   var root, current, next: ref;
      q,q1,q2: ref;
      i, k: integer;
      u, u1, u2, u3, u4: lineposition;

   function tree(i,j: index): ref;
      var p: ref;
   begin if i = j then p := nil else
      begin new(p);
         p↑.left := tree(i, r[i,j]-1);
         p↑.pos := trunc((lw-kln)*k/(n-1)) + (kln div 2); k := k+1;
         p↑.key := key[r[i,j]];
         p↑.right := tree(r[i,j], j)
      end ;
      tree := p
   end ;
```

Program 4.6 (Continued)

```
begin k := 0;  root := tree(0,n);
   current := root; root↑.link := nil;
   next := nil;
   while current ≠ nil do
   begin {proceed down; first write vertical lines}
      for i := 1 to 3 do
      begin u := 0; q := current;
         repeat u1 := q↑.pos;
            repeat write(' '); u := u+1
            until u = u1;
            write('|'); u := u+1; q := q↑.link
         until q = nil;
         writeln
      end ;
      {now print master line; descending from nodes on current list
       collect their descendants and form next list}
      q := current; u := 0;
      repeat unpack(q↑.key, buf, 1);
         {center key about pos}  i := kln;
         while buf[i] = ' ' do  i := i-1;
         u2 := q↑.pos - ((i-1) div 2); u3 := u2+i;
         q1 := q↑.left; q2 := q↑.right;
         if q1 = nil then u1 := u2 else
            begin u1 := q1↑.pos; q1↑.link := next; next := q1
            end ;
         if q2 = nil then u4 := u3 else
            begin u4 := q2↑.pos+1; q2↑.link := next; next := q2
            end ;
         i := 0;
         while u < u1 do begin write(' '); u := u+1 end ;
         while u < u2 do begin write('-'); u := u+1 end ;
         while u < u3 do begin i := i+1; write(buf[i]); u := u+1 end ;
         while u < u4 do begin write('-'); u := u+1 end ;
         q := q↑.link
      until q = nil;
      writeln;
      {now invert next list and make it current list}
      current := nil;
      while next ≠ nil do
         begin q := next; next := q↑.link;
            q↑.link := current; current := q
         end
   end
end {printtree} ;
```

Program 4.6 (Continued)

```
begin {initialize table of keys and counters}
   key[ 1] := 'ARRAY';     key[ 2] := 'BEGIN';
   key[ 3] := 'CASE';      key[ 4] := 'CONST';
   key[ 5] := 'DIV';       key[ 6] := 'DOWNTO';
   key[ 7] := 'DO';        key[ 8] := 'ELSE';
   key[ 9] := 'END';       key[10] := 'FILE';
   key[11] := 'FOR';       key[12] := 'FUNCTION';
   key[13] := 'GOTO';      key[14] := 'IF';
   key[15] := 'IN';        key[16] := 'LABEL';
   key[17] := 'MOD';       key[18] := 'NIL';
   key[19] := 'OF';        key[20] := 'PROCEDURE';
   key[21] := 'PROGRAM';   key[22] := 'RECORD';
   key[23] := 'REPEAT';    key[24] := 'SET';
   key[25] := 'THEN';      key[26] := 'TO';
   key[27] := 'TYPE';      key[28] := 'UNTIL';
   key[29] := 'VAR';       key[30] := 'WHILE';
   key[31] := 'WITH';
   for i := 1 to n do
      begin a[i] := 0;  b[i] := 0
      end ;
   b[0] := 0; k2 := kln;
   {scan input text and determine a and b}
   while ¬eof(input) do
   begin read(ch);
      if ch in ['A' .. 'Z'] then
      begin {identifier or key} k1 := 0;
         repeat if k1 < kln then
                  begin k1 := k1+1; buf[k1] := ch
                  end ;
               read(ch)
         until ¬ch in ['A' .. 'Z', '0' .. '9']);
         if k1 ≥ k2 then k2 := k1 else
         repeat buf[k2] := '  '; k2 := k2-1
         until k2 = k1;
         pack(buf,1,id);
         i := 1; j := n;
         repeat k := (i+j) div 2;
            if key[k] ≤ id then i := k+1;
            if key[k] ≥ id then j := k-1;
         until i > j;
         if key[k] = id then a[k] := a[k] + 1 else
            begin k := (i+j) div 2; b[k] := b[k]+1
            end
      end else
```

Program 4.6 (Continued)

```
        if ch = '''' then
           repeat read(ch) until ch = '''' else
        if ch = '{' then
           repeat read(ch) until ch = '}'
     end ;
     writeln ('KEYS AND FREQUENCIES OF OCCURRENCE:');
     suma := 0; sumb := b[0];
     for i := 1 to n do
     begin suma := suma+a[i]; sumb := sumb+b[i];
        writeln(b[i-1], a[i], '   ', key[i])
     end ;
     writeln(b[n]);
     writeln('    ———————      ———————');
     writeln(sumb, suma);
     {compute w from a and b}
     for i := 0 to n do
     begin w[i,i] := b[i];
        for j := i+1 to n do w[i,j] := w[i,j-1] + a[j] + b[j]
     end ;
     write('AVERAGE PATH LENGTH OF BALANCED TREE= ');
     writeln(baltree(0,n)/w[0,n]:6:3); printtree;
     opttree;
     write('AVERAGE PATH LENGTH OF OPTIMAL TREE= ');
     writeln(p[0,n]/w[0,n]:6:3);  printtree;
     {now consider keys only, setting b = 0}
     for i := 0 to n do
     begin w[i,i] := 0;
        for j := i+1 to n do w[i,j] := w[i,j-1] + a[j]
     end ;
     opttree;
     writeln('OPTIMAL TREE CONSIDERING KEYS ONLY');
     printtree
  end
```

Program 4.6 (Continued)

4.5. MULTIWAY TREES

So far, we have restricted our discussion to trees in which every node has at most two descendants, i.e., to binary trees. This is entirely satisfactory if, for instance, we wish to represent family relationship with a preference to the "pedigree view," in which every person is associated with his parents.

After all, no one has more than two parents! But what about someone who prefers the "posterity view"? He has to cope with the fact that some people have more than two children, and his trees will contain nodes with many branches. For lack of a better term, we shall call them *multiway trees.*

Of course, there is nothing special about such structures, and we have already encountered all the programming and data definition facilities to cope with such situations. If, for instance, an absolute upper limit on the number of children is given (which is admittedly a somewhat futuristic assumption), then one may represent the children as an array component of the record representing a person. If the number of children varies strongly among different persons, however, this may result in a poor utilization of available storage. In this case it will be much more appropriate to arrange the offspring as a linear list, with a pointer to the youngest (or eldest) offspring assigned to the parent. A possible type definition for this case is (4.80) and a possible data structure is shown in Fig. 4.43.

```
type person = record name: alfa;
                     sibling: ↑person;              (4.80)
                     offspring: ↑person
              end
```

We now realize that by tilting this picture 45° it will look like a perfect binary tree. But this view is misleading because functionally the two references have entirely different meanings. One usually doesn't treat a sibling as an offspring and get away unpunished, and hence one should not do so even in constructing data definitions. This example could also be easily extended into an even more complicated data structure by introducing more components in each person's record, thus being able to represent further family relationships. A likely candidate which cannot generally be derived from the sibling and offspring references is that of husband and wife, or even the inverse relationship of father and mother. Such a structure quickly grows into a complex "relational data bank," and it may be possible to map several trees into it. The algorithms operating on such structures are intimately tied to their data definitions, and it does not make sense to specify any general rules or widely applicable techniques.

However, there is a very practical area of application of multiway trees which *is* of general interest. This is the construction and maintenance of large-scale search trees in which insertions and deletions are necessary, but in which the primary store of a computer is not large enough or is too costly to be used for long-time storage.

Assume, then, that the nodes of a tree are to be stored on a secondary storage medium such as a disk store. Dynamic data structures introduced in this chapter are particularly suitable for incorporation of secondary storage

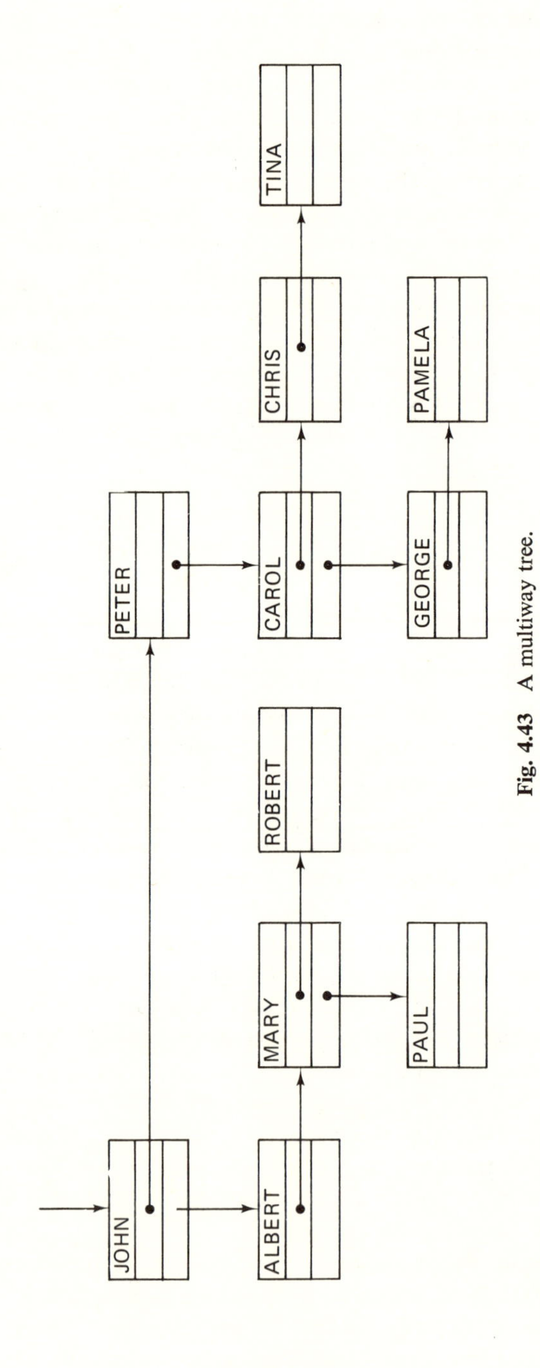

Fig. 4.43 A multiway tree.

media. The principal innovation is merely that pointers are represented by disk store addresses instead of main store addresses. Using a binary tree for a data set of, say, a million items, requires on the average approximately $\log_2 10^6 \cong 20$ search steps. Since each step now involves a disk access (with inherent latency time), a storage organization using fewer accesses will be highly desirable. The multiway tree is a perfect solution to this problem. If an item located on a secondary store is accessed, an entire group of items may also be accessed without much additional cost. This suggests that a tree be subdivided into subtrees and that the subtrees are represented as units which are accessed all together. We shall call these subtrees *pages*. Figure 4.44 shows a binary tree subdivided into pages, each page consisting of 7 nodes.

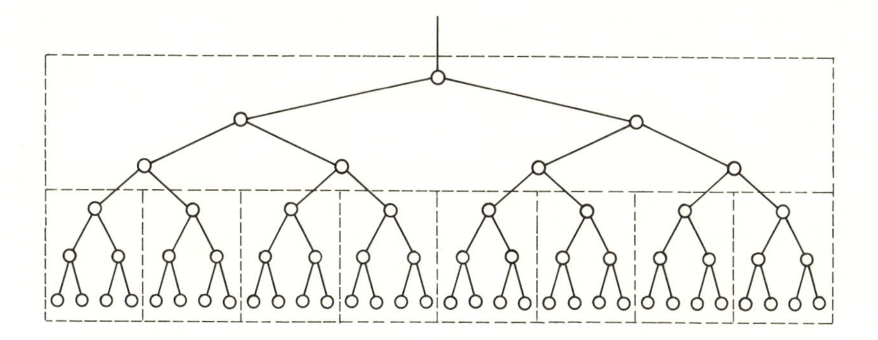

Fig. 4.44 A binary tree subdivided into "pages."

The savings in the number of disk accesses—each page access now involves a disk access—can be considerable. Assume that we choose to place 100 nodes on a page (this is a reasonable figure); then the million item search tree will on the average require only $\log_{100} 10^6 = 3$ page accesses instead of 20. But, of course, if the tree is left to grow "at random," then the worst case may still be as large as 10^4! It is plain that a schema for controlled growth is almost mandatory in the case of multiway trees.

4.5.1. B-Trees

If one is looking for a controlled growth criterion, the one requiring a perfect balance is quickly eliminated because it involves too much balancing overhead. The rules must clearly be somewhat relaxed. A very sensible criterion was postulated by R. Bayer [4.2] in 1970: every page (except one) contains between n and $2n$ nodes for a given constant n. Hence, in a tree with N items and a maximum page size of $2n$ nodes per page, the worst case requires $\log_n N$ page accesses—and page accesses clearly dominate the entire search effort. Moreover, the important factor of store utilization is at least 50% since pages are always at least half full. With all these advantages, the schema involves comparatively simple algorithms for search, insertion, and deletion. We will subsequently study them in detail.

The underlying data structures are called B-trees and have the following characteristics; n is said to be the *order* of the B-tree.

1. Every page contains at most $2n$ items (keys.)
2. Every page, except the root page, contains at least n items.
3. Every page is either a leaf page, i.e., has no descendants or it has $m + 1$ descendants, where m is its number of keys.
4. All leaf pages appear at the same level.

Figure 4.45 shows a B-tree of order 2 with 3 levels. All pages contain 2, 3,

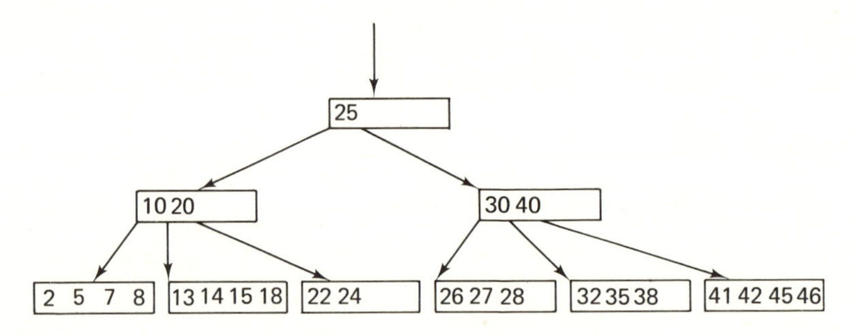

Fig. 4.45 B-tree of order 2.

or 4 items; the exception is the root which is allowed to contain a single item only. All leaf pages appear at level 3. The keys appear in increasing order from left to right if the B-tree is squeezed into a single level by inserting the descendants in between the keys of their ancestor page. This arrangement represents a natural extension of the organization of binary search trees, and it determines the method of searching an item with given key. Consider a page of the form shown in Fig. 4.46 and a given search argument x. Assuming that the page has been moved into the primary store, we may use conventional search methods among the keys $k_1 \dots k_m$. If m is sufficiently

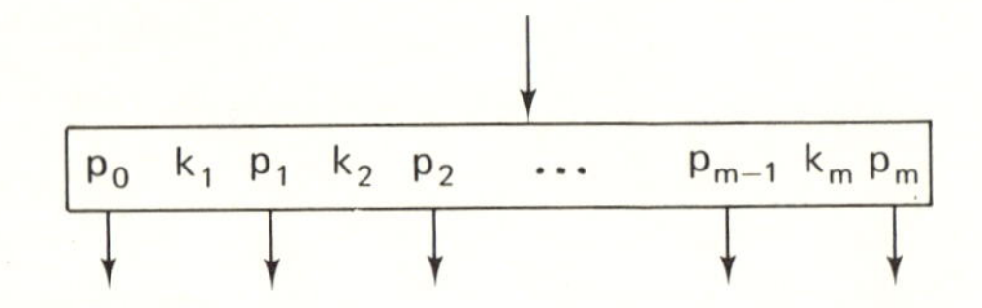

Fig. 4.46 B-tree page with m keys.

large, one may use binary search; if it is rather small, an ordinary sequential search will do. (Note that the time required for a search in main store is probably negligible compared to the time it takes to move the page from secondary into primary store.) If the search is unsuccessful, we are in one of the following situations:

1. $k_i < x < k_{i+1}$, for $1 \leq i < m$. We continue the search on page $p_i\uparrow$
2. $k_m < x$. The search continues on page $p_m\uparrow$.
3. $x < k_1$. The search continues on page $p_0\uparrow$.

If in some case the designated pointer is **nil**, i.e., if there is no descendant page, then there is no item with key x in the whole tree, and the search is terminated.

Surprisingly, insertion in a B-tree is comparatively simple too. If an item is to be inserted in a page with $m < 2n$ items, the insertion process remains constrained to that page. It is only insertion into an already full page that has consequences upon the tree structure and may cause the allocation of new pages. To understand what happens in this case, refer to Fig. 4.47, which illustrates the insertion of key 22 in a B-tree of order 2. It proceeds in the following steps:

1. Key 22 is found to be missing; insertion in page C is impossible because C is already full.
2. Page C is *split* into two pages (i.e., a new page D is allocated).
3. The $m + 1$ keys are equally distributed onto C and D, and the middle key is moved up one level into the ancestor page A.

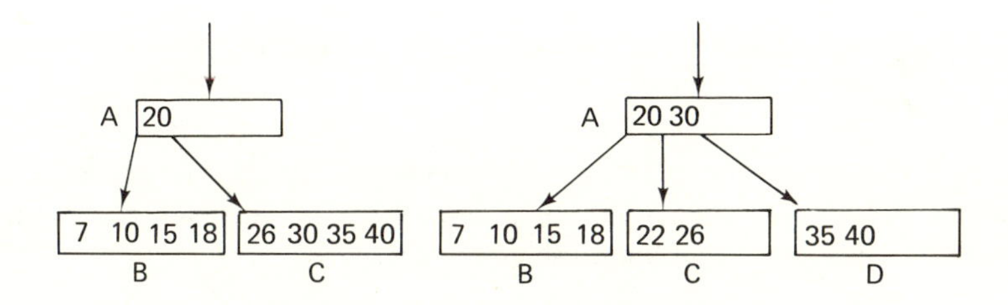

Fig. 4.47 Insertion of key 22 in B-tree.

This very elegant scheme preserves all the characteristic properties of B-trees. In particular, the split pages contain exactly n items. Of course, the insertion of an item in the ancestor page may again cause that page to overflow, thereby causing the *splitting to propagate*. In the extreme case it may propagate up to the root. This is, in fact, the only way that the B-tree may increase its height. The B-tree has thus a strange habit of growing: it grows from its leaves upward to the root.

We shall now develop a detailed program from these sketchy descriptions. It is already apparent that a recursive formulation will be most convenient because of the property of the splitting process to propagate back along the search path. The general structure of the program will therefore be similar to balanced tree insertion, although the details are different.

First of all, a definition of the page structure has to be formulated. We choose to represent the items in the form of an array.

```
type page = record m: index;
                   p0: ref;
                   e: array[1 .. nn] of item
            end                                          (4.81)
```

where

```
const nn = 2*n;
type ref = ↑page;
     index = 0 .. nn
```

and

```
type item = record key: integer;                         (4.82)
                   p: ref;
                   count: integer
            end
```

Again, the item component *count* stands for all kinds of other information that may be associated with each item, but it plays no role in the actual search process. Note that each page offers space for $2n$ items. The field m indicates how many item locations are actually used. As $m \geq n$ (except for the root page), a storage utilization of at least 50% is guaranteed.

The algorithm of B-tree search and insertion is part of Program 4.7, formulated as a procedure called *search*. Its main structure is straightforward, reminding one of the simple binary tree search, with the exception that the branching decision is not a binary choice. Instead, the "in-page search" is represented as a binary search upon the array e.

The insertion algorithm is formulated as a separate procedure merely for clarity. It is activated after *search* has indicated that an item is to be passed up on the tree (in the direction toward the root). This fact is indicated by the Boolean result parameter h; it assumes a similar role as in the algorithm for balanced tree insertion, where h indicates that the subtree had grown. If h is true, the second result parameter, u, represents the item being passed up. Note that insertions start in hypothetical pages, namely, the "special nodes" of Fig. 4.19; the new item is immediately handed up via the parameter u to the leaf page for true insertion. The scheme is sketched in (4.83).

```
procedure search(x: integer; a: ref; var h: boolean; var u: item);
begin if a = nil then
      begin {x is not in tree}
         Assign x to item u, set h to true, indicating that an
         item u is passed up in the tree
      end else
      with a↑ do
      begin {search x on page a↑}
         binary array search;
         if found then                                        (4.83)
           increment the relevant item's occurrence count else
         begin search(x, descendant, h, u);
            if h then {an item u is being passed up}
               if (no. items on a↑) < 2n then
                 insert u on page a↑ and set h to false
               else split page and pass middle item up
         end
      end
end
```

If the parameter *h* is true after the call of *search* in the main program, a split of the root page is indicated. Since the root page plays an exceptional role, this process has to be programmed separately. It consists merely of the allocation of a new (root) page and the insertion of the single item given by the parameter *u*. As a consequence, the new root page contains a single item only. The details can be gathered from Program 4.7.

Figure 4.48 shows the result of using Program 4.7 to construct a B-tree with the following insertion sequence of keys:

20; 40 10 30 15; 35 7 26 18 22; 5; 42 13 46
27 8 32; 38 24 45 25;

The semicolons designate the positions of the "snapshots" taken upon each page allocation. Insertion of the last key causes two splits and the allocation of three new pages.

Note the special significance of the **with** clause in this program. It is already evident in the sketch (4.83). In the first place it indicates that identifiers of page components automatically refer to page $a\uparrow$ within the statement prefixed by the clause. If, in fact, the pages are allocated on secondary store—as would certainly be necessary in a large data bank system—then the **with** clause may in addition be interpreted as implying the transfer of the designated page into primary store. Since each activation of *search* therefore implies one page allocation in main store, $k = \log_n N$ recursive calls are neces-

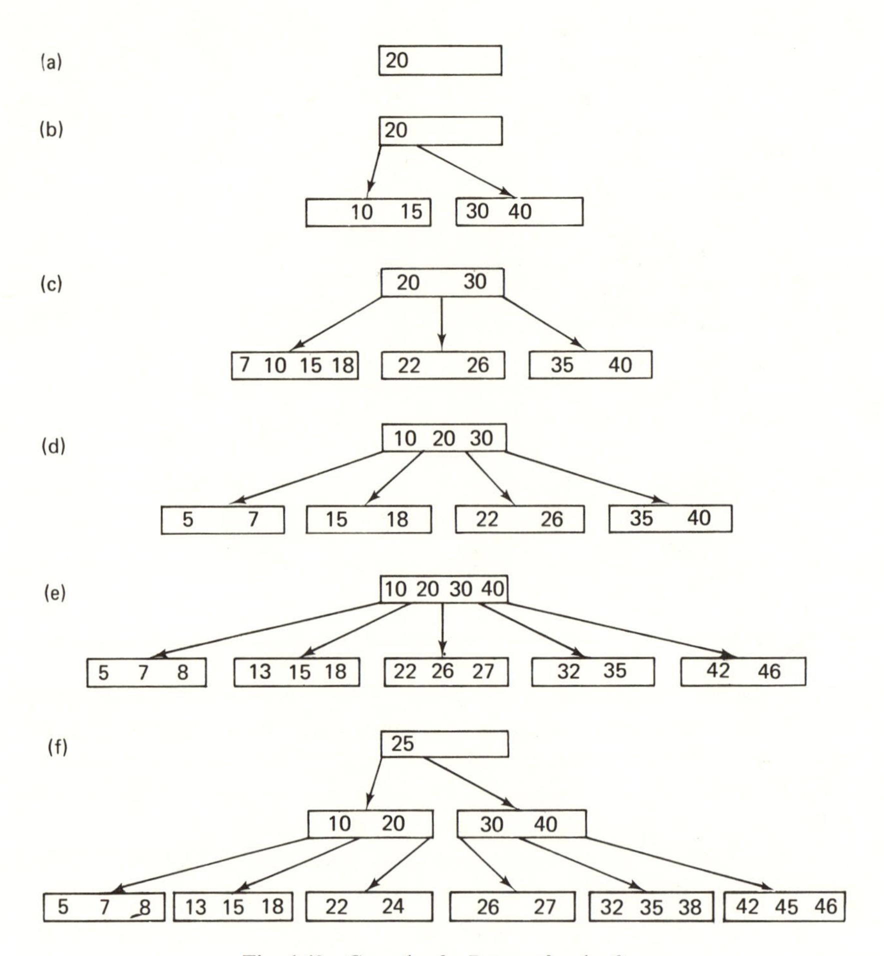

Fig. 4.48 Growth of a B-tree of order 2.

sary at most. Hence, if the tree contains N items, we must be capable of accommodating *k pages in main store*. This is one limiting factor on the page size $2n$. In fact, we need to accommodate even more than k pages, for insertion may cause page splitting. A corollary is that the root page is best allocated permanently in the primary store because each query proceeds necessarily from the root page.

Another positive quality of the B-tree organization is its suitability and economy in the case of purely sequential updating of the entire data bank. Every page is fetched into primary store exactly once.

Deletion of items from a B-tree is fairly straightforward in principle, but it is complicated in the details. We may distinguish two different circumstances:

1. The item to be deleted is on a leaf page; here its removal algorithm is plain and simple.
2. The item is not on a leaf page; it must be replaced by one of the two lexicographically adjacent items, which happen to be on leaf pages and can easily be deleted.

In case 2 finding the adjacent key is analogous to finding the one used in binary tree deletion. We descend along the rightmost pointers down to the leaf page P, replace the item to be deleted by the rightmost item on P, and then reduce the size of P by 1.

In any case, reduction of size must be followed by a check of the number of items m on the reduced page. For if $m < n$, the primary characteristic of B-trees would be violated. Some additional action has to be taken; this *underflow* condition is indicated by the Boolean variable parameter h.

The only recourse is to borrow, or to "annect," an item from one of the neighboring pages. Since this involves fetching page Q into main store—a relatively costly operation—one is tempted to make the best of this undesirable situation and to annect more than a single item at once. The usual strategy is to distribute the items on pages P and Q evenly on both pages. This is called *balancing*.

Of course, it may happen that there is no item left to be annected since Q has already reached its minimal size n. In this case the total number of items on pages P and Q is $2n - 1$; we may *merge* the two pages into one, adding the middle item from the ancestor page of P and Q, and then entirely dispose of page Q. This is exactly the inverse process of page splitting. The process may be visualized by considering the deletion of key 22 in Fig. 4.47.

Once again, the removal of the middle key in the ancestor page may cause its size to drop below the permissible limit n, thereby requiring that further special action (either balancing or merging) be undertaken at the next level. In the extreme case the page merging may *propagate* all the way up to the root. If the root is reduced to size 0, it is itself deleted, thereby causing a reduction in the height of the B-tree. It is, in fact, the only way that a B-tree may shrink in height.

Figure 4.49 shows the gradual decay of the B-tree of Fig. 4.48 upon the sequential deletion of the keys

25 45 24; 38 32; 8 27 46 13 42; 5 22 18 26; 7 35 15;

The semicolons again designate the places where the "snapshots" are taken, namely, where pages are being disposed. The deletion algorithm is included as a procedure in Program 4.7. The similarity of its structure to that of balanced tree deletion is particularly noteworthy.

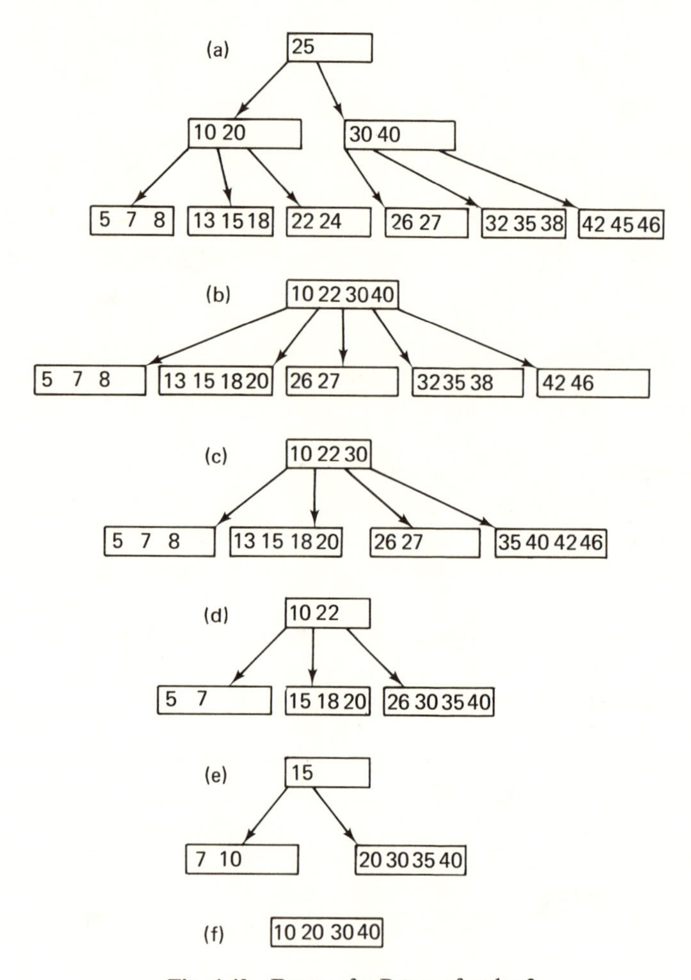

Fig. 4.49 Decay of a B-tree of order 2.

Program 4.7 B-Tree Search, Insertion, and Deletion.

```
program Btree(input,output);
{B-tree search, insertion and deletion}
const n = 2; nn = 4;  {page size}
type ref = ↑page;
   item = record key: integer;
              p: ref;
              count: integer;
          end ;
   page = record m: 0 . . nn;  {no. of items}
              p0: ref;
              e: array [1 . . nn] of item;
          end ;
var root, q: ref; x: integer;
   h: boolean;  u: item;
```

```
procedure search(x: integer; a: ref; var h: boolean; var v: item);
{Search key x on B-tree with root a; if found, increment counter.
 otherwise insert an item with key x and count 1 in tree. If an item
 emerges to be passed to a lower level, then assign it to v;
 h := "tree a has become higher"}
   var k,l,r: integer; q: ref; u: item;
   procedure insert;
      var i: integer; b: ref;
   begin {insert u to the right of a↑.e[r]}
      with a↑ do
      begin if m < nn then
         begin m := m+1; h := false;
            for i := m downto r+2 do e[i] := e[i-1];
            e[r+1] := u
         end else
         begin {page a↑ is full; split it and assign the emerging
                  item to v}      new(b);
            if r ≤ n then
            begin if r = n then v := u else
                  begin v := e[n];
                     for i := n downto r+2 do e[i] := e[i-1];
                     e[r+1] := u
                  end ;
               for i := 1 to n do b↑.e[i] := a↑.e[i+n]
            end else
            begin {insert u in right page} r := r-n; v := e[n+1];
               for i := 1 to r-1 do b↑.e[i] := a↑.e[i+n+1];
               b↑.e[r] := u;
               for i := r+1 to n do b↑.e[i] := a↑.e[i+n]
            end ;
            m := n; b↑.m := n; b↑.p0 := v .p; v .p := b
         end
      end {with}
   end {insert} ;
begin {search key x on page a↑; h = false}
   if a = nil then
   begin {item with key x is not in tree} h := true;
      with v do
      begin key := x; count := 1; p := nil
      end
   end else
```

Program 4.7 (Continued)

```
    with a↑ do
    begin l := 1; r := m;  {binary array search}
       repeat k := (l+r) div 2;
          if x ≤ e[k] .key then r := k−1;
          if x ≥ e[k] .key then l := k+1;
       until r < l;
       if l−r > 1 then
       begin {found} e[k] .count := e[k] .count + 1; h := false
       end else
       begin {item is not on this page}
          if r = 0 then q := p0 else q := e[r] .p;
          search(x,q,h,u); if h then insert
       end
    end
end {search} ;

procedure delete(x: integer; a: ref; var h: boolean);
{search and delete key x in b-tree a; if a page underflow is
 necessary, balance with adjacent page if possible, otherwise merge;
 h := "page a is undersize"}
    var i,k,l,r: integer; q: ref;
    procedure underflow(c,a: ref; s: integer; var h: boolean);
       {a = underflow page, c = ancestor page}
       var b: ref; i,k,mb,mc: integer;
    begin mc := c↑.m;  {h = true, a↑.m = n−1}
       if s < mc then
       begin {b := page to the right of a}  s := s+1;
          b := c↑.e[s].p; mb := b↑.m; k := (mb−n+1) div 2;
          {k = no. of items available on adjacent page b}
          a↑.e[n] := c↑.e[s]; a↑.e[n] .p := b↑.p0;
          if k > 0 then
          begin {move k items from b to a}
             for i := 1 to k−1 do a↑.e[i+n] := b↑.e[i];
             c↑.e[s] := b↑.e[k]; c↑.e[s] .p := b;
             b↑.p0 := b↑.e[k] .p; mb := mb−k;
             for i := 1 to mb do b↑.e[i] := b↑.e[i+k];
             b↑.m := mb; a↑.m := n−1+k; h := false
          end else
          begin {merge pages a and b}
             for i := 1 to n do a↑.e[i+n] := b↑.e[i];
             for i := s to mc−1 do c↑.e[i] := c↑.e[i+1];
             a↑.m := nn; c↑.m := mc−1; {dispose(b)}
          end
       end else
```

Program 4.7 (Continued)

```
        begin {b := page to the left of a}
           if s = 1 then b := c↑.p0 else b := c↑.e[s-1] .p;
           mb := b↑.m + 1; k := (mb-n) div 2;
           if k > 0 then
           begin {move k items from page b to a}
              for i := n-1 downto 1 do a↑.e[i+k] := a↑.e[i];
              a↑.e[k] := c↑.e[s]; a↑.e[k] .p := a↑.p0; mb := mb-k;
              for i := k-1 downto 1 do a↑.e [i]:= b↑.e[i+mb];
              a↑.p0 := b↑.e[mb] .p;
              c↑.e[s] := b↑.e[mb]; c↑.e[s] .p := a;
              b↑.m := mb-1; a↑.m := n-1+k; h := false
           end else
           begin {merge pages a and b}
              b↑.e[mb] := c↑.e[s]; b↑.e[mb] .p := a↑.p0;
              for i := 1 to n-1 do b↑.e[i+mb] := a↑.e[i];
              b↑.m := nn; c↑.m := mc-1; {dispose(a)}
           end
        end
    end {underflow} ;

    procedure del(p: ref; var h: boolean);
        var q: ref;       {global a,k}
    begin
        with p↑ do
        begin q := e[m] .p;
           if q ≠ nil then
           begin del(q,h); if h then underflow(p,q,m,h)
           end else
           begin p↑.e[m] .p := a↑.e[k] .p; a↑.e[k] := p↑.e[m];
              m := m-1; h := m<n
           end
        end
    end {del} ;

begin {delete}
    if a = nil then
    begin writeln ('KEY IS NOT IN TREE'); h := false
    end else
```

Program 4.7 (Continued)

```
    with a↑ do
    begin l := 1; r := m;  {binary array search}
       repeat k := (l+r) div 2;
          if x ≤ e[k] .key then r := k−1;
          if x ≥ e[k] .key then l := k+1;
       until l > r;
       if r=0 then q := p0 else q := e[r] .p;
       if l−r > 1 then
       begin {found, now delete e[k]}
          if q = nil then
          begin {a is a terminal page}  m := m−1; h := m<n;
             for i := k to m do e[i] := e[i+1];
          end else
          begin del(q,h); if h then underflow(a,q,r,h)
          end
       end else
       begin delete(x,q,h); if h then underflow(a,q,r,h)
       end
    end
end {delete} ;

procedure printtree(p: ref; l: integer);
    var i: integer;
begin if p ≠ nil then
    with p↑ do
    begin for i := 1 to l do write('      ');
          for i := 1 to m do write(e[i].key: 4);
          writeln;
          printtree(p0,l+1);
          for i := 1 to m do printtree(e[i] .p, l+1)
    end
end ;

begin root := nil; read(x);
    while x ≠ 0 do
    begin writeln('SEARCH KEY', x);
       search(x,root,h,u);
       if h then
       begin {insert new base page} q := root; new(root);
          with root↑ do
             begin m := 1; p0 := q; e[1] := u
             end
       end ;
       printtree(root,1); read(x)
    end ;
```

Program 4.7 (Continued)

```
    read(x);
    while x ≠ 0 do
    begin writeln('DELETE KEY', x);
        delete(x,root,h);
        if h then
        begin {base page size was reduced}
            if root↑.m = 0 then
            begin q := root; root := q↑.p0; {dispose(q)}
            end
        end ;
        printtree(root,1); read(x)
    end
end .
```

Program 4.7 (Continued)

Extensive analysis of B-tree performance has been undertaken and is reported in the referenced article (Bayer and McCreight). In particular, it includes a treatment of the question of optimal page size *n*, which strongly depends on the characteristics of the storage and computing system available.

Variations of the B-tree scheme are discussed in Knuth, Vol. 3, pp. 476–479. The one notable observation is that page splitting should be delayed in the same way that page merging is delayed, by first attempting to balance neighboring pages. Apart from this, the suggested improvements seem to yield marginal gains.

4.5.2. Binary B-Trees

The species of B-trees which seem to be least interesting are the first-order B-trees ($n = 1$). But sometimes it is worthwhile to pay attention even to these cases. It is plain, however, that first-order B-trees are not useful in representing large, ordered, indexed data sets involving secondary stores; approximately 50% of all pages will contain a single item only. Therefore, we shall forget secondary stores and again consider the problem of search trees involving a *one-level store* only.

A binary B-tree (BB-tree) consists of nodes (pages) with either one or two items. Hence, a page contains either two or three pointers to descendants; this suggested the term *2–3 tree*. According to the definition of B-trees, all leaf pages appear at the same level, and all non-leaf pages of BB-trees have either two or three descendants (including the root). Since we now are dealing with primary store only, an optimal economy of storage space is mandatory, and the representation of the items inside a node in the form of an array appears unsuitable. An alternative is the dynamic, linked allocation; that is, inside each node there exists a linked list of items of length 1 or 2. Since each

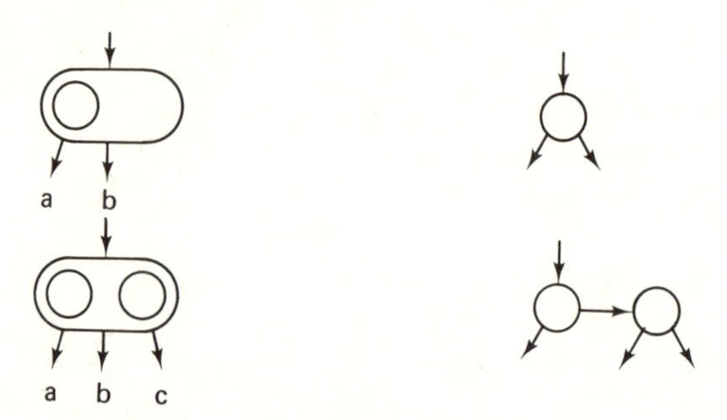

Fig. 4.50 Representation of BB-tree nodes.

node has at most three descendants and thus needs to harbor only up to three pointers, one is tempted to combine the pointers for descendants and pointers in the item list as shown in Fig. 4.50. The B-tree node thereby loses its actual identity, and the items assume the role of nodes in a regular binary tree. It remains necessary, however, to distinguish between pointers to descendants (vertical) and pointers to "siblings" on the same page (horizontal). Since only the pointers to the right may be horizontal, a single bit is sufficient to record this distinction. We therefore introduce the Boolean field h with the meaning "horizontal." The definition of a tree node based on this representation is given in (4.84). It was suggested and investigated by R. Bayer [4-3] in 1971 and represents a search tree organization guaranteeing a maximum path length $p = 2 \cdot \lceil \log N \rceil$.

```
type node = record key: integer;
                   . . . . . . . . . .
                   left,right: ref;                    (4.84)
                   h: boolean
            end
```

Considering the problem of key insertion, one must distinguish four possible situations that arise from a growth of the left or right subtrees. The four cases are illustrated in Fig. 4.51. Remember that B-trees have the characteristic of growing from the bottom toward the root and that the property of all leafs being at the same level must be maintained.

The simplest case (1) is when the *right* subtree of a node A grows and when A is the only key on its (hypothetical) page. Then, the descendant B merely becomes the sibling of A, i.e., the vertical pointer becomes a horizontal pointer. This simple "raising" of the right arm is not possible if A already has a sibling. Then we obtain a page with 3 nodes, and we have to split it (case 2). Its middle node B is passed up to the next higher level.

Now assume that the *left* subtree of a node B has grown in height. If B is again alone on a page (case 3), i.e., its right pointer refers to a descendant,

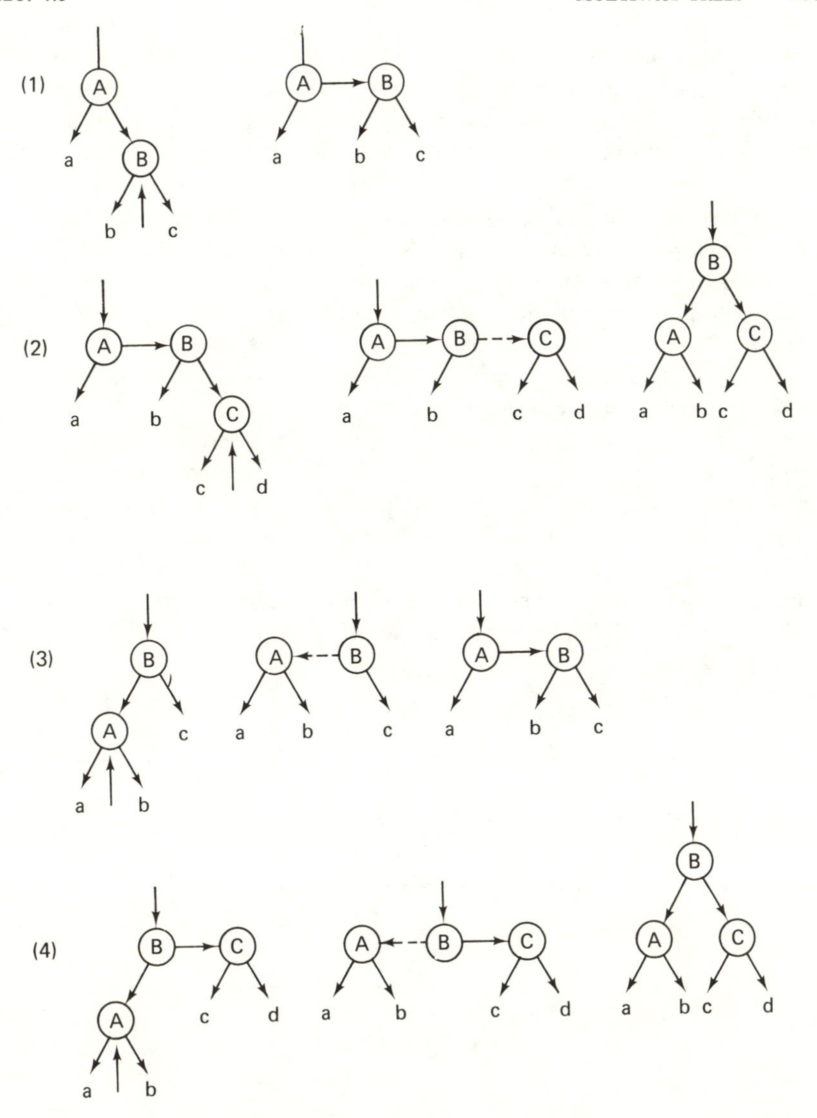

Fig. 4.51 Node insertion in BB-tree.

then the left subtree (A) is allowed to become B's sibling. (A simple rotation of pointers is necessary since the left pointer cannot be horizontal.) If, however, B already has a sibling, the raising of A yields a page with three members, requiring a split. This split is realized in a very straightforward manner: C becomes a descendant of B, which is raised to the next higher level (case 4).

It should be noted that upon searching a key, it makes no effective difference whether we proceed along a horizontal or a vertical pointer. It therefore appears artificial to worry about a left pointer in case 3 becoming horizontal, although its page still contains not more than two members. Indeed, the insertion algorithm reveals a strange asymmetry in handling the growth of left and right subtrees, and it lets the BB-tree organization appear rather artificial. There is no "proof" of strangeness of this organization; yet a healthy intuition tells us that something is "fishy" and that we should remove this asymmetry. It leads to the notion of the *symmetric binary B-tree* (SBB-tree) which was also investigated by Bayer [4-4] in 1972. On the average it leads to slightly more efficient search trees, but the algorithms for insertion and deletion are also slightly more complex. Moreover, each node now requires two bits (Boolean variables *lh* and *rh*) to indicate the nature of its two pointers.

Since we will restrict our detail considerations to the problem of insertion, we have once again to distinguish among four cases of grown subtrees. They are illustrated in Fig. 4.52, which makes the gained symmetry evident. Note that whenever a subtree of node A without siblings grows, the root of the subtree becomes the sibling of A. This case need not be considered any further.

The four cases considered in Fig. 4.52 all reflect the occurrence of a page overflow and the subsequent page split. They are labelled according to the directions of the horizontal pointers linking the three siblings in the

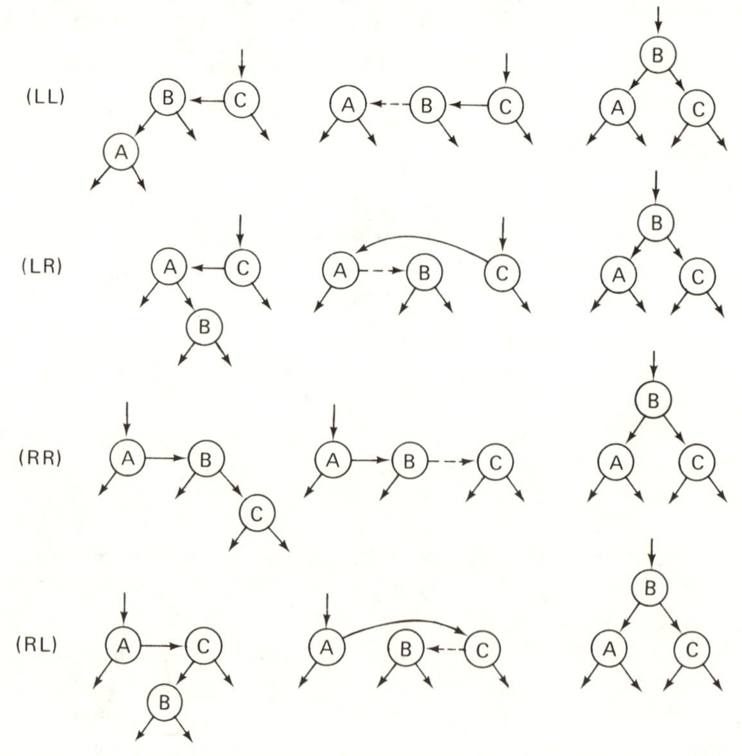

Fig. 4.52 Insertion in SBB-trees.

middle figures. The initial situation is shown in the left column; the middle column illustrates the fact that the lower node has been raised as its subtree has grown; the figures in the right column show the result of node re-arrangement (page split).

It is advisable to stick no longer to the notion of pages out of which this organization had developed, for all we are interested in is to bound the maximum path length to $2 \cdot \log N$. For this we need only to ensure that nowhere are there two successive horizontal pointers on any search path. However, there is no reason to forbid any nodes with horizontal pointers to the left *and* right. We will therefore define the SBB-tree as a tree that has the following properties:

1. Every node contains one key and at most two (pointers to) subtrees.
2. Every pointer is either horizontal or vertical. There are no two consecutive horizontal pointers on any search path.
3. All terminal nodes (nodes without descendants) appear at the same (terminal) level.

From this definition it follows that the longest search path is no longer than twice the height of the tree. Since no SBB-tree with N nodes can have a height larger than $\lceil \log N \rceil$, it follows immediately that $2\lceil \log N \rceil$ is an upper bound on the search path length.

In order to let the reader visualize how these trees grow, he is referred to Fig. 4.53. The lines represent snapshots taken during the insertion of the following sequences of keys, where every semicolon marks a snapshot.

$$\begin{array}{l} (1)\ \ 1\ \ 2;\ \ 3;\ \ 4\ \ 5\ \ 6;\ \ 7; \\ (2)\ \ 5\ \ 4;\ \ 3;\ \ 1\ \ 2\ \ 7\ \ 6; \\ (3)\ \ 6\ \ 2;\ \ 4;\ \ 1\ \ 7\ \ 3\ \ 5; \\ (4)\ \ 4\ \ 2\ \ 6;\ \ 1\ \ 7;\ \ 3\ \ 5; \end{array} \tag{4.85}$$

These pictures make the third property of B-trees particularly obvious: all terminal nodes appear on the same level. One is therefore inclined to compare these structures with garden hedges that have been recently trimmed with hedge scissors. We call these structures *hedges*.

The algorithm for the construction of hedge-trees is formulated in (4.87). It is based on a definition of the node type (4.86) with the two components *lh* and *rh* denoting horizontality of the left and right pointers.

```
type node = record key: integer;
                   count: integer;
                   left,right: ref;                    (4.86)
                   lh,rh: boolean
            end
```

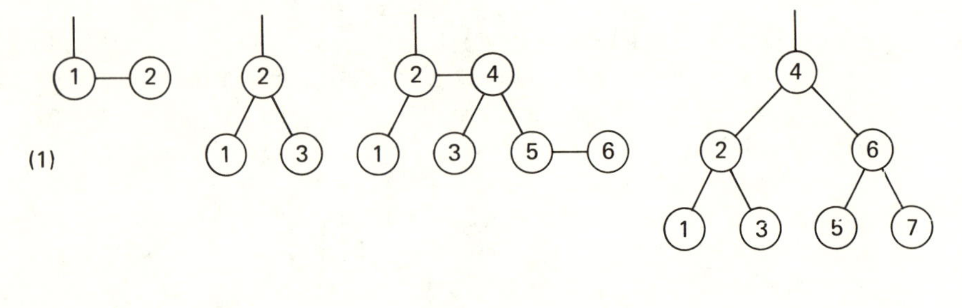

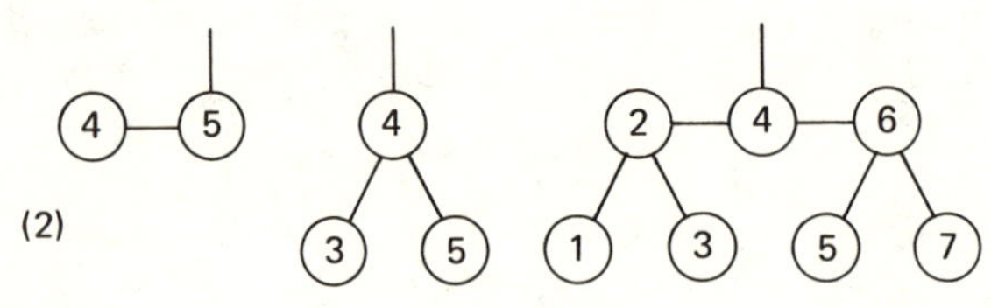

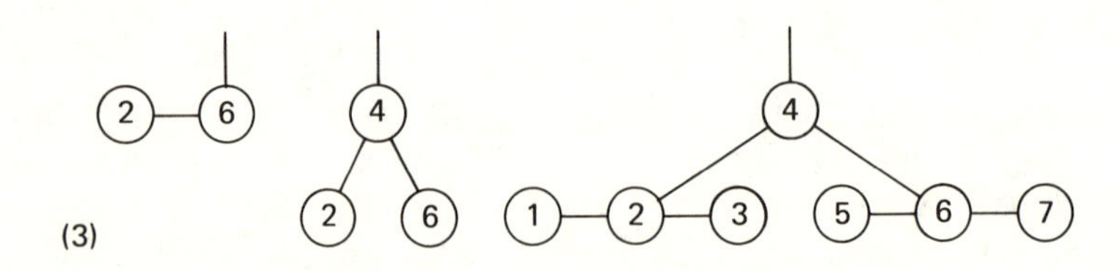

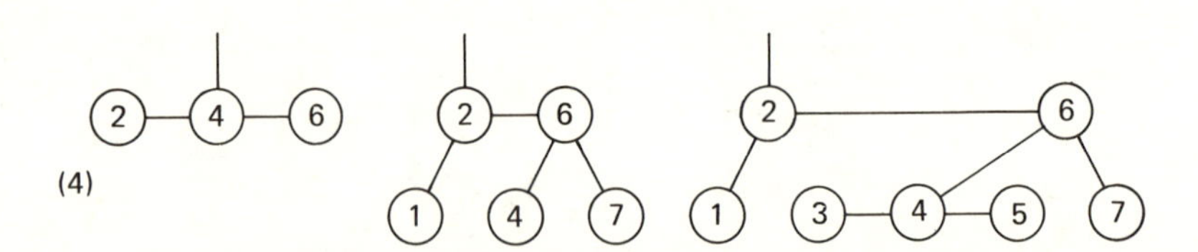

Fig. 4.53 The development of "hedge" trees with insertion sequences (4.85).

The recursive procedure *search* again follows the pattern of the basic binary tree insertion algorithm (see 4.87). A third parameter h is added; it indicates whether or not the subtree with root p has changed, and it corresponds directly to the parameter h of the B-tree search program. We must note, however, the consequence of representing "pages" as linked lists: a page is traversed by either one or two calls of the search procedure. We must distinguish between the case of a subtree (indicated by a vertical pointer) that has grown and a sibling node (indicated by a horizontal pointer) that has obtained another sibling and hence requires a page split. The problem is easily solved by introducing a three-valued h with the following meanings:

1. $h = 0$: the subtree p requires no changes of the tree structure.
2. $h = 1$: node p has obtained a sibling.
3. $h = 2$: the subtree p has increased in height.

```
procedure search(x: integer; var p: ref; var h: integer);
    var p1,p2: ref;
begin
    if p = nil then
    begin {word is not in tree; insert it}
        new(p); h := 2;
        with p↑ do
        begin key := x; count := 1; left := nil;
            right := nil; lh := false; rh := false
        end
    end else
    if x < p↑.key then
    begin search(x,p↑.left,h);
        if h ≠ 0 then
        if p↑.lh then
        begin p1 := p↑.left; h := 2; p↑.lh := false;
            if p1↑.lh then
            begin {LL} p↑.left := p1↑.right;
                p1↑.right := p; p1↑.lh := false; p := p1
            end else                                           (4.87)
            if p1↑.rh then
            begin {LR} p2 := p1↑.right; p1↑.rh := false;
                p1↑.right := p2↑.left; p2↑.left := p1;
                p↑.left := p2↑.right; p2↑.right := p; p := p2
            end
        end else
        begin h := h-1; if h ≠ 0 then p↑.lh := true
        end
    end else
    if x > p↑.key then
    begin search(x,p↑.right,h);
        if h ≠ 0 then
        if p↑.rh then
        begin p1 := p↑.right; h := 2; p↑.rh := false;
            if p1↑.rh then
            begin {RR} p↑.right := p1↑.left;
                p1↑.left := p; p1↑.rh := false; p := p1
            end else
```

```
                if p1↑.lh then
                begin {RL} p2 := p1↑.left; p1↑.lh := false;
                   p1↑.left := p2↑.right; p2↑.right := p1;
                   p↑.right := p2↑.left; p2↑.left := p; p := p2
                end
            end else
            begin h := h-1; if h ≠ 0 then p↑.rh := true
            end
        end else
        begin p↑.count := p↑.count + 1; h := 0
        end
    end {search}
```

Note that the actions to be taken for node re-arrangement very strongly resemble those developed in the balanced tree search algorithm (4.63). From (4.87) it is evident that all four cases can be implemented by simple pointer rotations: single rotations in the *LL* and *RR* cases, double rotations in the *LR* and *RL* cases. In fact, procedure (4.87) appears slightly simpler than (4.63). Clearly, the hedge-tree scheme emerges as an alternative to the AVL-balance criterion. A performance comparison is therefore both possible and desirable.

We refrain from involved mathematical analysis and concentrate on some basic differences. It can be proven that the *AVL-balanced trees are a subset of the hedge-trees*. Hence, the class of the latter is larger. It follows that their path length is on the average larger than in the AVL case. Note in this connection the "worst-case" tree (4) in Fig. 4.53. On the other hand, node re-arrangement will be called for less frequently. The balanced tree will therefore be preferred in those applications in which key retrievals are much more frequent than insertions (or deletions); if this quotient is moderate, the hedge-tree scheme may be preferred.

It is very difficult to say where the borderline lies. It strongly depends not only on the quotient between the frequencies of retrieval and structural change, but also on the characteristics of an implementation. This is particularly the case if the node records have a densely packed representation and consequently access to fields involves part word selection. Boolean fields (*lh*, *rh* in the case of hedge-trees) may be handled more efficiently on many implementations than three-valued fields (*bal* in the case of balanced trees).

4.6. KEY TRANSFORMATIONS (HASHING)

The general problem addressed in the last section and used to develop solutions demonstrating dynamic data allocation techniques is the following:

> Given a set S of items characterized by a key value upon which an ordering relation is defined, how is S to be organized so that

retrieval of an item with a given key k involves as little effort as possible.

Clearly, in a computer store each item is ultimately accessed by specifying a storage address a. Hence, the stated problem is essentially one of finding an appropriate mapping H of keys (K) into addresses (A):

$$H: K \longrightarrow A$$

In Sect. 4.5 this mapping was implemented in the form of various list and tree search algorithms based on different underlying data organizations. Here we present yet another approach that is basically simple and very efficient in many cases. The fact that it also has some disadvantages will be discussed subsequently.

The data organization used in this technique is the array structure. H is therefore a mapping transforming keys into array indices, which is the reason for the term *key transformation* that is generally used for this technique. It should be noted that we shall not need to rely on any dynamic allocation procedures because the array is one of the fundamental, static structures. This paragraph is thus somewhat misplaced under the chapter heading of dynamic information structures, but since it is often used in problem areas where tree structures are comparable competitors, this seems to be an appropriate place for its presentation.

The fundamental difficulty in using a key transformation is that the set of possible key values is very much larger than the set of available store addresses (array indices). A typical example is the use of alphabetical words with, say, up to 10 letters as keys for the identification of individuals in a set of, say, up to a thousand persons. Hence, there are 26^{10} possible keys, which are to be mapped onto 10^3 possible indices. The function H is therefore obviously a many-to-one function. Given a key k, the first step in a retrieval (search) operation is to compute its associated index $h = H(k)$, and the second —evidently necessary—step is to verify whether or not the item with the key k is indeed identified by h in the array (table) T, i.e., to check whether $T[H(k)]$.key $= k$. We are immediately confronted with two questions:

1. What kind of function H should be used?
2. How do we cope with the situation that H does not yield the location of the desired item?

The answer to question 2 is that some method must be used to yield an alternative location, say index h', and, if this is still not the location of the wanted item, yet a third index h'', and so on. The case in which a key other than the desired one is at the identified location is called a *collision*; the task of generating alternative indices is termed *collision handling*. In the following we shall discuss the choice of a transformation function and methods of collision handling.

4.6.1. Choice of a Transformation Function

A prerequisite of a good transformation function is that it distributes the keys as evenly as possible over the range of index values. Apart from satisfying this requirement, the distribution is not bound to any pattern, and it is actually desirable if it gives the impression that it is entirely at random. This property has given this method the somewhat unscientific name *hashing*, i.e., "chopping the argument up" or "making a mess," and H is called the *hash function*. Clearly, it should be efficiently computable, i.e., be composed of very few basic arithmetic operations.

Assume that a transfer function $ord(k)$ is available and denotes the ordinal number of the key k in the set of all possible keys. Assume, furthermore, that the array indices i range over the integers $0 \ldots N - 1$, where N is the size of the array. Then an obvious choice is

$$H(k) = ord(k) \textbf{ mod } N \tag{4.88}$$

It has the property that the key values are spread evenly over the index range, and it is therefore the basis of most key transformations. It is also extremely efficiently computable if N is a power of 2. But it is exactly this case that must be avoided if the keys are sequences of letters. The assumption that all keys are equally likely is in this case entirely erroneous. In fact, words which differ by only a few characters will then most likely map onto identical indices, thus effectively causing a most uneven distribution. In (4.88) it is therefore particularly recommended to let N be a *prime number* [4-7]. This has the consequence that a full division operation is needed that cannot be replaced by a mere masking of binary digits, but this is no serious drawback on most modern computers that feature a built-in division instruction.

Often, hash functions are used which consist of applying logical operations such as the "exclusive or" to some parts of the key represented as a sequence of binary digits. These operations may be faster than division on some computers, but they sometimes fail spectacularly to distribute the keys evenly over the range of indices. We therefore refrain from discussing such methods in further detail.

4.6.2. Collision Handling

If an entry in the table corresponding to a given key turns out to be not the desired item, then a collision is present, i.e., two items have keys mapping onto the same index. A second probe is necessary, one based on an index obtained in a deterministic manner from the given key. There exist several methods of generating secondary indices. An obvious and effective one is linking all entries with identical primary index $H(k)$ together as a linked list. This is called *direct chaining*. The elements of this list may either be in the primary table or not; in the latter case, storage in which they are allocated is usually called an *overflow area.* This method is quite effective, although it has the disadvantage that secondary lists must be maintained and that each entry must provide space for a pointer (or index) to its list of collided items.

An alternative solution for resolving collisions is to dispense with links entirely and instead simply look at other entries in the same table until the item is found or an open position is encountered, in which case one may assume that the specified key is not present in the table. This method is called *open addressing* [4-9]. Naturally, the sequence of indices of secondary probes must always be the same for a given key. The algorithm for a table lookup can then be sketched as follows:

```
h := H(k); i := 0;
repeat
   if T[h].key = k then item found else
   if T[h].key = free then item is not in table else
   begin {collision}
         i := i+1; h := H(k)+G(i)
   end
until found or not in table (or table full)
```
(4.89)

Various functions for resolving collisions have been proposed in the literature. A survey of the topic by Morris in 1968 [4-8] stimulated considerable activities in this field. The simplest method is to try for the next location —considering the table to be circular—until either the item with the specified key is found or an empty location is encountered. Hence, $G(i) = i$; the indices h_i used for probing in this case are

$$\begin{aligned} h_0 &= H(k) \\ h_i &= (h_0 + i) \ \mathbf{mod}\ N, \qquad i = 1 \dots N-1 \end{aligned} \tag{4.90}$$

This method is called *linear probing* and has the disadvantage that entries have a tendency to *cluster* around the primary keys (keys that had not collided upon insertion). Ideally, of course, a function G should be chosen which again spreads the keys uniformly over the remaining set of locations. In practice, however, this tends to be too costly, and methods which offer a compromise by being simple to compute and still superior to the linear function (4.90) are preferred. One of them consists of using a quadratic function such that the sequence of indices for probing is

$$\begin{aligned} h_0 &= H(k) \\ h_i &= (h_0 + i^2) \ \mathbf{mod}\ N \qquad (i > 0) \end{aligned} \tag{4.91}$$

Note that computation of the next index need not involve the operation of squaring if we use the recurrence relations (4.92) for $h_i = i^2$ and $d_i = 2i + 1$.

$$h_{i+1} = h_i + d_i \qquad (i > 0) \qquad (4.92)$$
$$d_{i+1} = d_i + 2$$

with $h_0 = 0$ and $d_0 = 1$. This is called *quadratic probing* and it essentially avoids primary clustering, although practically no additional computations are required. A very slight disadvantage is that in probing not all table entries are searched, that is, upon insertion one may not encounter a free slot although there are some left. In fact, in quadratic probing at least *half* the table is visited if its size N is a *prime number*. This assertion can be derived from the following deliberation. If the ith and the jth probes coincide upon the same table entry, we can express this by the equation

$$i^2 \ \mathbf{mod}\ N = j^2 \ \mathbf{mod}\ N$$

or

$$(i^2 - j^2) \equiv 0 \pmod N$$

Splitting the differences up into two factors, we obtain

$$(i + j)(i - j) \equiv 0 \pmod N$$

Since $i \neq j$, we realize that either i or j have to be *at least* $N/2$ in order to yield $i + j = cN$, with c being an integer.

In practice, the drawback is of no importance since having to perform $N/2$ secondary probes and collision evasions is extremely rare and occurs only if the table is already almost full.

As an application of the scatter storage technique, the Cross-Reference-Generator Program 4.5 is rewritten in the form of Program 4.8. The principal differences lie in the procedure *search* and in the replacement of the pointer type *wordref* by the table of words T. The hash function H is the modulus of the table size; quadratic probing was chosen for collision handling. Note that it is essential for good performance that the table size be a prime number.

Although the method of key transformation is most effective in this case—actually more efficient than tree organizations—it also has a disadvantage. After having scanned the text and collected the words, we wish to tabulate these words in alphabetical order. This is very straightforward when using a tree organization because its very basis is the ordered search tree. It is not, however, when key transformations are used. The full significance of the word *hashing* becomes apparent. Not only does the table printout process have to be preceded by a sort process (for simplicity Program 4.8 uses a straight selection sort), but it even turns out to be advantageous to keep track of inserted keys by linking them together in a special list. Hence, the superior performance of the hashing method considering the process of retrieval only is partly offset by additional operations required to complete the full task of generating an ordered cross-reference index.

Program 4.8 Cross Reference Generator Using Hash Table.

```
program crossref(f,output);
{cross reference generator using hash table}
label 13;
const c1 = 10;      {length of words}
      c2 = 8;      {numbers per line}
      c3 = 6;      {digits per number}
      c4 = 9999;      {max line number}
      p = 997;      {prime number}
      free = '          ';
type  index = 0..p;
      itemref = ↑item;
      word = record key: alfa;
                      first, last: itemref;
                      fol: index
                  end ;
      item = packed record
                      lno: 0..c4;
                      next: itemref
                  end ;
var i, top: index;
      k,k1: integer;
      n: integer;            {current line number}
      id: alfa;
      f: text;
      a: array [1..c1] of char;
      t: array [0..p] of word;            {hash table}
procedure search;
   var h,d,i: index;
      x: itemref; f: boolean;
   {global variables: t, id, top}
begin h := ord(id) mod p;
   f := false; d := 1;
   new(x); x↑.lno := n; x↑.next := nil;
   repeat
      if t[h].key = id then
         begin {found} f := true;
            t[h] .last↑.next := x; t[h].last := x
         end else
```

```
      if t[h].key = free then
         begin {new entry} f := true;
            with t[h] do
            begin key := id; first := x; last := x; fol := top
            end ;
            top := h
         end else
         begin {collision} h := h+d; d := d+2;
            if h ≥ p then h := h-p;
            if d = p then
               begin writeln('TABLE OVERFLOW'); goto 13
               end
         end
   until f
end {search} ;
procedure printtable;
   var i,j,m: index;
   procedure printword(w: word);
      var l: integer; x: itemref;
   begin write('   ', w.key);
      x := w.first; l := 0;
      repeat if l = c2 then
            begin writeln;
               l := 0; write('   ':c1+1)
            end ;
         l := l+1; write(x↑.lno:c3); x := x↑.next
      until x = nil;
      writeln
   end {printword} ;
begin i := top;
   while i ≠ p do
   begin {scan linked list and find minimal key}
      m := i; j := t[i].fol;
      while j ≠ p do
         begin if t[j].key < t[m].key then m := j;
            j := t[j].fol
       end ;
      printword(t[m]);
      if m ≠ i then
         begin t[m].key := t[i].key;
            t[m].first := t[i].first; t[m].last := t[i].last
         end ;
      i := t[i].fol
   end
end {printtable} ;
```

Program 4.8 (Continued)

```
begin n := 0; k1 := c1; top := p; reset(f);
   for i := 0 to p do t[i].key := free;
   while ¬eof(f) do
   begin if n = c4 then n := 0;
      n := n+1; write(n:c3);      {next line}
      write(' ');
      while ¬eoln(f) do
      begin {scan non-empty line}
         if f↑ in ['A' .. 'Z'] then
         begin k := 0;
            repeat if k < c1 then
               begin k := k+1;  a[k] := f↑;
               end ;
               write(f↑); get(f)
            until ¬(f↑ in ['A' .. 'Z', '0' .. '9']);
            if k ≥ k1 then k1 := k else
            repeat a[k1] := ' ';  k1 := k1−1
            until k1 = k;
            pack(a,1,id); search;
         end else
         begin {check for quote or comment}
            if f↑ = '''' then
               repeat write(f↑); get(f)
               until f↑ = '''' else
            if f↑ = '{' then
               repeat write(f↑); get(f)
               until f↑ = '}' ;
            write(f↑); get(f)
         end
      end ;
      writeln; get(f)
   end ;
13: page; printable
end .
```

Program 4.8 (Continued)

4.6.3. Analysis of Key Transformation

Insertion and retrieval by key transformation has evidently a miserable worst-case performance. After all, it is entirely possible that a search argument may be such that the probes hit exactly all occupied locations, missing consistently the desired (or free) ones. Actually, considerable confidence in the correctness of the laws of probability theory is needed by anyone using the hash technique. What we wish to be assured of is that on the *average* the number of probes is small. The following probabilistic argument reveals that it is even *very small.*

Let us once again assume that all possible keys are equally likely and that the hash function H distributes them uniformly over the range of table indices. Assume, then, that a key has to be inserted in a table of size n which already contains k items. The probability of hitting a free location the first time is then $1 - k/n$. This is also the probability p_1 that a single comparison only is needed. The probability that exactly one second probe is needed is equal to the probability of a collision in the first try times the probability of hitting a free location the next time. In general, we obtain the probability p_i of an insertion requiring i probes as listed in (4.93).

$$\begin{aligned} p_1 &= \frac{n-k}{n} \\ p_2 &= \frac{k}{n} \cdot \frac{n-k}{n-1} \\ p_3 &= \frac{k}{n} \cdot \frac{k-1}{n-1} \cdot \frac{n-k}{n-2} \\ &\cdots \\ p_i &= \frac{k}{n} \cdot \frac{k-1}{n-1} \cdot \frac{k-2}{n-2} \cdot \ldots \cdot \frac{k-i+2}{n-i+2} \cdot \frac{n-k}{n-i+1} \end{aligned} \tag{4.93}$$

The expected number of probes required upon insertion of the $k + 1$st key is therefore

$$\begin{aligned} E_{k+1} = \sum_{i=1}^{k+1} i \cdot p_i &= 1 \cdot \frac{n-k}{n} + 2 \frac{k}{n} \frac{n-k}{n-1} + \cdots + (k+1) \cdot \\ &\cdot \left(\frac{k}{n} \frac{k-1}{n-1} \frac{k-2}{n-2} \cdots \frac{1}{n-k+1} \right) = \frac{n+1}{n-k+1} \end{aligned} \tag{4.94}$$

Since the number of probes required to insert an item is identical with the number of probes needed to retrieve it, the result (4.94) can be used to compute the average number E of probes needed to access a random key in a table. Let the table size again be denoted by n, and let m be the number of keys actually in the table. Then

$$E = \frac{1}{m} \sum_{k=1}^{m} E_k = \frac{n+1}{m} \sum_{k=1}^{m} \frac{1}{n-k+2} = \frac{n+1}{m} (H_{n+1} - H_{n-m+1}) \tag{4.95}$$

where

$$H_n = 1 + \frac{1}{2} + \cdots + \frac{1}{n}$$

is the harmonic function. H_n can be approximated as $H_n \cong \ln(n) + \gamma$, where γ is Euler's constant. If, moreover, we substitute $\alpha = m/(n + 1)$, we obtain

$$E = \frac{1}{\alpha}(\ln(n+1) - \ln(n-m+1)) = \frac{1}{\alpha}\ln\frac{n+1}{n+1-m}$$
$$= \frac{-1}{\alpha}\ln(1-\alpha) \tag{4.96}$$

α is approximately the quotient of occupied and available locations, called the *load factor*; $\alpha = 0$ implies an empty table, $\alpha = n/(n+1)$ a full table. The expected number E of probes to retrieve or insert a randomly chosen key is listed in Table 4.6 as a function of the load factor α. The numerical results are indeed surprising, and they explain the exceptionally good performance of the key transformation method. Even if a table is 90% full, on the average only 2.56 probes are necessary to either locate the key or to find an empty location! Note in particular that this figure does not depend on the absolute number of keys present, but only on the load factor.

α	E
0.1	1.05
0.25	1.15
0.5	1.39
0.75	1.85
0.9	2.56
0.95	3.15
0.99	4.66

Table 4.6 Expected Number of Probes As a Function of the Load Factor.

The above analysis was based on the use of a collision handling method that spreads the keys uniformly over the remaining locations. Methods used in practice yield slightly worse performance. Detailed analysis for *linear probing* yields an expected number of probes as given by (4.97) [4-10].

$$E = \frac{1-\alpha/2}{1-\alpha} \tag{4.97}$$

Some numerical values for $E(\alpha)$ are listed in Table 4.7. The results obtained even for the poorest method of collision handling are so good that there is

α	E
0.1	1.06
0.25	1.17
0.5	1.50
0.75	2.50
0.9	5.50
0.95	10.50

Table 4.7 Expected Number of Probes for Linear Probing.

a temptation to regard key transformation (hashing) as the panacea for everything. This is particularly so because its performance is superior even to the most sophisticated tree organization discussed, at least on the basis of comparison steps needed for retrieval and insertion. It is therefore important to point out explicitly some of the drawbacks of hashing, even if they are obvious upon unbiased consideration.

Certainly the major disadvantage over techniques using dynamic allocation is that the *size of the table is fixed* and cannot be adjusted to actual demand. A fairly good *a priori* estimate of the number of data items to be classified is therefore mandatory if either poor storage utilization or poor performance (or even table overflow) is to be avoided. Even if the number of items is exactly known—an extremely rare case—the desire for good performance dictates to dimension the table slightly (say 10%) too large.

The second major deficiency of scatter storage techniques becomes evident if keys are not only to be inserted and retrieved, but if they are also to be deleted, for *deletion* of entries in a hash table is extremely *cumbersome* unless direct chaining in a separate overflow area is used. It is thus fair to say that tree organizations are still attractive, and actually to be preferred, if the volume of data is largely unknown, is strongly variable, and at times even decreases.

EXERCISES

4.1. Let us introduce the notion of a *recursive type*,

$$\textbf{rectype } T = T_0$$

as denoting the union of the set of values defined by the type T_0 and the single value **none**, i.e.,

$$T = T_0 \cup \{\textbf{none}\}.$$

The definition of the type *ped* [see (4.3)], for example, could then be simplified to

rectype *ped* = **record** *name: alfa*;
father, *mother*: *ped*
end

Which is the storage pattern of the recursive structure corresponding to Fig. 4.2?

Presumably, an implementation of such a feature would be based on a dynamic storage allocation scheme, and the fields named *father* and *mother* in the above example would contain pointers generated automatically but hidden from the programmer. What are the difficulties encountered in the realization of such a feature?

4.2. Define the data structure described in the last paragraph of Section 4.2 in terms of records and pointers. Is it also possible to represent this family constellation in terms of recursive types as proposed in the preceding exercise?

4.3. Assume that a first-in-first-out queue Q with elements of type T_0 is implemented as a linked list. Define a suitable data structure, procedures to insert and extract an element from Q, and a function to test whether or not the queue is empty. The procedures should contain their own mechanism for an economical re-use of storage.

4.4. Assume that the records of a linked list contain a key field of type *integer*. Write a program to sort the list in order of increasing value of the keys. Then construct a procedure to invert the list.

4.5. Circular lists (see Fig. 4.54) are usually set up with a so-called *list header*. What is the reason for introducing such a header? Write procedures for the insertion, deletion, and search of an element identified by a given key. Do this once assuming the existence of a header, once without header.

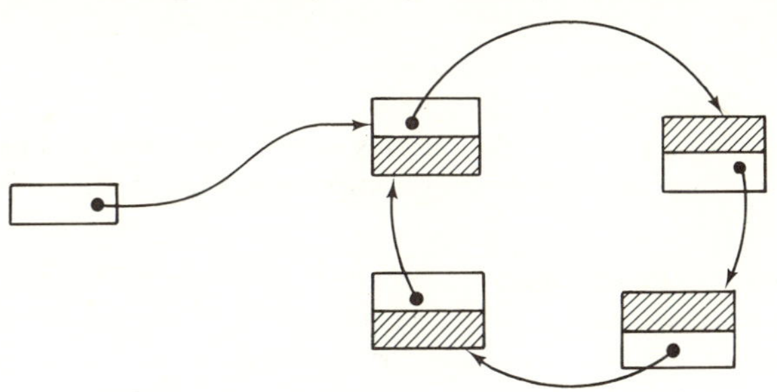

Fig. 4.54 Circular list.

4.6. *A bidirectional list* is a list of elements that are linked in both ways. (See Fig. 4.55.) Both links are originating from a header. Analogous to the preceding exercise, construct a package of procedures for searching, inserting, and deleting elements.

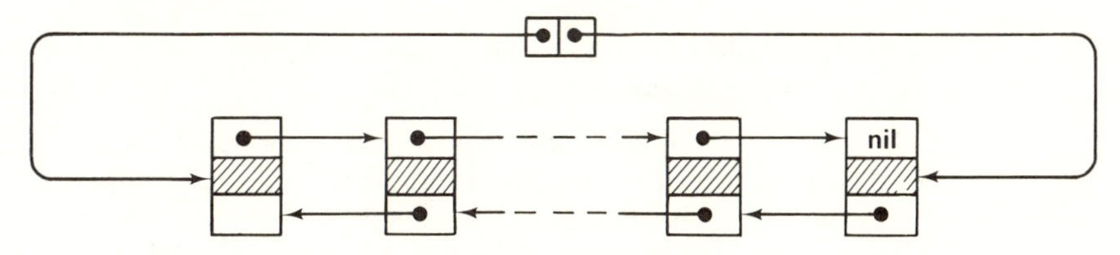

Fig. 4.55 Bidirectional list.

4.7. Does Program 4.2 work correctly if a certain pair $\langle x, y\rangle$ occurs more than once in the input?

4.8. The message "THIS SET IS NOT PARTIALLY ORDERED" in Program 4.2 is not very helpful in many cases. Extend the program so that it outputs a sequence of elements which form a loop if there exists a loop.

4.9. Write a program that reads a program text, identifies all procedure (subroutine) definitions and calls, and tries to establish a topological ordering among the subroutines. Let $P \prec Q$ hold whenever P is called by Q.

4.10. Draw the tree constructed by Program 4.3 if the input consists of the $n + 1$ numbers

$$n, 1, 2, 3, \ldots, n$$

4.11. What are the sequences of nodes encountered when traversing the tree of Fig. 4.23 in preorder, inorder, and postorder?

4.12. Find a composition rule for the sequence of n numbers which, if applied to Program 4.4, yields a perfectly balanced tree.

4.13. Consider the following two orders for traversing binary trees:

(a) (1) Traverse the right subtree.
(2) Visit the root.
(3) Traverse the left subtree.

(b) (1) Visit the root.
(2) Traverse the right subtree.
(3) Traverse the left subtree.

Are there any simple relationships between the sequences of nodes encountered following these orders and those generated by the three orders defined in the text?

4.14. Define a data structure to represent n-ary trees. Then write a procedure that traverses the n-ary tree and generates a binary tree containing the same elements. Assume that the key stored in an element occupies k words and that each pointer occupies one word of storage. What is the gain in storage when using a binary tree versus an n-ary tree?

4.15. Assume that a tree is built upon the following definition of a recursive data structure (see Exercise 4.1).

```
rectype tree = record x: integer;
                      left,right: tree
               end
```

Formulate a procedure to find an element with a given key x and to perform an operation P on this element.

4.16. In a file system a catalog of all files is organized as an ordered binary tree. Each node denotes a file and specifies the file name and, among other things, the date of its last access, encoded as an integer.

Write a program that traverses the tree and deletes all files whose last access was before a certain date.

4.17. In a tree structure the frequency of access of each element is measured empirically by attributing to each node an access count. At certain intervals of time, the tree organization is updated by traversing the tree and generating a new tree by using Program 4.4, and inserting the keys in the order of decreasing frequency count. Write a program that performs this reorganization. Is the average path length of this tree equal to, worse, or much worse than that of an optimal tree?

4.18. The method of analyzing the tree insertion algorithm described in Sect. 4.5 can also be used to compute the expected numbers C of comparisons and M of moves (exchanges) which are performed by Quicksort (Program 2.10) sorting N elements of an array, assuming that all $n!$ permutations of the n keys $\{1, 2, \ldots, n\}$ are equally likely. Find the analogy and determine C_n and M_n.

4.19. Draw the balanced tree with 12 nodes which has the maximum height of all 12-node balanced trees. In which sequence do the nodes have to be inserted so that procedure (4.63) generates this tree?

4.20. Find a sequence of n insertion keys so that procedure (4.63) performs each of the four rebalancing acts (*LL*, *LR*, *RR*, *RL*) at least once. What is the minimal length n for such a sequence?

4.21. Find a balanced tree with keys $1 \ldots n$ and a permutation of these keys so that, when applied to the deletion procedure (4.64), this procedure performs each of the four rebalancing routines at least once. What is the sequence with minimal length n?

4.22. What is the average path length of the Fibonacci-tree T_n?

4.23. Write a program that generates a nearly optimal tree according to the algorithm based on the selection of a centroid as root (4.78).

4.24. Assume that the keys 1, 2, 3, . . . are inserted into an empty B-tree of order 2 (Program 4.7). Which keys cause page splits to occur? Which keys cause the height of the tree to increase?

If the keys are deleted in the same order, which keys cause pages to be merged (and disposed) and which keys cause the height to decrease? Answer the question for (a) a deletion scheme using balancing (as in Program 4.7) and (b) a scheme without balancing (upon underflow, a single item is fetched from a neighboring page).

4.25. Write a program for the search, insertion, and deletion of keys in a binary B-tree. Use the node type definition (4.84). The insertion scheme is shown in Fig. 4.51.

4.26. Find a sequence of insertion keys which, starting from the empty symmetric binary B-tree, causes procedure (4.87) to perform all four rebalancing acts (*LL*, *LR*, *RR*, *RL*) at least once. What is the shortest such sequence?

4.27. Write a procedure for the deletion of elements in a symmetric binary B-tree. Then find a tree and a short sequence of deletions causing all four rebalancing situations to occur at least once.

4.28. Compare the performances of the insertion and deletion algorithm of binary trees, of AVL-balanced trees, and of symmetric binary B-trees on your computer. In particular, investigate the effect of data packing, i.e., of choosing an economical data representation using only 2 bits for the balance information in each node.

4.29. Modify the printing algorithm of Program 4.6 in such a way that it can be used to display symmetric binary B-trees with horizontal and vertical edges.

4.30. If the amount of information associated with each key is relatively large (compared to the key itself), this information should not be stored in the hash table. Explain why and propose a scheme for representing such a set of data.

4.31. Consider the proposal to solve the clustering problem by constructing overflow trees instead of overflow lists, i.e., of organizing those keys which collided as tree structures. Hence, each entry of the scatter (hash) table can be considered as the root of a (possibly empty) tree (tree-hashing).

4.32. Devise a scheme that performs insertions and *deletions* in a hash table using quadratic increments for collision resolution. Compare this scheme experimentally with the straight binary tree organization by applying random sequences of keys for insertion and deletion.

4.33. The primary drawback of the hash table technique is that the size of the table has to be fixed at a time when the actual number of entries is not known. Assume that your computer system incorporates a dynamic storage allocation mechanism that allows obtaining storage at any time. Hence, when the hash table H is full (or nearly full), a larger table H' is generated, and all keys in H are transferred to H', whereafter the store for H can be returned to the mechanism's disposal. This is called *rehashing*. Write a program that performs a rehash of a table H of size n.

4.34. Very often keys are not integers but sequences of letters. These words may greatly vary in length, and therefore they cannot conveniently and economically be stored in key fields of fixed size. Write a program that operates with a hash table and *variable length keys.*

REFERENCES

4-1. ADELSON-VELSKII, G. M. and LANDIS, E. M., *Doklady Akademia Nauk SSSR*, **146**, (1962), 263-66; English translation in Soviet Math, 3, 1259–63.

4-2. BAYER, R. and MCCREIGHT, E., "Organization and Maintenance of Large Ordered Indexes," *Acta Informatica*, **1**, No. 3 (1972), 173–89.

4-3. _____, "Binary B-trees for Virtual Memory," *Proc. 1971 ACM SIGFIDET Workshop*, San Diego, Nov. 1971, pp. 219–35.

4-4. _____, "Symmetric Binary B-trees: Data Structure and Maintenance Algorithms," *Acta Informatica*, **1**, No. 4 (1972), 290–306.

4-5. HU, T. C. and TUCKER, A. C., *SIAM J. Applied Math*, **21**, No. 4 (1971) 514–32.

4-6. KNUTH, D. E., "Optimum Binary Search Trees," *Acta Informatica*, **1**, No. 1 (1971), 14–25.

4-7. MAURER, W. D., "An Improved Hash Code for Scatter Storage," *Comm. ACM*, **11**, No. 1 (1968), 35–38.

4-8. MORRIS, R., "Scatter Storage Techniques," *Comm. ACM*, **11**, No. 1 (1968), 38–43.

4-9. PETERSON, W. W., "Addressing for Random-access Storage," *IBM J. Res. & Dev.*, **1**, (1957), 130–46.

4-10. SCHAY, G. and SPRUTH, W., "Analysis of a File Addressing Method," *Comm. ACM*, **5**, No. 8 (1962), 459–62.

4-11. WALKER, W. A. and GOTLIEB, C. C., "A Top-down Algorithm for Constructing Nearly Optimal Lexicographic Trees," in *Graph Theory and Computing* (New York: Academic Press, 1972), pp. 303–23.

5 LANGUAGE STRUCTURES AND COMPILERS

In this chapter we are aiming at developing a compiler (translator) for a simple, rudimentary programming language. This compiler program may serve as an example for the systematic, well-structured development of a program of non-trivial complexity and size. In this respect, it constitutes a welcome application of the program and data structuring disciplines exposed and elaborated in the preceding chapters, but in addition to this, the aim is to present a general introduction to the structure and operation of compilers. Knowledge and insight on this subject will both enhance the general understanding of the art of programming in terms of high-level languages and will make it easier for a programmer to develop his own systems appropriate for specific purposes and areas of application. Since it is well recognized that the discipline of compiler engineering is a complicated and wide subject, the chapter's character in this latter respect will necessarily be introductory and expository. Perhaps the most important single point is that the structure of language is mirrored in the structure of its compiler and that its complexity—or simplicity—intimately determines the complexity of its compiler. We shall therefore start by describing language composition and will then concentrate exclusively on simple structures that lead to simple, modular translators. Language constructs of this kind of structural simplicity are, as it turns out, adequate for virtually all genuine needs arising in practical programming languages.

5.1. LANGUAGE DEFINITION AND STRUCTURE

Every language is based on a *vocabulary*. Its elements are ordinarily called words; in the realm of formal languages, however, they are called (basic) *symbols*. It is characteristic of languages that some sequences of words are recognized as correct, well-formed *sentences* of the language and that others are said to be incorrect or ill-formed. What is it that determines whether a sequence of words is a correct sentence or not? It is the grammar, syntax, or structure of the language. In fact, we define the *syntax* as the set of rules or formulas which defines the set of (formally correct) sentences. More importantly, however, such a set of rules not only allows us to decide whether or not a given sequence of words is a sentence, but it also provides the sentences with a structure which is instrumental in the recognition of a sentence's meaning. Hence, it is clear that syntax and *semantics* (= meaning) are intimately connected. The structural definitions are therefore always to be considered as auxiliary to a higher purpose. This, however, must not prevent us from initially studying structural aspects exclusively, ignoring the issues of meaning and interpretation.

Take, for example, the sentence, "Cats sleep." The word "cats" is the subject and "sleep" is the predicate. This sentence belongs to the language that may, for instance, be defined by the following syntax.

$$\begin{aligned} \langle\text{sentence}\rangle &::= \langle\text{subject}\rangle\,\langle\text{predicate}\rangle \\ \langle\text{subject}\rangle &::= \mathit{cats} \mid \mathit{dogs} \\ \langle\text{predicate}\rangle &::= \mathit{sleep} \mid \mathit{eat} \end{aligned}$$

The meaning of these three lines is

1. A sentence is formed by a subject followed by a predicate.
2. A subject consists of either the single word "cats" or the word "dogs."
3. A predicate consists of either the word "sleep" or the word "eat."

The idea then is that a sentence may be derived from the *start symbol* ⟨sentence⟩ by repeated application of *replacement rules*.

The formalism or notation in which these rules are written is called *Backus-Naur-Form* (BNF). It was first used in the definition of ALGOL 60 [5-7]. The sentential constructs ⟨sentence⟩, ⟨subject⟩, and ⟨predicate⟩ are called *non-terminal symbols*; the words *cats*, *dogs*, *sleep*, and *eat* are called *terminal symbols*, and the rules are called *productions*. The symbols $::=$ and $\mid$ are called *meta-symbols* of the BNF notation. If, for the sake of brevity, we use single capital letters for non-terminal symbols and lower case letters for terminal symbols, then the example can be rewritten as

EXAMPLE 1

$$\begin{aligned} S &::= AB \\ A &::= x \mid y \\ B &::= z \mid w \end{aligned} \tag{5.1}$$

and the language defined by this syntax consists of the four sentences xz, yz, xw, yw.

To be more precise, we present the following mathematical definitions:

1. Let a language $L = L(T, N, P, S)$ be specified by
   (a) A vocabulary T of terminal symbols.
   (b) A set N of non-terminal symbols (grammatical categories).
   (c) A set P of productions (syntactical rules).
   (d) A symbol S (from N), called the start symbol.
2. The language $L(T, N, P, S)$ is the set of sequences of terminal symbols ξ that can be generated from S according to rule 3 below.

$$L = \{\xi \mid S \xrightarrow{*} \xi \quad \text{and} \quad \xi \in T^*\} \tag{5.2}$$

   (we use Greek letters to denote sequences of symbols.) T^* denotes the set of all sequences of symbols from T.
3. A sequence σ_n can be *generated* from a sequence σ_0 if and only if there exist sequences $\sigma_1, \sigma_2, \ldots, \sigma_{n-1}$ such that every σ_i can be directly generated from σ_{i-1} according to rule 4 below:

$$(\sigma_0 \xrightarrow{*} \sigma_n) \longleftrightarrow ((\sigma_{i-1} \longrightarrow \sigma_i) \quad \text{for } i = 1 \ldots n) \tag{5.3}$$

4. A sequence η can be *directly generated* from a sequence ξ if and only if there exist sequences $\alpha, \beta, \xi', \eta'$ such that
   (a) $\xi = \alpha\xi'\beta$
   (b) $\eta = \alpha\eta'\beta$
   (c) P contains the production $\xi' ::= \eta'$

Note: We use $\alpha ::= \beta_1 \mid \beta_2 \mid \ldots \mid \beta_n$ as a short form for the set of productions $\alpha ::= \beta_1, \quad \alpha ::= \beta_2, \ldots, \quad \alpha ::= \beta_n$.

For instance, the sequence xz of Example 1 can be generated by the following sequence of direct generating steps: $S \longrightarrow AB \longrightarrow xB \longrightarrow xz$; hence $S \xrightarrow{*} xz$, and since $xz \in T^*$, xz is a sentence of the language, i.e., $xz \in L$. Note that the non-terminal symbols A and B occur in non-terminating steps only, whereas the terminating step must lead to a sequence that contains *terminal* symbols only. The grammatical rules are called productions because they determine how new forms may be generated or *produced*.

A language is said to be *context free* if and only if it can be defined in terms of a context free production set. A set of productions is context free if and only if all its members have the form

$$A ::= \xi \qquad (A \in N, \quad \xi \in (N \cup T)^*)$$

i.e., if the left side consists of a single non-terminal symbol and can be replaced by ξ regardless of the context in which A occurs. If a production has the form

$$\alpha A \beta ::= \alpha \xi \beta,$$

then it is said to be *context sensitive* because the replacement of A by ξ may take place only in the context of α and β. We shall subsequently restrict our attention to context free systems.

Example 2 shows how through recursion an infinity of sentences can be generated by a finite set of productions.

EXAMPLE 2

$$\begin{aligned} S &::= xA \\ A &::= z \mid yA \end{aligned} \tag{5.4}$$

The following sentences can be generated from the start symbol S.

xz
xyz
$xyyz$
$xyyyz$
$\ldots\ldots$

5.2. SENTENCE ANALYSIS

The task of language translators or processors is primarily not the generation but the *recognition* of sentences and sentence structure. This implies that the generating steps which lead to a sentence must be reconstructed upon reading the sentence, and that its generation steps must be retraced. This is generally a very complicated and sometimes even impossible task. Its complexity intimately depends on the kind of production rules used to define the language. It is the task of the theory of *syntax analysis* to develop recognizing algorithms for languages with rather complicated structural rules. Here, however, our goal is to outline a method for constructing recognizers that are sufficiently simple and efficient to serve in practice. This implies nothing less than that the computational effort to analyze a sentence must be a linear function of the length of the sentence; in the very worst case the dependency function may be $n \cdot \log n$, where n is the sentence length. Clearly, we cannot be bothered with the problem of finding a recognition algorithm for any given language, but we will work pragmatically in the reverse direction: define an efficient algorithm and then determine the class of languages that can be treated by it [5-3].

A first consequence of the basic efficiency requirement is that the choice of every analysis step must depend only on the present state of computation and on a single next symbol being read. Another most important requirement is that no step will have to be revoked later on. These two requirements are commonly known under the technical term *one-symbol-lookahead without backtracking*.

The basic method to be explained here is called *top-down* parsing because it consists of trying to reconstruct the generating steps (which in general form a structural tree) from their start symbol to the final sentence, from the top down [5-5 and 5-6]. Let us start by revisiting Example 1: We are given the sentence, *Dogs eat*, and we must determine whether or not it belongs to the language. This, by definition, is only the case if it can be generated from the start symbol ⟨sentence⟩. From the grammatical rules it is evident that it can only be a sentence if it is a subject followed by a predicate. We now divide the remaining task; first, we determine whether or not some initial part of the sentence may be generated from the symbol ⟨subject⟩. This is indeed so since *dogs* can be directly generated; the symbol *dogs* is checked off in the input sentence (i.e., we advance our reading position), and we proceed to the second task: checking whether or not the remaining part can be generated from the symbol ⟨predicate⟩. Since this is again the case, the result of the analysis process is affirmative. We may visualize this process by the following trace, showing on the left the tasks still ahead and on the right the part of the input still remaining unread.

⟨sentence⟩	*dogs eat*
⟨subject⟩ ⟨predicate⟩	*dogs eat*
dogs ⟨predicate⟩	*dogs eat*
⟨predicate⟩	*eat*
eat	*eat*
—	—

A second example shows the trace of the analysis process of the sentence *xyyz* according to the productions of Example 2.

S	$xyyz$
xA	$xyyz$
A	yyz
yA	yyz
A	yz
yA	yz
A	z
z	z
—	—

Since the process of retracing the generating steps of a sentence is called *parsing*, what is described above is a *parsing algorithm*. In the two examples the individual replacement steps could decidedly be taken upon inspection of the single next symbol in the input sequence. Unfortunately, this is not always possible, as is evident from the following example:

EXAMPLE 3

$$\begin{aligned} S &::= A \mid B \\ A &::= xA \mid y \\ B &::= xB \mid z \end{aligned} \tag{5.5}$$

We try to parse the sentence *xxxz*

S	$xxxz$
A	$xxxz$
xA	$xxxz$
A	xxz
xA	xxz
A	xz
xA	xz
A	z

and get stuck. The difficulty arises in the very first step in which the decision about replacing S by A or B cannot be taken by looking at the first symbol only. A possible solution lies in simply proceeding according to one of the possible options and to retreat along the taken path of pursued subgoals if no further progress is possible. This action is called *backtracking*. In the language of Example 3, there is no limit to the number of steps that may have to be "undone." This situation is clearly most undesirable; hence, those language structures that lead to backtracking should be identified and avoided in practical applications. Consequently, we shall decree that we will only consider grammar systems satisfying the following restriction that specifies that the initial symbols of alternative right parts of productions be distinct.

RULE 1

Given the production

$$A ::= \xi_1 | \xi_2 | \ldots | \xi_n$$

the sets of initial symbols of all sentences that can be generated from the ξ_i's must be disjoint, i.e.,

$$first(\xi_i) \cap first(\xi_j) = \varnothing \qquad \text{for all } i \neq j.$$

The set $first(\xi)$ is the set of all terminal symbols that can appear in the first position of sentences derived from ξ. Let this set be computed according to the following rules:

1. The first symbol of the argument is terminal:

$$first(a\xi) = \{a\}$$

2. The first symbol is a non-terminal symbol with the derivation rule

$$A ::= \alpha_1 | \alpha_2 | \ldots | \alpha_n$$

   Then

$$first(A\xi) = first(\alpha_1) \cup first(\alpha_2) \cup \ldots \cup first(\alpha_n)$$

In Example 3, we notice that $x \in first(A)$ and $x \in first(B)$. Hence Rule 1 is violated by the first production. It is indeed trivial to find a syntax for the language of Example 3 that satisfies Rule 1. The solution lies in delaying the factoring until all x's have been dealt with. The following productions are *equivalent* with those of (5.5) in the sense that they generate the same set of sentences:

$$\begin{aligned} S &::= C \mid xS \\ C &::= y \mid z \end{aligned} \tag{5.5a}$$

Unfortunately, Rule 1 is not strong enough to shield us from further trouble. Consider

EXAMPLE 4

$$\begin{aligned} S &::= Ax \\ A &::= x \mid \epsilon \end{aligned} \tag{5.6}$$

Here, ϵ denotes the null sequence of symbols. As we try to parse the sentence x, we may proceed into the following "dead alley":

S	x
Ax	x
xx	x
x	—

The trouble arose because we should have followed the production $A ::= \epsilon$ instead of $A ::= x$. This situation is called the *null string problem*, and it arises only for non-terminal symbols that can generate the empty sequence. In order to avoid it, we postulate

RULE 2

For every symbol $A \in N$ which generates the empty sequence ($A \xrightarrow{*} \epsilon$), the set of its initial symbols must be disjoint from the set of symbols that may follow any sequence generated from A, i.e.,

$$\mathit{first}(A) \cap \mathit{follow}(A) = \varnothing$$

The set $\mathit{follow}(A)$ is computed by considering every production P_i of the form

$$X ::= \xi A \eta$$

and taking the set $S_i = \mathit{first}(\eta_i)$. $\mathit{follow}(A)$ is the union of all such sets S_i. If at least one η_i is capable of generating the empty sequence, then the set $\mathit{follow}(X)$ has to be included in $\mathit{follow}(A)$ as well. In Example 4, Rule 2 is violated for the symbol A since

$$\mathit{first}(A) = \mathit{follow}(A) = \{x\}$$

The usual way of expressing a repeated pattern of symbols is by using a recursive definition of a sentential construct. For example, the production

$$A ::= B \mid AB$$

describes the set of sequences $B, BB, BBB, \ldots$. Its use, however, is now prohibited by Rule 1 because

$$first(B) \cap first(AB) = first(B) \neq \varnothing$$

If we replace the production by the slightly modified version

$$A ::= \epsilon \mid AB$$

generating the sequences ϵ, B, BB, BBB, . . . , we violate Rule 2 because

$$first(A) = first(B)$$

and therefore

$$first(A) \cap follow(A) \neq \varnothing$$

The two restrictive rules obviously prohibit the use of left recursive definitions. A simple method used to avoid these forms is by either using right recursion

$$A ::= \epsilon \mid BA$$

or by extending the symbolism of BNF to allow to express replication explicitly; we shall do so by letting $\{B\}$ denote the set of sequences

$$\epsilon, B, BB, BBB, \ldots$$

Of course, one must be aware that every such construct is capable of generating the empty sequence. (The brackets { and } are meta-symbols of extended BNF.)

From the preceding argument and from the transformation of the productions (5.5) into (5.5a) it may appear that the "trick" of transforming grammars might be the panacea to all problems of syntax analysis. We must, however, keep in mind that sentential structure is instrumental in defining sentential meaning, that explanations of the meaning of a sentential construct are usually expressed in terms of the meaning of the sentential components. Take, for example, the language of expressions consisting of operands a, b, c and the minus sign meaning subtraction.

$$\begin{aligned} S &::= A \mid S-A \\ A &::= a \mid b \mid c \end{aligned}$$

According to this grammar, the sentence $a-b-c$ has a structure that can be expressed by using parentheses as follows: $((a-b)-c)$. However, if the grammar is transformed into the syntactically equivalent but left recursion free form

$$\begin{aligned} S &::= A \mid A-S \\ A &::= a \mid b \mid c \end{aligned}$$

then the same sentence would be given another structure, namely, the one expressed as $(a-(b-c))$. Considering the conventional meaning of subtraction, we see that the two forms are not at all semantically equivalent.

The lesson, then, is that when defining a language with an inherent meaning, one must always be aware of the *semantic structure* when devising its syntactic structure because the latter must reflect the former.

5.3. CONSTRUCTING A SYNTAX GRAPH

In the previous paragraph, a top-down recognition algorithm was presented that is applicable to grammars satisfying the restrictive Rules 1 and 2. We now turn to the problem of converting this algorithm into a concrete program. There are two essentially different techniques that can be applied. One is to design a general top-down parsing program valid for all possible grammars (satisfying Rules 1 and 2). In this case, particular grammars are to be supplied in the form of some data structure, on the basis of which the program operates. This general parser is in some sense controlled by the data structure; the program is then called *table driven.* The other technique is to develop a top-down parsing program which is specific for the given language and to construct it systematically according to a set of rules which map a given syntax into a sequence of statements, i.e., into a program. Both techniques have their advantages and disadvantages; both will be introduced subsequently. In the development of a compiler for a given programming language the high degree of flexibility and parametrization of the general parser are hardly needed, whereas the specific parser approach usually leads to more efficient and more easily manageable systems and is therefore preferable. In both cases it is advantageous to represent the given syntax by a so-called *recognition-* or *syntaxgraph.* This graph reflects the flow of control during the process of parsing a sentence.

It is a characteristic of the top-down approach that the *goal* of the parsing process is known at the start. The goal is to recognize a sentence, i.e., a sequence of symbols generatable from the start symbol. The application of a production, that is, the replacement of a single symbol by a sequence of symbols, corresponds to the splitting up of a single goal into a number of subgoals to be pursued in specified order. The top-down method is therefore also called *goal-oriented* parsing. In constructing a parser it is easy to take advantage of this obvious correspondence of non-terminal symbols and goals: we construct a subparser for each non-terminal symbol. Each subparser has the goal of recognizing a subsentence generatable from its corresponding non-terminal symbol. Since we wish to construct a graph to represent the total parser, each non-terminal will be mapped into a subgraph. This leads us to the following rules for constructing a recognizer graph.

RULES OF GRAPH CONSTRUCTION:

A1. Each nonterminal symbol A with corresponding production set

$$A ::= \xi_1 | \xi_2 | \ldots | \xi_n$$

is mapped into a recognition graph A, whose structure is determined by the righthand side of the production according to Rules A2 through A6.

A2. Every occurrence of a *terminal* symbol x in a ξ_i corresponds to a recognizing statement for this symbol and the advancing of the reader to the next symbol of the input sentence. This is represented in the graph by an edge labelled x enclosed in a circle.

A3. Every occurrence of a *non-terminal* symbol B in a ξ_i corresponds to an activation of the recognizer B. This is represented in the graph by an edge labelled B:

A4. A production having the form

$$A ::= \xi_1 | \ldots | \xi_n$$

is mapped into the graph

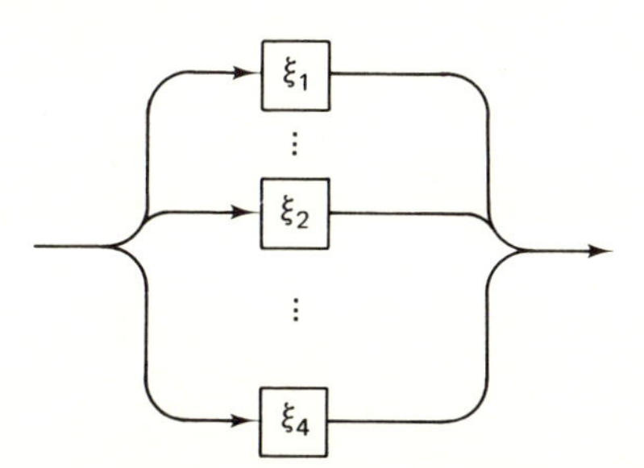

where every $\boxed{\xi_i}$ is obtained by applying construction Rules A2 through A6 to ξ_i.

A5. A ξ having the form

$$\xi = \alpha_1\alpha_2 \ldots \alpha_m$$

is mapped into the graph

where every $\boxed{\alpha_i}$ is obtained by applying construction Rules A2 through A6 to α_i.

A6. A ξ having the form

$$\xi = \{\alpha\}$$

is mapped into the graph

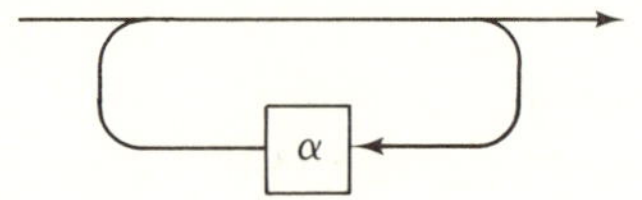

where $\boxed{\alpha}$ is obtained by applying construction Rules A2 through A6 to α.

EXAMPLE 5

$$\begin{aligned} A &::= x \mid (B) \\ B &::= AC \\ C &::= \{+A\} \end{aligned} \tag{5.7}$$

Here, +, x, (, and) are the terminal symbols, whereas { and } belong to the extended BNF and, hence, are meta-symbols. The language generatable from A consists of expressions with operands x, operator +, and parentheses. Examples of sentences are

$$\begin{array}{l} x \\ (x) \\ (x+x) \\ ((x)) \\ \cdots\cdots\cdot \end{array}$$

The graphs resulting from the application of the six construction rules are shown in Fig. 5.1. Note that it is possible to reduce this system of graphs into a single graph by suitable substitution of C in B and of B in A (see Fig. 5.2).

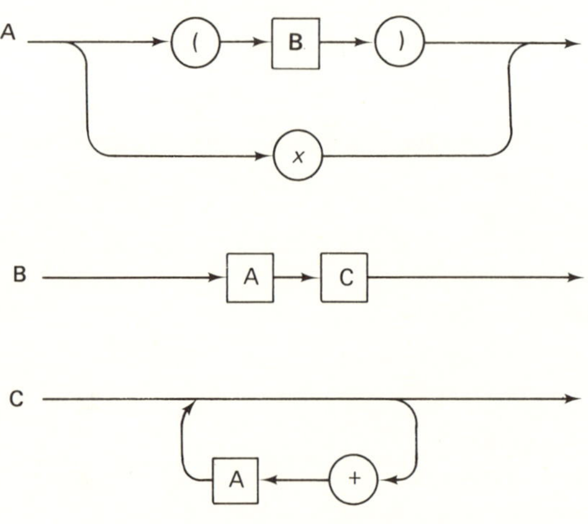

Fig. 5.1 Syntax graphs according to Example 5.

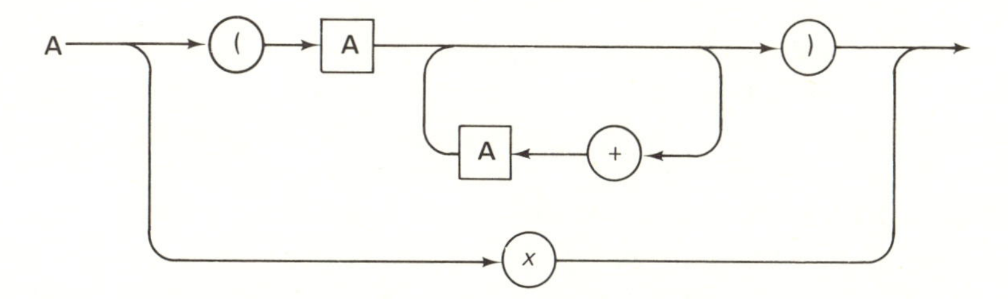

Fig. 5.2 Reduced syntax graph corresponding to Example 5.

The recognition graph is an equivalent representation of the language grammar; it can be used instead of the set of productions in BNF. It is a very convenient form and in many (if not most) instances preferable to BNF. It certainly gives a clearer and more concise picture of a language structure and also conveys a more direct understanding of the parsing process. *The graph is an appropriate form for the designer of a language to start from.* Examples of syntax specifications of entire languages are shown in Sect. 5.7 for PL/0, and in Appendix B for PASCAL.

Restrictive Rules 1 and 2 were imposed in order to allow for deterministic parsing with only one symbol lookahead. How are these rules manifested in the graph representation? It is in this respect that the clarity of the graph becomes most obvious:

1. Rule 1 translates into the requirement that at every fork the branch to be pursued must be selectable by looking only at the next symbol on this branch. This implies that no two branches must start with the same next symbol.
2. Rule 2 translates into the requirement that if any graph *A* can be traversed without reading an input symbol at all, then this "null branch" must be labelled with all symbols that may follow *A*. (This will affect the decision to be made upon entering this branch).

It is simple to verify, whether or not a system of graphs satisfies these two adapted rules, without resorting to a BNF representation of the grammar. As an auxiliary step, the sets *first*(*A*) and *follow*(*A*) are determined for each graph *A*. Application of Rules 1 and 2 is then immediate. We call a system of graphs that satisfies these two rules a *deterministic syntax graph.*

5.4. CONSTRUCTING A PARSER FOR A GIVEN SYNTAX

A program which accepts and parses a language is readily derived from its deterministic syntax graph (if such a graph exists). The graph essentially represents the flowchart of the program. In developing this program, however, one is well-advised to strictly follow a given set of translation rules similar to those that may have led from a BNF to a graph representation of the syntax in the first place. These rules are listed below. They are applicable in a specific framework. This framework consists of a main program in which the procedures corresponding to the various subgoals are embedded and of a routine to proceed to the next symbol.

For the sake of simplicity, let us assume that the sentence to be parsed is represented by the file *input* and that terminal symbols are individual characters. We now postulate the existence within this framework of a character variable *ch* that always represents the next symbol being read. Stepping to the next symbol is then expressed by the statement

read(*ch*)

The main program now consists of an initial statement to read the first character, followed by a statement activating the main parsing goal. The individual routines corresponding to parsing goals or graphs are obtained by obeying the following rules. Let the statement obtained by translating the graph S be denoted by $T(S)$.

RULES OF GRAPH TO PROGRAM TRANSLATION:

B1. Reduce the system of graphs to as few individual graphs as possible by appropriate substitutions.

B2. Translate each graph into a procedure declaration according to the subsequent rules B3 through B7.

B3. A *sequence* of elements

$\longrightarrow \boxed{S_1} \longrightarrow \boxed{S_2} \longrightarrow \cdots \longrightarrow \boxed{S_n} \longrightarrow$

is translated into the compound statement

begin $T(S_1)$; $T(S_2)$; . . .; $T(S_n)$ **end**

B4. A *choice* of elements

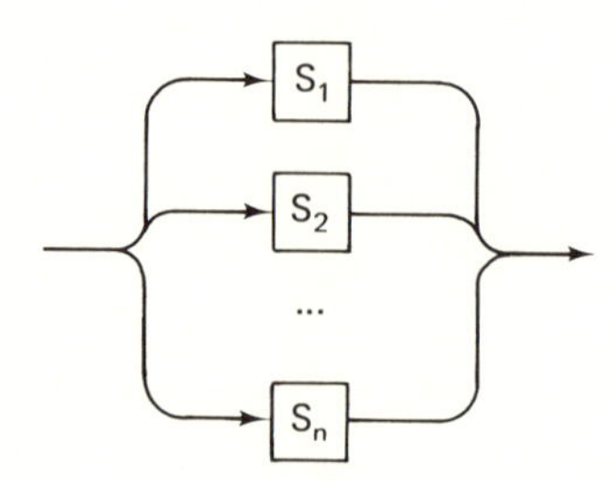

is translated into the selective or conditional statement

```
case ch of
L1 : T(S1);
L2 : T(S2);
  ......
Ln : T(Sn)
end
```

```
if ch in L1 then T(S1) else
if ch in L2 then T(S2) else
  .........
if ch in Ln then T(Sn) else
error
```

where L_i denotes the set of initial symbols of the construct $S_i(L_i = first(S_i))$. *Note*: if L_i consists of a single symbol a, then of course "ch **in** L_i" should be expressed as "$ch = a$".

B5. A loop of the form

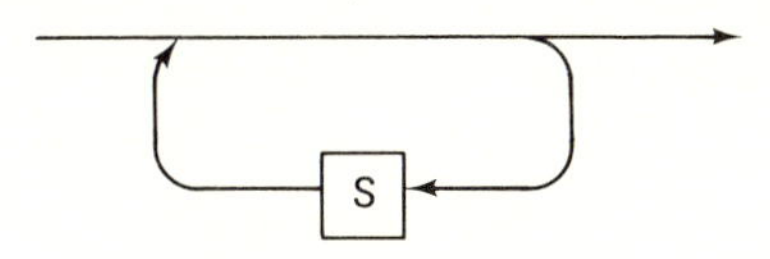

is translated into the statement

```
while ch in L do T(S)
```

where $T(S)$ is the translation of S according to rules B3 through B7, and L is the set $L = first(S)$ (see preceding note).

B6. An element of the graph denoting another graph A

is translated into the procedure call statement A.

B7. An element of the graph denoting a terminal symbol x

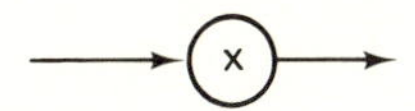

is translated into the statement

```
if ch = x then read(ch) else error
```

where *error* is a routine called when an ill-formed construct is encountered.

The application of these rules is now demonstrated by translating the reduced graph of Example 5 (Fig. 5.2) into a recognizer program (Program 5.1).

```
program parse (input, output);
  var ch: char;
  procedure A;
  begin if ch = 'x' then read(ch) else
          if ch = '(' then
          begin read(ch); A;
            while ch = '+' do
              begin read(ch); A
              end ;
            if ch = ')' then read(ch) else error
          end else error
  end ;
begin read(ch); A
end
```

Program 5.1 Parsing Program for Grammar of Example 5.

During this translation some obvious programming rules have been freely applied in order to simplify the program. A literal translation would have resulted, for example, in the fourth line reading

```
if ch = 'x' then
      if ch = 'x' then read(ch) else error
else . . . .
```

which can obviously be reduced into the simpler form presented in the program. Also the read statements in the fifth and seventh lines resulted from similar reductions.

It seems sensible to find out where such reductions are possible in general and then to represent them directly in terms of the graph. The two relevant cases are covered by the following additional rules:

B4a

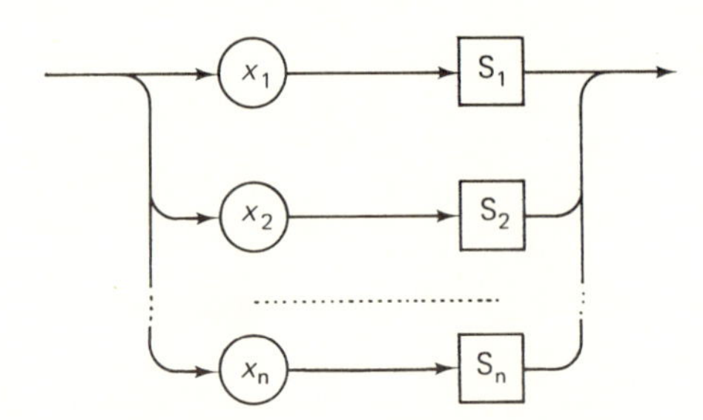

if *ch* = 'x_1' **then begin** *read*(*ch*); $T(S_1)$ **end else** **if** *ch* = 'x_2' **the begin** *read*(*ch*); $T(S_2)$ **end else** . . . . **if** *ch* = 'x_n' **then begin** *read*(*ch*); $T(S_n)$ **end else** *error*

B5a

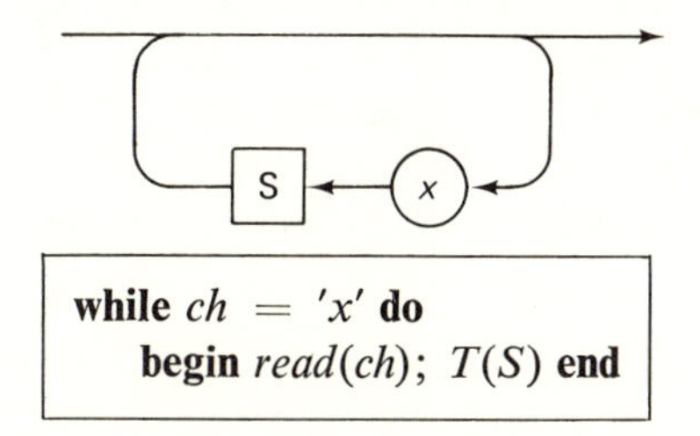

while *ch* = '*x*' **do**   **begin** *read*(*ch*); $T(S)$ **end**

In addition, the frequently occurring construct

> *read*(*ch*); $T(S)$;
> **while** B **do**
>   **begin** *read*(*ch*); $T(S)$ **end**

can of course be expressed by the shorter form

repeat *read*(*ch*); $T(S)$ **until** B (5.8)

The procedure *error* has so far been left unspecified on purpose. Since we are now only interested in finding out whether an input sequence is well- or ill-formed, we may think of this procedure as a program terminator. Naturally, in practice, more refined principles of coping with ill-formed sentences have to be used. This will be the subject of Sect. 5.9.

5.5. CONSTRUCTING A TABLE-DRIVEN PARSING PROGRAM

Instead of composing a specific program according to the rules given in the preceding chapter for each language and syntax that arises, one may construct a single, general parsing program. Individual language grammars are then fed to the general program in the form of initial data preceding the sentences that are to be parsed. The general program strictly follows the rules of the simple top-down parsing method, and it is straightforward if the underlying syntax graph is deterministic, that is, if the grammar is such that sentences can be parsed with one symbol of lookahead and without backtracking.

Hence, the grammar, which we assume to be represented in the form of a deterministic set of syntax graphs, is translated into an appropriate data structure instead of into a program structure [5-2]. The natural technique for representing a graph is by introducing a node for each symbol and by connecting these nodes by pointers. Hence, the "table" is not a simple array structure. The rules which guide the translation are given below and are self-evident. The nodes of the data structure are records of two variants: one for terminal and the other for non-terminal symbols. The former are identified by the terminal symbol for which they stand, the latter by a pointer to the data structure representing the corresponding non-terminal symbol. Both variants contain two pointers, one designating the symbol that follows, the *successor*, and the other forming the list of possible *alternatives*. The resulting data type definition is given in (5.9), and within graphs we will depict a node as

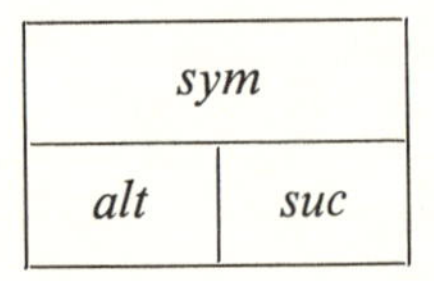

At it turns out, we also need an element to represent the empty sequence, the null symbol. We shall denote it by a terminal element, called *empty*.

```
type pointer = ↑node;
   node =
   record suc,alt: pointer;
      case terminal: boolean of                    (5.9)
         true: (tsym: char);
         false: (nsym: hpointer)
   end
```

The translation rules from graphs into data structures are analogous to Rules B1 through B7.

RULES OF GRAPH TO DATA STRUCTURE TRANSLATION:

C1. Reduce the system of graphs to as few individual graphs as possible by suitable substitution.

C2. Translate each graph into a data structure according to the subsequent rules C3 through C5.

C3. A sequence of elements (see picture of Rule B3) is translated into the following list of data nodes:

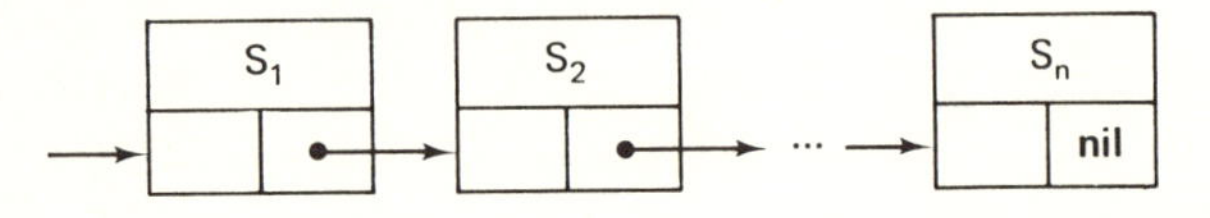

C4. The list of alternatives (see picture of Rule B4) is translated into the data structure

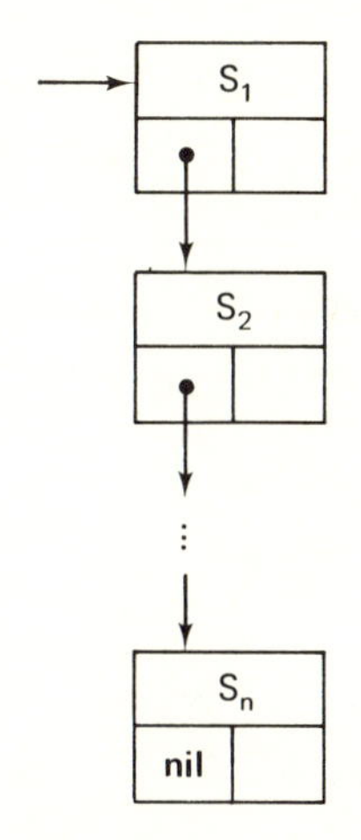

C5. A loop (see picture in Rule B5) is translated into the structure

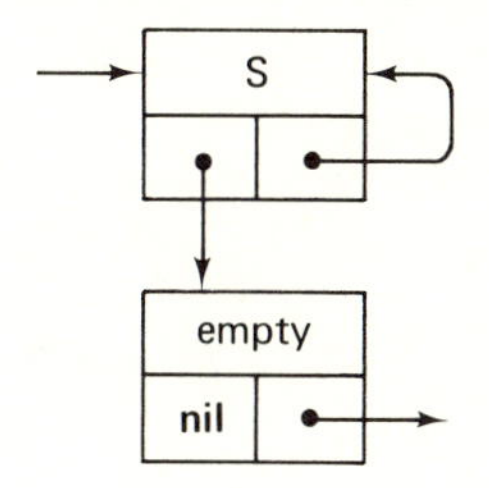

As an example, the graph corresponding to the syntax of Example 5 (Fig. 5.2) results in the structure in Fig. 5.3. The data structure is identified by a *header* node that contains the name of the non-terminal symbol (goal) for which the structure stands. This header is so far unnecessary, for the

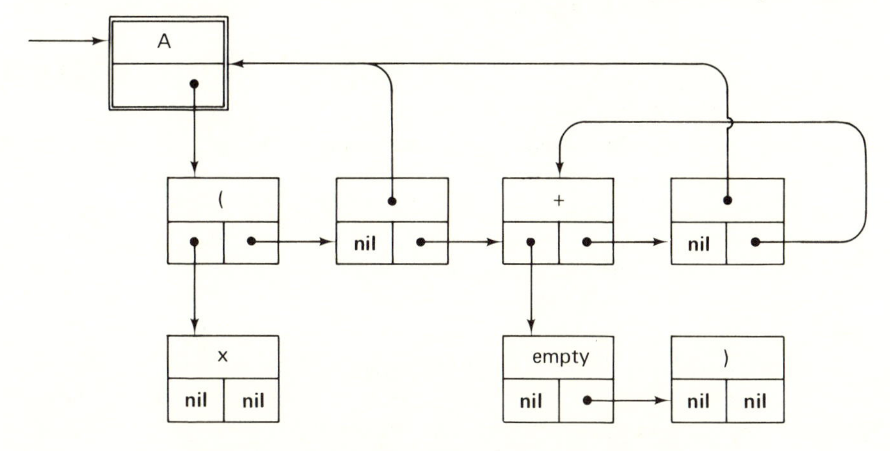

Fig. 5.3 Data structure representing graph of Fig. 5.2.

pointer of the goal field could as well be pointing directly at the "entrance" of the appropriate structure. The header may be used, however, to carry a printable name of the structure.

```
type hpointer = ↑header;
    header =
        record entry: pointer;                    (5.10)
            sym: char
        end
```

The program to parse a sentence—represented as a sequence of characters on the input file—now consists of a repeated statement describing the transition from one node to a next node. The program is expressed as a procedure describing the interpretation of a graph; if a node representing a non-terminal symbol is encountered, then the interpretation of that graph precedes the completion of the interpretation of the present graph. Hence, the interpretation procedure is activated *recursively*. If the current symbol (*sym*) in the input file matches the symbol in the current node of the data structure, then the *suc* field is selected to indicate the next step, otherwise the *alt* field.

```
procedure parse(goal: hpointer; var match: boolean);
    var s : pointer;
begin s := goal↑.entry;
    repeat
        if  s↑.terminal then
            begin if s↑.tsym = sym then
                    begin match := true; getsym
                    end                                   (5.11)
                else match := (s↑.tsym = empty)
            end
        else parse(s↑.nsym, match);
        if match then s := s↑.suc else s := s↑.alt
    until s = nil
end
```

The parsing program (5.11) has the property of immediately pursuing a new subgoal G whenever one appears, without first inspecting whether or not the current symbol is contained in the set of initial symbols *first*(G). This implies that the underlying syntax graph must be void of choices between several alternative non-terminal elements. In particular, if a non-terminal symbol is capable of generating the empty sequence, then none of its right parts must start with a non-terminal symbol.

More sophisticated table-driven parsers may readily be derived from (5.11) which operate on less restrictive classes of grammars. Only slight modifications will also enable it to perform backtracking with, however, notable loss of effectiveness.

Representing a syntax by a graph has one decisive disadvantage: computers cannot directly read graphs. But the data structure that drives the parser must somehow be constructed before parsing can start. It is in this respect that the BNF-representation of grammars appears ideal—as input form for general parsing programs. The next section is therefore devoted to the design of a program which reads a sequence of BNF-productions and transforms them according to rules B1 through B6 into an internal data structure, upon which the parser (5.11) can operate [5-8].

5.6. A TRANSLATOR FROM BNF INTO PARSER-DRIVING DATA STRUCTURES

A translator accepting BNF-productions, converting them into some other representation, is a genuine example of a program whose input data can be regarded as sentences belonging to a language. In fact, it is most appropriate to consider BNF as a language itself, characterized by its own syntax that may, of course, once again be specified in terms of BNF-productions. As a consequence, this translator may serve as a further example of the construction of a recognizer that is, moreover, extended into a translator, or, in general, a processor of its input. Therefore, we shall proceed in the following manner:

Step 1. Define a syntax of the meta language, called EBNF (for Extended BNF).

Step 2. Construct a recognizer for EBNF according to the rules given in Sect. 5.4.

Step 3. Extend the recognizer into a translator, combining it with the table-driven parser.

Let the meta-language—the language of syntax productions—be described by the following productions:

⟨production⟩ ::= ⟨symbol⟩ = ⟨expression⟩.
⟨expression⟩ ::= ⟨term⟩ {, ⟨term⟩}
⟨term⟩ ::= ⟨factor⟩ { ⟨factor⟩}
⟨factor⟩ ::= ⟨symbol⟩ | [⟨term⟩] (5.12)

Note that symbols different from the usual BNF meta-symbols have been used to denote exactly these symbols in the production input language. There are two reasons for this:

1. To distingish meta-symbols and language symbols in (5.12).
2. To use characters more commonly available on computing equipment and, in particular, to be able to use the single character = instead of ::=.

The correspondences of usual BNF with our input form are shown in Table 5.1. In addition, each production is terminated by an explicit period.

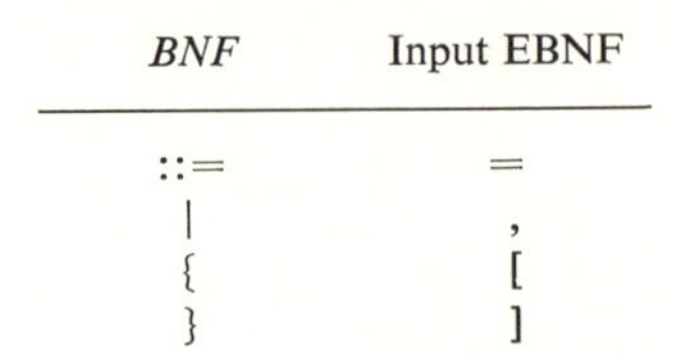

BNF	Input EBNF
::=	=
|	,
{	[
}	]

Table 5.1 Meta and Language Symbols.

Using this input language to describe the syntax of Example 5 (5.7), we obtain

$$\begin{aligned} A &= x, (B). \\ B &= AC. \\ C &= [+A]. \end{aligned} \tag{5.13}$$

In order to simplify the translator to be constructed, we postulate that terminal symbols be *single letters* and that each production be written on a separate line. This includes the possibility of using blanks in the input (to make it more readable) and of ignoring these blanks by the translator. However, the statement *read*(*ch*) in Rule B7 must now be replaced by a call to a routine that obtains the next relevant character. This is a very simple case of what is generally called a lexical scanner, or simply a *scanner*. The purpose of a scanner is to extract the next *symbol*—as defined by language representation rules—from the input sequence of *characters*. So far we have considered symbols to be identical to characters; this, however, is a special case and is rarely done in practice.

As a last rule, in the input-BNF we postulate that non-terminal symbols be represented by the letters A through H and terminal symbols by the letters I through Z. This is merely a rule of convenience that has no deeper reasons. But it makes it unnecessary to list the vocabularies of terminal and non-terminal symbols prior to the list of productions.

By proceeding strictly according to the parser construction rules B1 through B7, and after having verified that (5.12) satisfies the Restrictive Rules 1 and 2, we obtain Program 5.2 as a recognizer for the language specified by (5.12). Note that the scanner is called *getsym*.

```
program parser(input, output);
label 99;
const empty = '*';
var sym: char;

procedure getsym;
begin
   repeat read(sym); write(sym) until sym ≠ ' '
end {getsym} ;

procedure error;
begin writeln;
   writeln (' INCORRECT INPUT'); goto 99
end {error} ;

procedure term;
   procedure factor;
   begin
      if sym in ['A' . . 'Z', empty] then getsym else
      if sym = '[' then
      begin getsym; term;
         if sym = ']' then getsym else error
      end else error
   end {factor} ;
begin factor;
   while sym in ['A' . . 'Z', '[', empty] do factor
end {term} ;

procedure expression;
begin term;
   while sym = ',' do
   begin getsym; term
   end
end {expression} ;

begin {main program}
   while ¬eof(input) do
   begin getsym;
      if sym in ['A' . . 'Z'] then getsym else error;
      if sym = '=' then getsym else error;
      expression;
      if sym ≠ '.' then error;
      writeln; readln;
   end ;
99: end .
```

Program 5.2 Parser of Language (5.13).

Step 3 in the development of the translator is concerned with constructing the desired data structure that represents the BNF productions read and that can be interpreted by the parsing procedure (5.11). Unfortunately, this step is not susceptible to formalization as was the step concerned with the recognizer construction. Lacking a formal approach, we describe once again the structures that are desired to represent each language construct by a picture. The resulting structures are then passed as result parameters of the corresponding recognition procedures, which by then have been augmented into translation procedures. It is natural to return as results not the data structures themselves, but pointers p, q, r referring to the structures instead.

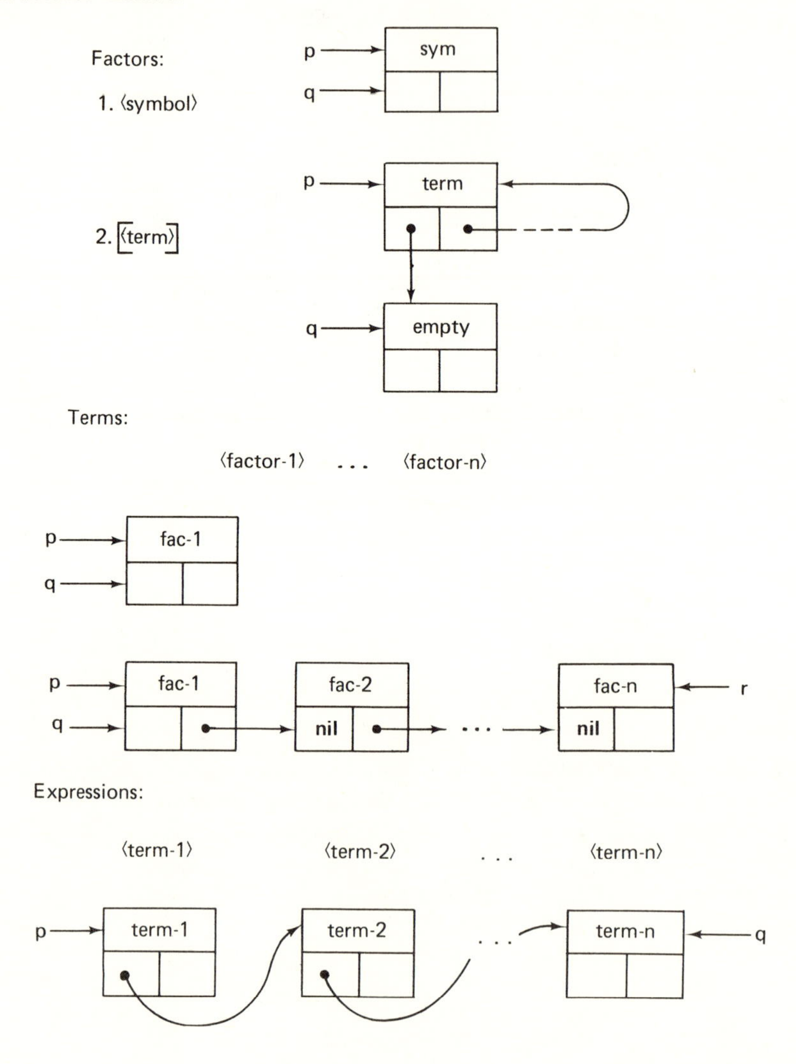

Clearly, it is the task of the procedure *factor* to generate new elements of the data structure; the task of the remaining two procedures is to link them together into a linear list in which *term* uses the *suc* field and *expression* uses the *alt* field for chaining. The details are evident from Program 5.3.

The technique of processing non-terminal symbols needs further clarification. It is possible for a non-terminal symbol to appear as a factor before it appears as a left part in a production. A procedure *find*(*sym*, *h*) is used to locate the symbol *sym* in a linear list in which all headers representing the non-terminal symbols are collected. If a symbol is located, its reference is assigned to *h*; if it is not yet present in the list, it is added to the list. Procedure *find* uses the sentinel technique discussed in detail in Chap. 4.

Program 5.3 consists of three parts, each corresponding to a section of input. Part 1 is concerned with the processing of *productions* into corresponding data structures. Part 2 reads and identifies a single symbol, namely, the one specified as the *start symbol* which generates sentences of the language. (It is preceded by a $ sign delimiting Parts 1 and 2 of the input data.) Part 3 is the parsing program (5.11) reading *input sentences* under control of the data structure generated in Part 1.

It is noteworthy that Program 5.3 has been developed by merely inserting further statements into the *unchanged* Program 5.2. The existing program deals exclusively with the recognition of correctly formed sentences, and it can be used as a framework for the extended program that not only recognizes but also processes or translates the accepted sentences. This method of constructing language processors by *stepwise refinement*, or rather stepwise *enrichment*, is highly recommended. It permits the designer to deal exclusively with a selected aspect of language processing before taking into account other aspects, and hence it facilitates the task of verifying the correctness of a translator program, or at least of maintaining a high confidence level throughout the program's development. In this rather simple example this development consists of only two steps. More complicated languages and more complicated translation tasks require a considerably higher number of individual enrichment steps. A highly similar development in three steps will be the subject of Sects. 5.8 through 5.11.

As evidenced by the development of Program 5.3, the *syntax table-driven* —or rather data structure-driven—approach to parsing provides a degree of freedom and flexibility not presented in the scheme of the specific parser program. This additional flexibility, although not required in general, is the very essence of compilers for so-called *extensible languages*. An extensible language can be extended by further syntactic constructs, more or less at the discretion of the programmer. Analogous to the input of Program 5.3, the input of an extensible language compiler consists of a section specifying the language extensions used in the subsequent program. A more ambitious scheme even allows altering the language during the process of compilation by intermixing parts of the program to be translated with sections of new language specifications.

As appealing or exciting as these ideas may seem, efforts to realize such compilers have been marked by a notable lack of success. The reason is that the aspect of syntax and of sentence recognition is but a part of the whole task of translation, and in fact even the minor part. It is also the part that is most easily formalized and is therefore most readily represented by a systematized table structure. The much harder part to formalize is the *meaning* of the language, that is, the output or result of translation. This problem has so far not been solved even nearly satisfactorily, which explains why compiler designers tend to be much more enthusiastic about extensible languages before than after their first completed assignment. We conclude our lesson by devoting the remainder of this chapter to developing a modest compiler for one specific, small programming language.

Program 5.3 Translator of Language (5.13).

```
program generalparser (input, output);
label 99;
const empty = '*';
type pointer = ↑node;
      hpointer = ↑header;
    node = record suc, alt: pointer;
                   case terminal: boolean of
                      true: (tsym: char);
                      false: (nsym: hpointer)
                end ;
    header = record sym: char;
                      entry: pointer;
                      suc: hpointer
                  end ;
var list, sentinel, h: hpointer;
    p: pointer;
    sym: char;
    ok: boolean;

procedure getsym;
begin
    repeat read(sym); write(sym) until sym ≠ ' '
end {getsym} ;
```

```
procedure find(s: char; var h: hpointer);
{locate nonterminal symbol s in list. if not present, insert it}
   var h1: hpointer;
begin h1 := list; sentinel↑.sym := s;
   while h1↑.sym ≠ s do h1 := h1↑.suc;
   if h1 = sentinel then
      begin {insert} new (sentinel);
         h1↑.suc := sentinel; h1↑.entry := nil
      end ;
   h := h1
end {find} ;

procedure error;
begin writeln;
   writeln ('INCORRECT SYNTAX'); goto 99
end {error} ;

procedure term (var p,q,r: pointer);
   var a,b,c: pointer;
   procedure factor (var p,q: pointer);
      var a,b: pointer; h: hpointer;

   begin if sym in ['A' . . 'Z', empty] then
      begin {symbol}  new(a);
         if sym in ['A' . . 'H'] then
         begin {nonterminal}  find(sym,h);
            a↑.terminal := false; a↑.nsym := h
         end else
         begin {terminal}
            a↑.terminal := true; a↑.tsym := sym
         end ;
         p := a; q := a; getsym
      end else
      if sym = '[' then
      begin getsym; term(p,a,b); b↑.suc := p;
         new(b); b↑.terminal := true; b↑.tsym := empty;
         a↑.alt := b; q := b;
         if sym = ']' then getsym else error
      end else error
   end {factor} ;

begin factor(p,a); q := a;
   while sym in ['A' . . 'Z', '[', empty] do
   begin factor(a↑.suc, b); b↑.alt :=  ; a nil:= b
   end ;
   r := a
end {term} ;
```

Program 5.3 (Continued)

```
procedure expression (var p,q: pointer);
    var a,b,c: pointer;
begin term(p,a,c): c↑.suc := nil;
    while sym = ',' do
    begin getsym;
        term(a↑.alt, b, c): c↑.suc := nil; a := b
    end ;
    q := a
end {expression} ;

procedure parse (goal: hpointer; var match: boolean);
    var s: pointer;
begin s := goal↑.entry;
    repeat
        if s↑.terminal then
        begin if s↑.tsym = sym then
                    begin match := true; getsym
                    end
                else match := (s↑.tsym = empty)
        end
        else parse(s↑.nsym, match); .
        if match then s := s↑.suc else s := s↑.alt
    until s = nil
end {parse} ;

begin {productions}
    getsym; new(sentinel); list := sentinel;
    while sym ≠ '$' do
    begin find(sym,h);
        getsym; if sym = '=' then getsym else error;
        expression (h↑.entry, p); p↑.alt := nil;
        if sym ≠ '.' then error;
        writeln; readln; getsym
    end ;
    h := list; ok := true;  {check whether all symbols are defined}
    while h ≠ sentinel do
    begin if h↑.entry = nil then
            begin writeln(' UNDEFINED SYMBOL ', h↑.sym);
                ok := false
            end ;
        h := h↑.suc
    end ;
```

Program 5.3 (Continued)

```
    if ¬ok then goto 99;
{goal symbol}
    getsym; find(sym,h); readln; writeln;
{sentences}
    while ¬eof(input) do
        begin write(' '); getsym; parse(h,ok);
            if ok ∧ (sym='.') then writeln (' CORRECT')
                                else writeln (' INCORRECT');
            readln
        end ;
99: end .
```

Program 5.3 (Continued)

5.7. THE PROGRAMMING LANGUAGE PL/0

The remaining sections of this chapter are devoted to the development of a compiler for a language to be called PL/0. The necessity of keeping this compiler reasonably small in order to fit into the framework of this book and the desire to be able to expose the most fundamental concepts of compiling high-level languages constitute the boundary conditions for the design of this language. There is no doubt that either an even simpler or a much more complicated language could have been chosen; PL/0 is one possible compromise between sufficient simplicity to make the exposition transparent and sufficient complexity to make the project worthwhile. A considerably more complicated language is PASCAL, whose compiler was developed using the same techniques, and whose syntax is shown in Appendix B.

As far as program structures are concerned, PL/0 is relatively complete. It features, of course, the assignment statement as the basic construct on the statement level. The structuring concepts are those of sequencing, conditional execution and repetition, represented by the familiar forms of **begin/end-**, **if-**, and **while** statements. PL/0 also features the subroutine concept and, hence, contains a procedure declaration and a procedure call statement.

In the realm of data types, however, PL/0 adheres to the demand for simplicity without compromise: integers are its only data type. It is possible to declare constants and variables of this type. Of course, PL/0 features the conventional arithmetic and relational operators.

The presence of procedures, that is, of more or less "self-contained" partitions of a program offers the opportunity to introduce the concept of *locality* of objects (constants, variables, and procedures). PL/0 therefore features declarations in the heading of each procedure, implying that these objects are understood to be local to the procedure in which they are declared.

This brief introduction and overview provide the necessary intuition to understand the syntax of PL/0. This syntax is presented in Fig. 5.4 in the form of seven diagrams. The task of transforming the diagrams into a set of equivalent BNF-productions is left to the interested reader. Fig. 5.4 is a convincing example of the expressive power of these diagrams which allow formulation of the syntax of an entire programming language in such a concise and readable form.

The following PL/0 program may demonstrate the use of some features that are included in this mini-language. The program contains the familiar algorithms for multiplication, division, and finding the greatest common divisor (gcd) of two natural numbers.

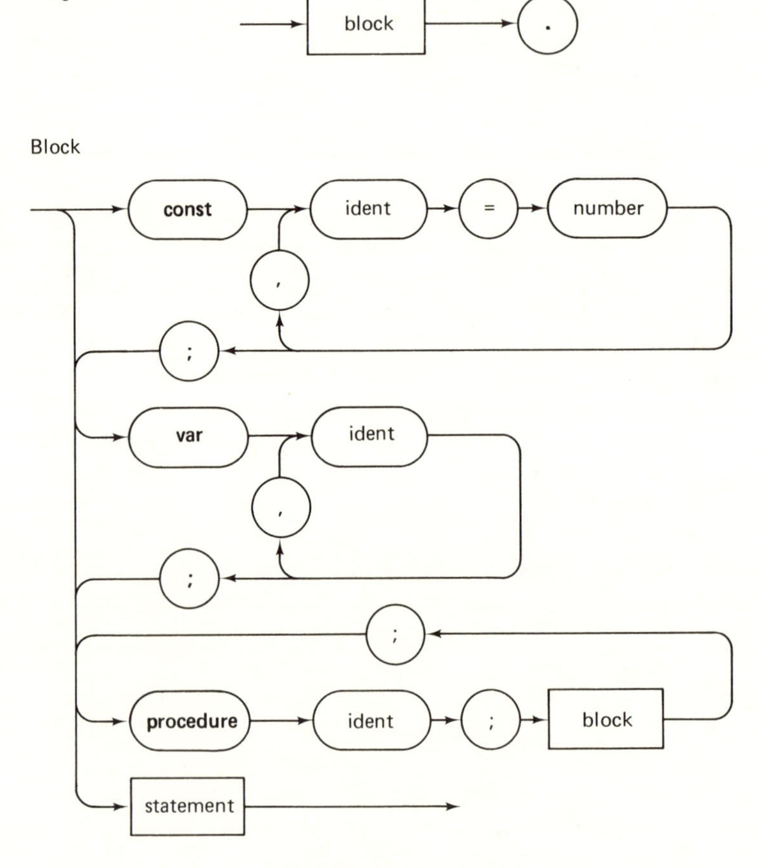

Fig. 5.4 Syntax of PL/0.

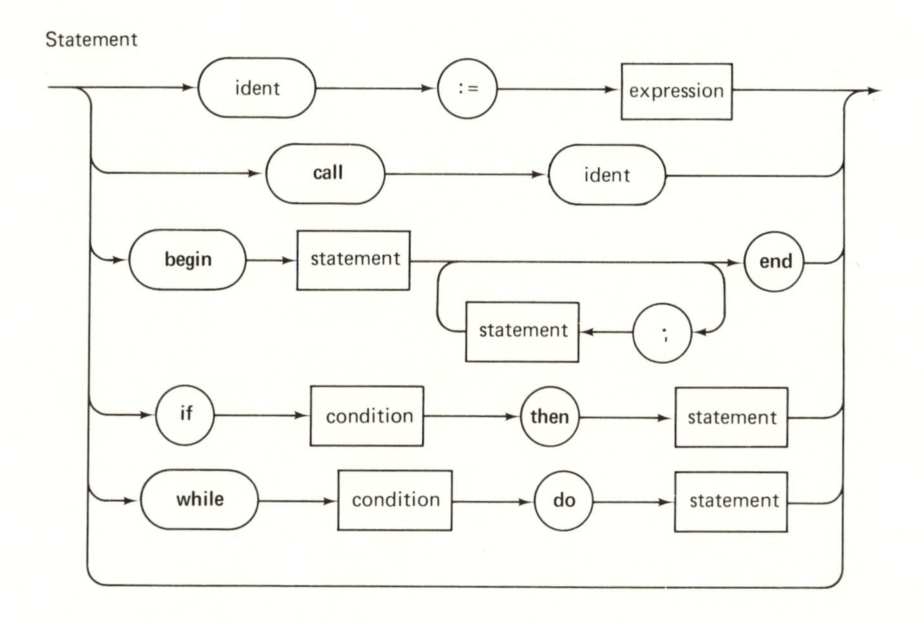

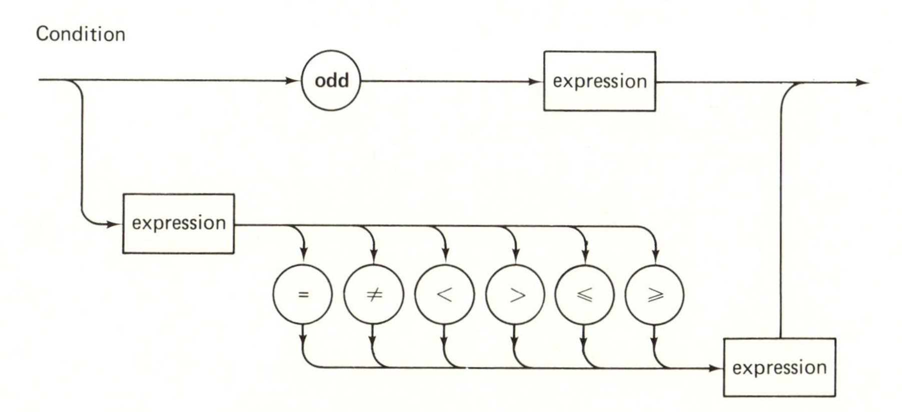

Fig. 5.4 (Continued)

Expression

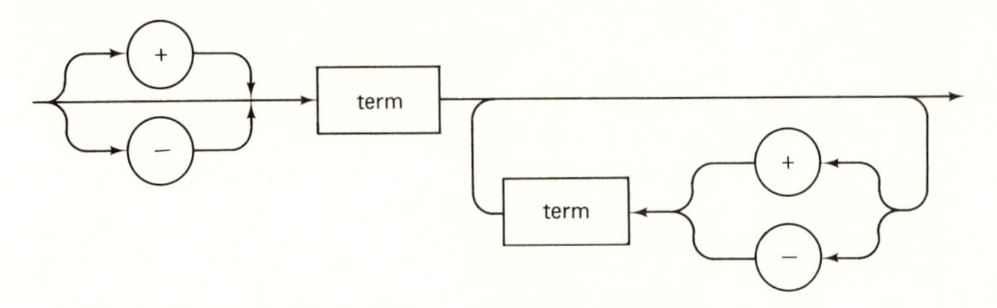

Term

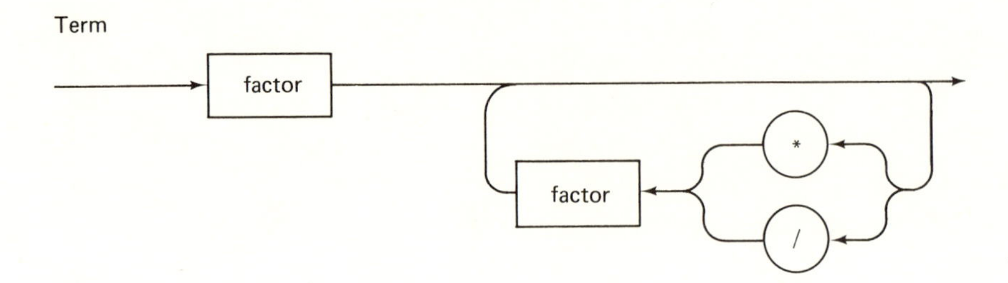

Factor

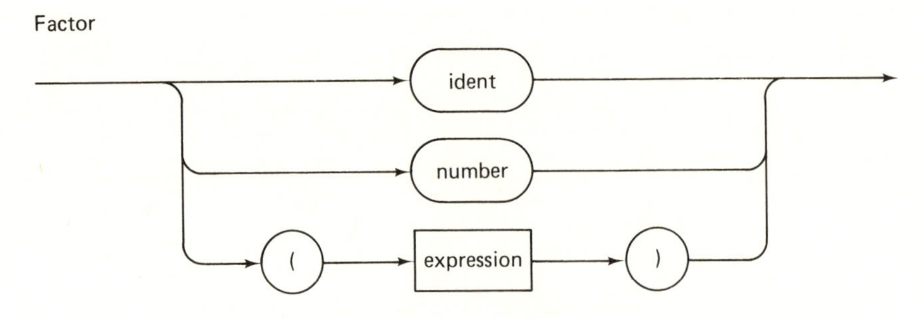

Fig. 5.4 (Continued)

```
const m = 7, n = 85;
var x,y,z,q,r;
procedure multiply;
    var a,b;
begin a := x; b := y; z := 0;
        while b > 0 do
        begin                                    (5.14)
            if odd b then z := z + a;
            a := 2*a; b := b/2;
        end
end ;
```

```
procedure divide;
    var w;
begin r := x; q := 0; w := y;
    while w ≤ r do w := 2*w;
    while w > y do
        begin q := 2*q; w := w/2;                    (5.15)
            if w ≤ r then
                begin r := r-w; q := q+1
                end
        end
end ;

procedure gcd;
    var f,g;
begin f := x; g := y;
    while f ≠ g do
        begin if f < g then g := g-f;                (5.16)
              if g < f then f := f-g;
        end ;
    z := f
end ;

begin
    x := m; y := n; call multiply;
    x := 25; y := 3; call divide;
    x := 84; y := 36; call gcd;
end .
```

5.8. A PARSER FOR PL/0

As a first step toward the PL/0 compiler a parser is being developed. This can be done strictly according to the parser Construction Rules B1 through B7 outlined in Sect. 5.4. This method, however, is only applicable if the Restrictive Rules 1 and 2 are satisfied by the underlying syntax. We are therefore obliged to verify this condition, as formulated for their application to syntax graphs.

Rule 1 specifies that every branch emanating from a fork point must lead toward a distinct first symbol. This is very simple to verify on the syntax diagrams of Fig. 5.4. Rule 2 applies to all graphs that can be traversed without reading any symbol. The only such graph in the PL/0 syntax is the one describing statements. Rule 2 demands that all first symbols that may follow a statement must be disjoint from initial symbols of statements. Since later on it will be useful to know the sets of initial and following symbols for all graphs, we shall determine these sets for all seven non-terminal symbols (graphs) of the PL/0 syntax (except for "program"). Table 5.2 provides the

Non-terminal Symbol S	Initial Symbols $L(S)$	Follow Symbols $F(S)$
Block	**const var** **procedure** *ident* **if call begin while**	. ;
Statement	*ident* **call** **begin if while**	. ; **end**
Condition	**odd** + − (*ident number*	**then do**
Expression	+ − (*ident number*	. ;) R **end then do**
Term	*ident number* (	. ;) R + − **end then do**
Factor	*ident number* (	. ;) R + − * / **end then do**

Table 5.2 Initial and Follow Symbols in PL/0.

desired assurance, namely, that the sets of initial and following symbols of statements do not intersect. Application of the parser Construction Rules B1 through B7 is thereby legalized.

The careful reader will have noticed that the basic symbols of PL/0 are no longer single characters as in the preceding examples. Instead, the basic symbols are themselves sequences of characters, such as BEGIN, or :=. As in Program 5.3, a so-called scanner is used to take care of the merely representational or lexical aspects of the input sequence of symbols. The scanner is conceived as a procedure *getsym* whose task is to get the next symbol. The scanner serves the following purposes:

1. It skips separators (blanks).
2. It recognizes reserved words, such as BEGIN, END, etc.
3. It recognizes non-reserved words as identifiers. The actual identifier is assigned to a global variable called *id*.
4. It recognizes sequences of digits as numbers. The actual value is assigned to a global variable *num*.
5. It recognizes pairs of special characters, such as :=.

In order to scan the input sequence of characters, *getsym* uses a local procedure *getch* whose task is to get the next character. Apart from this main purpose, *getch* also

1. Recognizes and suppresses line end information.
2. Copies the input onto the output file, thus generating a program listing.
3. Prints a line number or location counter at the beginning of each line.

The scanner constitutes the necessary one-symbol lookahead. Moreover, the auxiliary procedure *getch* represents an additional lookahead of one character. Therefore, the total lookahead of this compiler is one symbol plus one character.

The details of these routines are evident from Program 5.4 which represents the complete parser for PL/0. In fact, this parser is already extended in the sense that it collects the declared identifiers denoting constants, variables, and procedures in a *table*. The occurrence of an identifier within a statement then causes a search of this table to determine whether or not the identifier had been properly declared. The lack of such a declaration may duly be regarded as a syntactic error since it is a formal error in the composition of the program text because of the use of an "illegal" symbol. The fact that this error can only be detected by retaining information in a table is a consequence of the inherent *context dependence* of the language, manifest in the rule that all identifiers have to be declared in the appropriate context. Indeed, practically all programming languages are context sensitive in this sense; nevertheless, the context-free syntax is a most helpful model for these languages and greatly aids in the systematic construction of their recognizers. The framework thus obtained can then very easily be extended to take care of the few context sensitive elements of the language, as witnessed by the introduction of the identifier table in the present parser.

Before constructing the individual parser procedures corresponding to the individual syntax graphs, it is useful to determine how these graphs depend on each other. To this end, a so-called *dependence diagram* is constructed; it displays the relationships of the individual graphs, i.e., it lists for each graph G all those graphs $G_1 \ldots G_n$ in terms of which G is defined. Correspondingly, it shows those procedures that will be called by other procedures. The dependence graph for PL/0 is shown in Fig. 5.5.

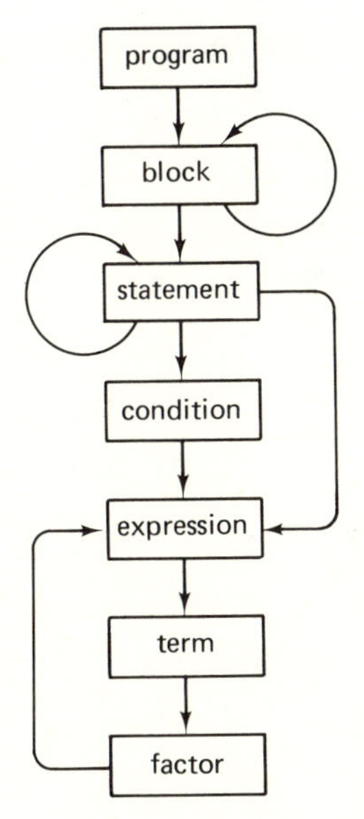

Fig. 5.5 Dependence diagram for PL/0.

The loops in Fig. 5.5 indicate instances of recursion. It is therefore essential that a language in which the PL/0 compiler is implemented is not burdened by prohibition of recursion. In addition, the dependence diagram also allows drawing conclusions on the hierarchical organization of the parser program. For instance, all routines may be contained in (be declared local to) the routine that parses the construct ⟨program⟩ (which is therefore the main program part of the parser). Furthermore, all routines below ⟨block⟩ may be defined locally to the routine representing the parsing goal ⟨block⟩. Naturally, all of these routines call upon the scanner *getsym*, which in turn calls upon *getch*.

Program 5.4 PL/0 Parser.

```
program PL0 (input, output);
{PL/0 compiler, syntax analysis only}
label 99;
const norw = 11;        {no. of reserved words}
    txmax = 100;        {length of indentifier table}
    nmax = 14;          {max. no of digits in numbers}
    al = 10;            {length of identifiers}
type symbol =
    (nul, ident, number, plus, minus, times, slash, oddsym,
     eql, neq, lss, leq, gtr, geq, lparen, rparen, comma, semicolon,
     period, becomes, beginsym, endsym, ifsym, thensym,
     whilesym, dosym, callsym, constsym, varsym, procsym);
    alfa = packed array [1 .. al] of char;
    object = (constant, variable, procedure);
var ch: char;           {last character read}
    sym: symbol;        {last symbol read}
    id: alfa;           {last identifier read}
    num: integer;       {last number read}
    cc: integer;        {character count}
    ll: integer;        {line length}
    kk: integer;
    line: array [1 .. 81] of char;
    a: alfa;
    word: array [1 .. norw] of alfa;
    wsym: array [1 .. norw] of symbol;
    ssym: array [char] of symbol;
    table: array [0 .. txmax] of
              record name: alfa;
                     kind: object
              end ;
```

```
procedure error (n: integer);
begin writeln (' ':cc, '↑', n:2); goto 99
end {error} ;

procedure getsym;
    var i,j,k: integer;

    procedure getch;
    begin if cc = ll then
        begin if eof(input) then
                    begin write (' PROGRAM INCOMPLETE'); goto 99
                    end ;
            ll := 0; cc := 0; write(' ');
            while ¬eoln(input) do
                begin ll := ll+1; read(ch); write(ch); line[ll] := ch
                end ;
            writeln; ll := ll+1; read(line[ll])
        end ;
        cc := cc+1; ch := line[cc]
    end {getch} ;

begin {getsym}
while ch = ' ' do getch;
    if ch in ['A' . . 'Z'] then
    begin {identifier or reserved word}  k := 0;
        repeat if k < al then
            begin k := k+1; a[k] := ch
            end ;
            getch
        until ¬(ch in ['A' . . 'Z', '0' . . '9']);
        if k ≥ kk then kk := k else
            repeat a[kk] := ' '; kk := kk−1
            until kk = k;
        id := a; i := 1; j := norw;
        repeat k := (i+j) div 2;
            if id ≤ word[k] then j := k−1;
            if id ≥ word[k] then i := k+1
        until i > j;
        if i−1 > j then sym := wsym[k] else sym := ident
    end else
    if ch in ['0' . . '9'] then
    begin {number} k := 0; num := 0; sym := number;
        repeat num := 10*num + (ord(ch)−ord('0'));
            k := k+1; getch
        until ¬(ch in ['0' . . '9']);
        if k > nmax then error (30)
    end else
```

Program 5.4 (Continued)

```
    if ch = ':' then
    begin getch;
       if ch = '=' then
       begin sym := becomes; getch
       end else sym := nul;
    end else
    begin sym := ssym[ch]; getch
    end
end {getsym} ;

procedure block (tx: integer);
    procedure enter (k: object);
    begin {enter object into table}
       tx := tx + 1;
       with table[tx] do
       begin name := id; kind := k;
       end
    end {enter} ;

    function position (id: alfa): integer;
       var i: integer;
    begin {find identifier id in table}
       table[0].name := id; i := tx;
       while table[i].name ≠ id do i := i-1;
       position := i
    end {position} ;

    procedure constdeclaration;
    begin if sym = ident then
       begin getsym;
          if sym = eql then
          begin getsym;
             if sym = number then
                begin enter (constant); getsym
                end
             else error (2)
          end else error (3)
       end else error (4)
    end {constdeclaration} ;

    procedure vardeclaration;
    begin if sym = ident then
          begin enter (variable); getsym
          end else error (4)
    end {vardeclaration} ;
```

Program 5.4 (Continued)

```
procedure statement;
    var i: integer;
    procedure expression;
        procedure term;
            procedure factor;
                var i: integer;
            begin
                if sym = ident then
                begin i := position(id);
                    if i = 0 then error (11) else
                    if table[i] .kind = procedure then error (21);
                    getsym
                end else
                if sym = number then
                begin getsym
                end else
                if sym = lparen then
                begin getsym; expression;
                    if sym = rparen then getsym else error (22)
                end
                else error (23)
            end {factor} ;
            begin {term} factor;
                while sym in [times, slash] do
                    begin getsym; factor
                    end
            end {term} ;
        begin {expression}
            if sym in [plus, minus] then
                begin getsym; term
                end else term;
            while sym in [plus, minus] do
                begin getsym; term
                end
        end {expression} ;
        procedure condition;
        begin
            if sym = oddsym then
            begin getsym; expression
            end else
            begin expression;
                if ¬(sym in [eql, neq, lss, leq, gtr, geq]) then
                    error (20) else
                begin getsym; expression
                end
            end
        end {condition} ;
```

```
        begin {statement}
        if sym = ident then
        begin i := position(id);
            if i = 0 then error (11) else
            if table [i] .kind ≠ variable then error (12);
            getsym; if sym = becomes then getsym else error (13);
            expression
        end else
        if sym = callsym then
        begin getsym;
            if sym ≠ ident then error (14) else
                begin i := position(id);
                    if i = 0 then error (11) else
                    if table[i] .kind ≠ procedure then error (15);
                    getsym
                end
        end else
        if sym = ifsym then
        begin getsym; condition;
            if sym = thensym then getsym else error (16);
            statement;
        end else
        if sym = beginsym then
        begin getsym; statement;
            while sym = semicolon do
                begin getsym; statement
                end ;
            if sym = endsym then getsym else error (17)
        end else
        if sym = whilesym then
        begin getsym; condition;
            if sym = dosym then getsym else error (18);
            statement
        end
    end {statement} ;

begin {block}
    if sym = constsym then
    begin getsym; constdeclaration;
        while sym = comma do
            begin getsym; constdeclaration
            end ;
        if sym = semicolon then getsym else error (5)
    end ;
```

Program 5.4 (Continued)

```
        if sym = varsym then
        begin getsym; vardeclaration;
           while sym = comma do
              begin getsym; vardeclaration
              end ;
           if sym = semicolon then getsym else error (5)
        end ;
        while sym = procsym do
        begin getsym;
           if sym = ident then
              begin enter (procedure); getsym
              end
           else error (4);
           if sym = semicolon then getsym else error (5);
           block (tx);
           if sym = semicolon then getsym else error (5);
        end ;
        statement
     end {block} ;
     begin {main program}
        for ch := 'A' to ';' do ssym[ch] := nul;
        word[ 1] := 'BEGIN ';     word[ 2] := 'CALL      ';
        word[ 3] := 'CONST';      word[ 4] := 'DO        ';
        word[ 5] := 'END   ';     word[ 6] := 'IF        ';
        word[ 7] := 'ODD   ';     word[ 8] := 'PROCEDURE';
        word[ 9] := 'THEN  ';     word[10] := 'VAR       ';
        word[11] := 'WHILE ';
        wsym[ 1] := beginsym;     wsym[ 2] := callsym;
        wsym[ 3] := constsym;     wsym[ 4] := dosym;
        wsym[ 5] := endsym;       wsym[ 6] := ifsym;
        wsym[ 7] := oddsym;       wsym[ 8] := procsym;
        wsym[ 9] := thensym;      wsym[10] := varsym;
        wsym[11] := whilesym;
        ssym['+'] := plus;        ssym['-'] := minus;
        ssym['*'] := times;       ssym['/'] := slash;
        ssym['('] := lparen;      ssym[')'] := rparen;
        ssym['='] := eql;         ssym[','] := comma;
        ssym['.'] := period;      ssym['≠'] := neq;
        ssym['<'] := lss;         ssym['>'] := gtr;
        ssym['≤'] := leq;         ssym['≥'] := geq;
        ssym[';'] := semicolon;
        page(output);
        cc := 0; ll := 0; ch := ' '; kk := al; getsym;
        block (0);
        if sym ≠ period then error (9);
     99: writeln
     end .
```

5.9. RECOVERING FROM SYNTACTIC ERRORS

Up to this point the parser had only the modest task of determining whether or not an input sequence of symbols belonged to a language. As a side product, the parser also discovered the inherent structure of a sentence. But as soon as an ill-formed construct was encountered, the parser's task was achieved, and the program could as well terminate. For practical compilers, this is of course no tenable proposition. Instead, a compiler must issue an appropriate error diagnostic and be able to continue the parsing process—probably to find further mistakes. A continuation is only possible either by making some likely assumption about the nature of the error and the intention of the author of the ill-formed program or by skipping over some subsequent part of the input sequence, or both. The art of choosing an assumption with a high likelihood of correctness is rather intricate. It has so far eluded any kind of successful formalization because formalizations of syntax and parsing do not take into account the many factors that strongly influence the human mind. For instance, it is a common error to omit interpunctuation symbols such as the semicolon (not only in programming!), whereas it is highly improbable that one forgets to write a + operator in an arithmetic expression. The semicolon and plus symbol are merely terminal symbols without further distinction for the parser; for the human programmer, the semicolon has hardly a meaning and appears redundant at the end of a line, whereas the significance of an arithmetic operator is obvious beyond doubt. There are many more such considerations that have to go into the design of an adequate recovery system, and they all depend on the individual language and cannot be generalized in the framework of all context-free languages.

Nevertheless, there are some rules and hints that can be postulated and that have validity beyond the scope of a single language such as PL/0. Characteristically, perhaps, they are concerned equally much with the initial conception of a language as with the design of the recovery mechanism of its parser. First of all, it is abundantly clear that sensible recovery is much facilitated, or even made possible, only by a *simple language structure.* In particular, if upon diagnosing an error some part of the subsequent input is to be skipped (ignored), then it is mandatory that the language contains *key words* that are highly unlikely to be misused, and that may therefore serve to bring the parser back into step. PL/0 notably follows this rule: every structured statement begins with an unmistakable keyword such as **begin**, **if**, **while**, and the same holds for declarations; they are headed by **var**, **const**, or **procedure**. We shall therefore call this rule the *keyword rule.*

The second rule concerns the construction of the parser more directly. It is the characteristic of top-down parsing that goals are split up into subgoals and that parsers call upon other parsers to tackle their subgoals. The second rule specifies that if a parser detects an error, it should not merely refuse to continue and report the happening back to its master parser. Instead, it should itself continue to scan text up to a point where some plausible analysis can be resumed. We shall therefore call this the *don't panic rule*. The programmatic consequence of this rule is that there will be no exit from a parser except through its regular termination point.

A possible strict interpretation of the don't panic rule consists of skipping input text upon detecting an illegal formation up to the next symbol that may correctly follow the currently parsed sentential construct. This implies that every parser know the set of its follow-symbols at the place of its present activation.

In the first refinement (or enrichment) step we shall therefore provide every parsing procedure with an explicit parameter *fsys* that specifies the possible follow-symbols. At the end of each procedure an explicit test is included to verify that the next symbol of the input text is indeed among those follow-symbols (if this condition is not already asserted by the logic of the program).

It would, however, be very shortsighted of us to skip the input text up to the next occurrence of such a follow-symbol under all circumstances. After all, the programmer may have mistakenly omitted exactly one symbol (say a semicolon); ignoring the entire text up to the next follow-symbol may be disastrous. We therefore augment these sets of symbols that terminate a possible skip by keywords that specifically mark the beginning of a construct not to be overlooked. The symbols passed as parameters to the parsing procedures are therefore *stopping symbols* rather than follow-symbols only. We may regard the sets of stopping symbols as being initialized by distinct key symbols and being gradually supplemented by legal follow-symbols upon penetration of the hierarchy of parsing subgoals. For flexibility, a general routine called *test* is introduced to perform the described verification. This procedure (5.17) has three parameters:

1. The set *s*1 of admissible next symbols; if the current symbol is not among them, an error is at hand.
2. A set *s*2 of additional stopping symbols whose presence is definitely an error, but which should in no case be ignored and skipped.
3. The number *n* of the pertinent error diagnostic.

```
procedure test (s1, s2: symset; n: integer);
begin if ¬(sym in s1) then
      begin error(n); s1 := s1+s2;
            while ¬(sym in s1) do getsym
      end
end                                                      (5.17)
```

Procedure (5.17) may also be conveniently used at the *entrance* of parsing procedures to verify whether or not the current symbol is an admissible initial symbol. This is recommended in all cases in which a parsing procedure X is called unconditionally, such as in the statement

$$\textbf{if } sym = a_1 \textbf{ then } S_1 \textbf{ else}$$

$$\cdots\cdots$$

$$\textbf{if } sym = a_n \textbf{ then } S_n \textbf{ else } X$$

which is the result of translation of the production

$$A ::= a_1S_1 | \ldots | a_nS_n | X \tag{5.18}$$

In these instances the parameter *s*1 must be equal to the set of initial symbols of X, whereas *s*2 is chosen as the set of the follow-symbols of A (see Table 5.2). The details of this procedure are given in Program 5.5, which represents the enriched version of Program 5.4. For the reader's convenience, the entire parser is listed again, with the exception of initializations of global variables and of the procedure *getsym*, all of which remain unchanged.

The scheme presented so far has the property of trying to recover, to fall back into step, by ignoring one or more symbols in the input text. This is an unfortunate strategy in all cases in which an error is caused by *omission* of a symbol. Experience shows that such errors are virtually restricted to symbols which have merely syntactic functions and do not represent an action. An example is the semicolon in PL/0. The fact that the follow-symbol sets are augmented by certain key words actually causes the parser to stop skipping symbols prematurely, thereby behaving as if a missing symbol had been inserted. This can be seen from the program part that parses compound statements shown in (5.19). It effectively "inserts" missing semicolons in front of key words. The set called *statbegsys* is the set of initial symbols of the construct "statement."

```
if sym = beginsym then
begin getsym;
   statement([semicolon, endsym]+fsys);
   while sym in [semicolon]+statbegsys do
      begin
         if sym = semicolon then getsym else error;        (5.19)
         statement([semicolon, endsym]+fsys)
      end;
   if sym = endsym then getsym else error
end
```

The degree of success with which this program diagnoses syntactic errors and recovers from unusual situations can be estimated by considering the PL/0 program (5.20). The listing represents an output delivered by Program

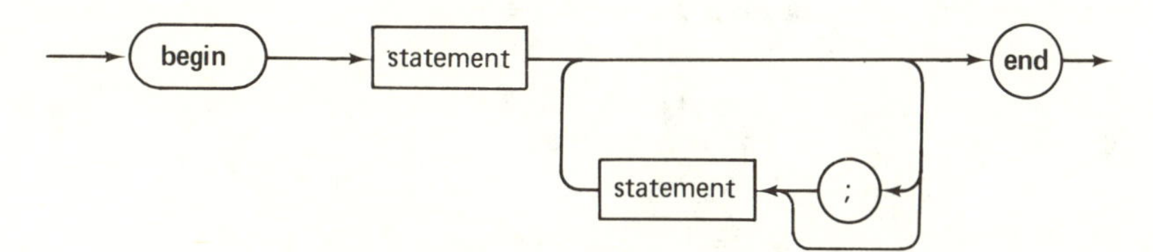

Fig. 5.6 Modified compound statement syntax.

1. Use = instead of :=.
2. = must be followed by a number.
3. Identifier must be followed by =.
4. **const, var, procedure** must be followed by an identifier.
5. Semicolon or comma missing.
6. Incorrect symbol after procedure declaration.
7. Statement expected.
8. Incorrect symbol after statement part in block.
9. Period expected.
10. Semicolon between statements is missing.
11. Undeclared identifier.
12. Assignment to constant or procedure is not allowed.
13. Assignment operator := expected.
14. **call** must be followed by an identifier.
15. Call of a constant or a variable is meaningless.
16. **then** expected.
17. Semicolon or **end** expected.
18. **do** expected.
19. Incorrect symbol following statement.
20. Relational operator expected.
21. Expression must not contain a procedure identifier.
22. Right parenthesis missing.
23. The preceding factor cannot be followed by this symbol.
24. An expression cannot begin with this symbol.
30. This number is too large.

Table 5.3 Error Messages of PL/0 Compiler.

5.5, and Table 5.3 lists a set of possible diagnostic messages corresponding to the error numbers in Program 5.5.

The following program (5.20) was obtained by the introduction of syntactic errors in (5.14) through (5.16).

```
const m = 7, n = 85
var x,y,z,q,r;
    ↑ 5
            ↑ 5
procedure multiply;
    var a,b
```

```
      begin a := u; b := y; z := 0
      ↑ 5
             ↑11
         while b > 0 do
             ↑10
         begin
             if odd b do z := z + a;
                         ↑16
                         ↑19
             a := 2a; b := b/2;
                   ↑23
         end
      end ;
   procedure divide
      var w;
         ↑ 5
      const two = 2, three := 3;
         ↑ 7
                          ↑ 1
   begin r = x; q := 0; w := y;                         (5.20)
            ↑13
            ↑24
      while w ≤ r do w := two*w;
      while w > y
         begin q := (2*q; w := w/2);
             ↑18
                     ↑22
                              ↑23
             if w ≤ r then
                 begin r := r—w q := q+1
                              ↑23
                 end
         end
   end ;
   procedure gcd;
      var f,g;
   begin f := x; g := y
      while f ≠ g do
         ↑17
         begin if f < g then g := g—f;
                  if g < f then f := f—g;
      z := f
   end ;
```

```
begin
   x := m; y := n; call multiply;
   x := 25; y := 3; call divide;
   x := 84; y := 36; call gcd;
   call x; x := gcd; gcd =x
      ↑15
                ↑21
                          ↑12
                            ↑13
                            ↑24
end .
    ↑17
    ↑ 5
    ↑ 7
PROGRAM INCOMPLETE
```

It should be clear that no scheme that reasonably efficiently translates correct sentences will also be able to handle all possible incorrect constructions in a sensible way. And why should it! Every scheme implemented with reasonable effort will fail, that is, will inadequately handle some misconstructions. The important characteristics of a good compiler, however, are that

1. No input sequence will cause the compiler to collapse.
2. All constructs that are illegal according to the language definition are detected and marked.
3. Errors that occur reasonably frequently and are true programmer's mistakes (caused by oversight or misunderstanding) are diagnosed correctly and do not cause any (or many) further stumblings of the compiler (so-called *spurious* error messages).

The presented scheme performs satisfactorily, although there is always room for improvement. Its merit is that it is built according to a few ground rules in a systematic fashion. The ground rules are merely supplemented by some choices of parameters based on heuristics and experience with actual use of the language.

Program 5.5 PL/0 Parser with Error Recovery.

```
program PL0 (input ,output);
{PL/0 compiler with syntax error recovery}
label 99;
const norw = 11;          {no. of reserved words}
   txmax = 100;           {length of identifier table}
   nmax = 14;             {max. no. of digits in numbers}
   al = 10;               {length of identifiers}
type symbol =
   (nul, ident, number, plus, minus, times, slash, oddsym,
   eql, neq, lss, leq, gtr, geq, lparen, rparen, comma, semicolon,
   period, becomes, beginsym, endsym, ifsym, thensym,
   whilesym, dosym, callsym, constsym, varsym, procsym);
   alfa = packed array [1 .. al] of char;
   object = (constant, variable, procedure);
   symset = set of symbol;
var ch: char;              {last character read}
   sym: symbol;            {last symbol read}
   id: alfa;               {last identifier read}
   num: integer;           {last number read}
   cc: integer;            {character count}
   ll: integer;            {line length}
   kk: integer;
   line: array [1 .. 81] of char;
   a: alfa;
   word: array [1 .. norw] of alfa;
   wsym: array [1 .. norw] of symbol;
   ssym: array [char] of symbol;
   declbegsys, statbegsys, facbegsys: symset;
   table: array [0 .. txmax] of
          record name: alfa;
                 kind: object
          end ;
procedure error (n: integer);
begin writeln(' ':cc, '↑', n: 2);
end {error} ;

procedure test (s1,s2: symset; n: integer);
begin if ¬(sym in s1) then
      begin error(n); s1 := s1 + s2;
         while ¬(sym in s1) do getsym
      end
end {test} ;
```

```
procedure block (tx: integer; fsys: symset);
    procedure enter (k: object);
    begin {enter object into table}
        tx := tx + 1;
        with table[tx] do
        begin name := id; kind := k;
        end
    end {enter} ;
    function position (id: alfa): integer;
        var i: integer;
    begin {find identifier id in table}
        table[0] .name := id; i := tx;
        while table[i] .name ≠ id do i := i-1;
        position := i
    end {position} ;

    procedure constdeclaration;
    begin if sym = ident then
        begin getsym;
            if sym in [eql, becomes] then
            begin if sym = becomes then error (1);
                getsym;
                if sym = number then
                    begin enter (constant); getsym
                    end
                else error (2)
            end else error (3)
        end else error (4)
        end {constdeclaration} ;

    procedure vardeclaration;
    begin if sym = ident then
            begin enter (variable); getsym
            end else error (4)
    end {vardeclaration} ;
procedure statement (fsys: symset);
var i: integer;
procedure expression (fsys: symset);
    procedure term (fsys: symset);
        procedure factor (fsys: symset);
            var i: integer;
```

Program 5.5 (Continued)

```
            begin test (facbegsys, fsys, 24);
               while sym in facbegsys do
               begin
                  if sym = ident then
                  begin i := position (id);
                     if i = 0 then error (11) else
                     if table[i] .kind = procedure then error (21);
                     getsym
                  end else
                  if sym = number then
                  begin getsym;
                  end else
                  if sym = lparen then
                  begin getsym; expression ([rparen]+fsys);
                     if sym = rparen then getsym else error (22)
                  end ;
                  test(fsys, [lparen], 23)
               end
            end {factor} ;
         begin {term} factor (fsys+[times, slash]);
            while sym in [times, slash] do
               begin getsym; factor(fsys+[times, slash])
               end
         end {term} ;
      begin {expression}
         if sym in [plus ,minus] then
            begin getsym; term(fsys+[plus, minus])
            end else term(fsys+[plus, minus]);
         while sym in [plus, minus] do
            begin getsym; term(fsys+[plus, minus])
            end
      end {expression} ;
      procedure condition(fsys: symset);
      begin
         if sym = oddsym then
         begin getsym; expression(fsys);
         end else
         begin expression ([eql, neq, lss, gtr, leq, geq]+fsys);
            if ¬(sym in [eql, neq, lss, leq, gtr, geq]) then
               error (20) else
            begin getsym; expression (fsys)
            end
         end
      end {condition} ;
```

Program 5.5 (Continued)

```
begin {statement}
   if sym = ident then
   begin i := position(id);
      if i = 0 then error (11) else
      if table[i] .kind ≠ variable then error (12);
      getsym; if sym = becomes then getsym else error (13);
      expression(fsys);
   end else
   if sym = callsym then
   begin getsym;
      if sym ≠ ident then error (14) else
         begin i := position(id);
            if i = 0 then error (11) else
            if table[i] .kind ≠ procedure then error (15);
            getsym
         end
   end else
   if sym = ifsym then
   begin getsym; condition ([thensym, dosym]+fsys);
      if sym = thensym then getsym else error (16);
      statement(fsys)
   end else
   if sym = beginsym then
   begin getsym; statement([semicolon, endsym]+fsys);
      while sym in [semicolon]+statbegsys do
      begin
         if sym = semicolon then getsym else error (10);
         statement([semicolon ,endsym]+fsys)
      end ;
      if sym = endsym then getsym else error (17)
   end else
   if sym = whilesym then
   begin getsym; condition([dosym]+fsys);
      if sym = dosym then getsym else error (18);
      statement(fsys);
   end ;
   test(fsys, [ ], 19)
end {statement} ;
```

Program 5.5 (Continued)

```
begin {block}
    repeat
        if sym = constsym then
        begin getsym;
            repeat constdeclaration;
                while sym = comma do
                    begin getsym; constdeclaration
                    end ;
                if sym = semicolon then getsym else error (5)
            until sym ≠ ident
        end ;
        if sym = varsym then
        begin getsym;
            repeat vardeclaration;
                while sym = comma do
                    begin getsym; vardeclaration
                    end ;
                if sym = semicolon then getsym else error (5)
            until sym ≠ ident;
        end ;
        while sym = procsym do
        begin getsym;
            if sym = ident then
                begin enter (procedure); getsym
                end
            else error (4);
            if sym = semicolon then getsym else error (5);
            block (tx, [semicolon]+fsys);
            if sym = semicolon then
                begin getsym; test(statbegsys+[ident, procsym], fsys, 6)
                end
            else error (5)
        end ;
        test(statbegsys+[ident], declbegsys, 7)
    until ¬(sym in declbegsys);
    statement([semicolon, endsym]+fsys);
    test(fsys, [ ], 8);
end {block} ;
begin {main program}
    . . . Initialization (see Program 5.4) . . .
    cc := 0; ll := 0; ch := ' '; kk := al; getsym;
    block (0, [period]+declbegsys+statbegsys);
    if sym ≠ period then error (9);
99: writeln
end .
```

5.10. A PL/0 PROCESSOR

It is indeed remarkable that the PL/0 compiler was so far developed without any knowledge of the machine for which it was supposed to generate code. But why should the structure of an object machine influence the parsing and error recovery scheme of a compiler! In fact, it *must not* do so. Instead, the proper scheme for code generation for any computer should be superimposed on the existing parser by the method of stepwise refinement of the existing program. Since we are about to do this, it becomes necessary to select a processor for which to compile.

In order to keep the description of the compiler reasonably simple and free from extraneous considerations of peculiar properties of a real, existing processor, we shall postulate a computer of our own choice, specifically tailored to the needs of PL/0. Since this processor does not really exist (in hardware), it is a hypothetical processor; it will be called the *PL/0 machine*.

It is not the aim of this section to explain the detailed reasoning that led to the choice of exactly this kind of machine architecture. Instead, it is to serve as a descriptive manual consisting of an intuitive introduction, followed by a detailed definition of the processor in the form of an algorithm. This formalization may serve as an example for accurate and detailed algorithmic descriptions of actual processors. The algorithm interprets PL/0 instructions sequentially, and is called an *interpreter*.

The PL/0 machine consists of two stores: an instruction register and three address registers. The *program store*, called *code*, is loaded by the compiler and remains unchanged during interpretation of the code. It can then be considered as a read-only store. The *data store S* is organized as a *stack*, and all arithmetic operators operate on the two elements on top of the stack, replacing their operands by a result. The top element is addressed (indexed) by the *top stack register T*. The *instruction register I* contains the instruction that is currently being interpreted. The *program address register P* designates the next instruction to be fetched for interpretation.

Every procedure in PL/0 may contain local variables. Since procedures may be activated recursively, storage for these local variables may not be allocated before the actual procedure call. Hence, the data segments for individual procedures are stacked up consecutively in the stack store *S*. Since procedure activations strictly obey the first-in-last-out scheme, the stack is the appropriate storage allocation strategy. Every procedure owns some internal information of its own, namely, the program address of its call (the so-called *return address*), and the address of the data segment of its caller. These two addresses are needed for proper resumption of program execution after termination of the procedure. They can be understood as internal or implicit local variables allocated in the procedure's data segment.

We call them the *return address RA* and the *dynamic link DL*. The origin of the dynamic link, that is, the address of the most recently allocated data segment, is retained in the *base address register B*.

Since the actual allocation of storage takes place during execution (interpretation) time, the compiler cannot equip the generated code with absolute addresses. Since it can only determine the location of variables within a data segment, it is capable of providing *relative addresses* only. The interpreter has to add to this so-called *displacement* to the base address of the appropriate data segment. If a variable is local to the procedure currently being interpreted, then this base address is given by the *B* register. Otherwise, it must be obtained by descending the chain of data segments. The compiler, however, can only know the static depth of an access path, whereas the dynamic link chain maintains the dynamic history of procedure activations. Unfortunately, these two access paths are not necessarily the same.

For example, assume that a procedure *A* calls a procedure *B* declared local to *A*, *B* calls *C* declared local to *B*, and *C* calls *B* (recursively). We say that *A* is declared at level 1, *B* at level 2, *C* at level 3 (see Fig. 5.7). If a variable *a* declared in *A* is to be accessed in *B*, then the compiler knows that there exists a *level difference* of 1 between *B* and *A*. Descending one step along the dynamic link chain, however, would result in an access to a variable local to *C*!

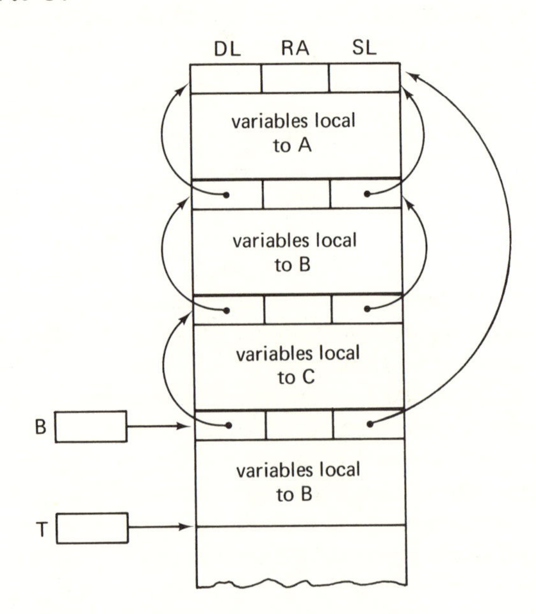

Fig. 5.7 Stack of PL/0 machine.

Hence, it is plain that a second link chain has to be provided that properly links data segments in the way the compiler can see the situation. We call this the *static link SL.*

Addresses are therefore generated as pairs of numbers indicating the static level difference and the relative displacement within a data segment. We assume that each location of the data store is capable of holding an address or an integer.

The instruction set of the PL/0 machine is tuned to the requirements of the PL/0 language. It includes the following orders:

1. An instruction to load numbers (literals) onto the stack (LIT).
2. An instruction to fetch variables onto the top of the stack (LOD).
3. A store instruction corresponding to assignment statements (STO).
4. An introduction to activate a subroutine corresponding to a procedure call (CAL).
5. An instruction to allocate storage on the stack by incrementing the stack pointer T (INT).
6. Instructions for unconditional and conditional transfer of control, used in if- and while statements (JMP, JPC).
7. A set of arithmetic and relational operators (OPR).

The format of instructions is determined by the need for three components, namely, an operation code f and a parameter consisting of one or two parts (see Fig. 5.8). In the case of operators the parameter a determines the identity of the operator; in the other cases it is either a number (LIT, INT), a program address (JMP, JPC, CAL), or a data address (LOD, STO).

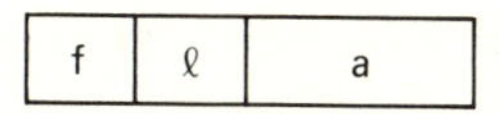

Fig. 5.8 Instruction format.

The details of operation of the PL/0 machine should be evident from the procedure called *interpret* that is part of Program 5.6, which combines the completed compiler with the interpreter into a system that translates and subsequently executes PL/0 programs. The modification of this program to generate code for an existing computer is left as an excercise for the interested reader. The resulting expansion of the compiler program may be taken as a measure of the appropriateness of the chosen computer for the present task.

There is no doubt that the presented PL/0 computer could be expanded into a more sophisticated organization in order to make certain operations more efficient. One instance is the chosen addressing mechanism. The presented solution was chosen because of its inherent simplicity and because all improvements must essentially be based on it and derived from it.

5.11. CODE GENERATION

In order to be able to assemble an instruction, the compiler must know its operation code and its parameter, which is a literal number or an address. These values are associated by the compiler itself with the respective identifiers. This association is performed upon processing the declaration of constants, variables, and procedures. For this purpose, the table containing the identifiers is expanded to contain the attributes associated with each identifier. If an identifier denotes a constant, its attribute is the constant value; if the identifier denotes a variable, the attribute is its address, consisting of a displacement and a level; and if the identifier denotes a procedure, then its attributes are the procedure's entry address and its level. The corresponding extension of the declaration of the variable *table* is shown in Program 5.6. It is a noteworthy example of a stepwise refinement (or enrichment) of a data declaration progressing simultaneously with the refinement of the statement part.

Whereas the constant values are provided by the program text, it is the compiler's task to determine addresses on its own. PL/0 is sufficiently simple to make sequential allocation of variables and code the obvious choice. Hence, every variable declaration is processed by incrementing a data allocation index by 1 (since each variable occupies by definition of the PL/0 machine exactly one storage cell). The data allocation index *dx* is to be initialized upon starting the compilation of a procedure, reflecting the fact that its data segment starts empty. [Actually, *dx* is given the initial value 3 since each data segment contains at least the three internal variables *RA*, *DL*, *SL* (see preceding section).] The appropriate computations to determine the identifiers' attributes are included in the procedure *enter* which is used to enter new identifiers into the table.

With this information about operands at hand, generating the actual code is a rather simple affair. Because of the convenient stack organization of the PL/0 machine, there exists practically a one-to-one correspondence between operands and operators in the source language and instructions in the target code. The compiler has merely to perform a suitable resequencing into *postfix* form. By "postfix form" is meant that operators always follow their operands instead of being embedded between the operands as in the conventional *infix* form. The postfix form is sometimes also called Polish form (after its originator Lukasciewicz) or *parenthesis-free* form since it makes parentheses superfluous. Some correspondences between infix and postfix forms of expressions are shown in Table 5.4 (see also Sect. 4.4.2).

The very simple technique of performing this transformation is shown by the procedures *expression* and *term* in Program 5.6. It is merely a matter

Infix Form	Postfix Form
$x + y$	$xy+$
$(x - y) + z$	$xy - z+$
$x - (y + z)$	$xyz+-$
$x*(y + z)*w$	$xyz + *w*$

Table 5.4 Expressions in Infix and Postfix Form.

of delaying the transmission of the arithmetic operator. At this point the reader should verify that the presented arrangement of parsing procedures also takes care of an appropriate interpretation of the conventional priority rules among the various operators.

A slightly less trivial matter is the translation of conditional and repetitive statements. In this case the generation of jump instructions is necessary, for which at times the destination address is still unknown. If one insists on a strictly sequential production of instructions in the form of an output file, then a *two-pass* compiler scheme is necessary. The second pass then assumes the task of supplementing the incomplete jump instructions with their destination addresses. An alternative solution adopted by the present compiler is to place the instructions into an array and essentially retaining them in directly accessible store. This method allows supplementing the missing addresses as soon as they become known. This operation is commonly called a *fixup*.

The only additional operation that has to be performed when issuing such a forward jump is to retain its location, i.e., its index in the program store. This address is then used to locate the incomplete instruction at the time of the fixup. The details are again evident from Program 5.6 (see routines processing **if**- and **while** statements). The patterns of code generated for the **if**- and **while** statements are as follows (*L*1 and *L*2 stand for code addresses):

	if *C* **then** *S*		**while** *C* **do** *S*
	code for condition *C*	L1:	code for *C*
	JPC L1		JPC L2
	code for statement *S*		code for *S*
L1:	...		JMP L1
		L2:	...

For convenience, an auxiliary procedure called *gen* is introduced. Its purpose is to assemble and emit an instruction according to its three parameters. It automatically increments the code index *cx* which designates the location of the next instruction to be issued.

As an example, the code emitted by compiling the multiplication routine (5.14) is listed below in mnemonic form. The comments on the right-hand side are merely added for explanatory purposes.

```
 2   INT   0,5    allocate space for links and local variables
 3   LOD   1,3    x
 4   STO   0,3    a
 5   LOD   1,4    y
 6   STO   0,4    b
 7   LIT   0,0    0
 8   STO   1,5    z
 9   LOD   0,4    b
10   LIT   0,0    0
11   OPR   0,12   >
12   JPC   0,29
13   LOD   0,4    b
14   OPR   0,7    odd
15   JPC   0,20
16   LOD   1,5    z
17   LOD   0,3    a
18   OPR   0,2    +
19   STO   1,5    z
20   LIT   0,2    2
21   LOD   0,3    a
22   OPR   0,4    *
23   STO   0,3    a
24   LOD   0,4    b
25   LIT   0,2    2
26   OPR   0,5    /
27   STO   0,4    b
28   JMP   0,9
29   OPR   0,0    return
```

Code corresponding to PL/0 procedure 5.14.

Many tasks in compiling programming languages are considerably more complex than the ones presented in the PL/0 compiler for the PL/0 machine [5-4]. Most of them are much more resistant to being neatly organized. The reader trying to extend the presented compiler in either direction toward a more powerful language or a more conventional computer will soon realize the truth of this statement. Nevertheless, the basic approach toward designing a complex program presented here retains its validity, and even increases its value when the task grows more complicated and more sophisticated. It has, in fact, been successfully used in the construction of large compilers [5-1 and 5-9].

Program 5.6 PL/0 Compiler.

```
program PL0(input,output);
{PL/0 compiler with code generation}
label 99;
const norw = 11;        {no. of reserved words}
   txmax = 100;         {length of identifier table}
   nmax = 14;           {max. no. of digits in numbers}
   al = 10;             {length of identifiers}
   amax = 2047;         {maximum address}
   levmax = 3;          {maximum depth of block nesting}
   cxmax = 200;         {size of code array}
type symbol =
   (nul, ident, number, plus, minus, times, slash, oddsym,
    eql, neq, lss, leq, gtr, geq, lparen, rparen, comma, semicolon,
    period, becomes, beginsym, endsym, ifsym, thensym,
    whilesym, dosym, callsym, constsym, varsym, procsym);
   alfa = packed array [1 .. al] of char;
   object = (constant, variable, procedure);
   symset = set of symbol;
   fct = (lit, opr, lod, sto, cal, int, jmp, jpc);      {functions}
   instruction = packed record
                    f: fct;              {function code}
                    l: 0 .. levmax;      {level}
                    a: 0 .. amax;        {displacement address}
                 end ;
{   LIT 0,a  :  load constant a
    OPR 0,a  :  execute operation a
    LOD l,a  :  load variable l,a
    STO l,a  :  store variable l,a
    CAL l,a  :  call procedure a at level l
    INT 0,a  :  increment t-register by a
    JMP 0,a  :  jump to a
    JPC 0,a  :  jump conditional to a    }
var ch: char;           {last character read}
   sym: symbol;         {last symbol read}
   id: alfa;            {last identifier read}
   num: integer;        {last number read}
   cc: integer;         {character count}
   ll: integer;         {line length}
   kk, err: integer;
   cx: integer;         {code allocation index}
```

```
    line: array [1 .. 81] of char;
    a: alfa;
    code: array [0 .. cxmax] of instruction;
    word: array [1 .. norw] of alfa;
    wsym: array [1 .. norw] of symbol;
    ssym: array [char] of symbol;
    mnemonic: array [fct] of
                  packed array [1 .. 5] of char;
    declbegsys, statbegsys, facbegsys: symset;
    table: array [0 .. txmax] of
            record name: alfa;
               case kind: object of
               constant: (val: integer);
               variable, procedure: (level, adr: integer)
            end ;
procedure error(n: integer);
begin writeln(' ****', ' ': cc-1, '↑', n: 2); err := err+1
end {error} ;

procedure getsym;
    var i,j,k: integer;

    procedure getch;
    begin if cc = ll then
       begin if eof(input) then
                begin write(' PROGRAM INCOMPLETE'); goto 99
                end ;
          ll := 0; cc := 0; write(cx: 5, ' ');
          while ¬eoln(input) do
             begin ll := ll+1; read(ch); write(ch); line[ll] := ch
             end ;
          writeln; ll := ll+1; read(line[ll])
       end ;
       cc := cc+1; ch := line[cc]
    end {getch} ;

begin {getsym}
    while ch = ' ' do getch;
    if ch in ['A' .. 'Z'] then
    begin {identifier or reserved word}  k := 0;
       repeat if k < al then
          begin k := k+1; a[k] := ch
          end ;
```

Program 5.6 (Continued)

```
        getch
      until ¬(ch in ['A' . . 'Z','0' . . '9']);
      if k ≥ kk then kk := k else
        repeat a[kk] := ' '; kk := kk-1
        until kk = k;
      id := a; i := 1; j := norw;
      repeat k := (i+j) div 2;
        if id ≤ word[k] then j := k-1;
        if id ≥ word[k] then i := k+1
      until i > j;
      if i-1 > j then sym := wsym[k] else sym := ident
    end else
    if ch in ['0' . . '9'] then
    begin {number} k := 0; num := 0; sym := number;
      repeat num := 10*num + (ord(ch)-ord('0'));
        k := k+1; getch
      until ¬(ch in ['0' . . '9']);
      if k > nmax then error (30)
    end else
    if ch = ':' then
    begin getch;
      if ch = '=' then
      begin sym := becomes; getch
      end else sym := nul;
    end else
    begin sym := ssym[ch]; getch
    end
end {getsym} ;

procedure gen(x: fct; y,z: integer);
begin if cx > cxmax then
        begin write(' PROGRAM TOO LONG'); goto 99
        end ;
   with code[cx] do
      begin f := x; l := y; a := z
      end ;
   cx := cx + 1
end {gen} ;

procedure test(s1,s2: symset; n: integer);
begin if ¬(sym in s1) then
       begin error(n); s1 := s1 + s2;
          while ¬(sym in s1) do getsym
       end
end {test} ;
```

Program 5.6 (Continued)

```
procedure block(lev,tx: integer; fsys: symset);
    var dx: integer;       {data allocation index}
        tx0: integer;      {initial table index}
        cx0: integer;      {initial code index}
    procedure enter(k: object);
    begin {enter object into table}
        tx := tx + 1;
        with table[tx] do
        begin name := id; kind := k;
            case k of
            constant: begin if num > amax then
                                    begin error (30); num := 0 end ;
                          val := num
                      end ;
            variable: begin level := lev; adr := dx; dx := dx+1;
                      end ;
            procedure: level := lev
            end
        end
    end {enter} ;

    function position(id: alfa): integer;
        var i: integer;
    begin {find indentifier id in table}
        table[0] .name := id; i := tx;
        while table[i] .name ≠ id do i := i-1;
        position := i
    end {position} ;

    procedure constdeclaration;
    begin if sym = ident then
        begin getsym;
            if sym in [eql ,becomes] then
            begin if sym = becomes then error(1);
                getsym;
                if sym = number then
                    begin enter(constant); getsym
                    end
                else error (2)
            end else error (3)
        end else error (4)
    end {constdeclaration} ;

procedure vardeclaration;
begin if sym = ident then
        begin enter(variable); getsym
        end else error (4)
end {vardeclaration} ;
```

Program 5.6 (Continued)

```
procedure listcode;
    var i: integer;
begin {list code generated for this block}
    for i := cx0 to cx−1 do
        with code[i] do
            writeln(i, mnemonic[f]:5, l:3, a:5)
end {listcode} ;

procedure statement(fsys: symset);
    var i,cx1,cx2: integer;
    procedure expression(fsys: symset);
        var addop: symbol;
        procedure term(fsys: symset);
            var mulop: symbol;
            procedure factor(fsys: symset);
                var i: integer;
            begin test(facbegsys, fsys, 24);
                while sym in facbegsys do
                begin
                    if sym = ident then
                    begin i := position(id);
                        if i = 0 then error (11) else
                        with table[i] do
                        case kind of
                          constant: gen(lit, 0, val);
                          variable: gen(lod, lev-level, adr);
                          procedure: error (21)
                        end ;
                        getsym
                    end else
                    if sym = number then
                    begin if num > amax then
                            begin error (30); num := 0
                            end ;
                        gen(lit, 0, num); getsym
                    end else
                    if sym = lparen then
                    begin getsym; expression([rparen]+fsys);
                        if sym = rparen then getsym else error (22)
                    end ;
                    test(fsys, [lparen], 23)
                end
            end {factor} ;
```

Program 5.6 (Continued)

```
        begin {term} factor(fsys+[times, slash]);
           while sym in [times, slash] do
              begin mulop := sym; getsym; factor(fsys+[times, slash]);
                 if mulop = times then gen(opr,0,4) else gen(opr,0,5)
              end
        end {term} ;
     begin {expression}
        if sym in [plus, minus] then
           begin addop := sym; getsym; term(fsys+[plus, minus]);
              if addop = minus then gen(opr,0,1)
           end else term(fsys+[plus, minus]);
        while sym in [plus, minus] do
           begin addop := sym; getsym; term(fsys+[plus, minus]);
              if addop = plus then gen(opr,0,2) else gen(opr,0,3)
           end
     end {expression} ;

     procedure condition(fsys: symset);
        var relop: symbol;
     begin
        if sym = oddsym then
        begin getsym; expression(fsys); gen(opr,0,6)
        end else
        begin expression([eql, neq, lss, gtr, leq, geq]+fsys);
           if ¬(sym in [eql, neq, lss, leq, gtr, geq]) then
              error (20) else
           begin relop := sym; getsym; expression(fsys);
              case relop of
                eql: gen(opr,0, 8);
                neq: gen(opr,0, 9);
                lss: gen(opr,0,10);
                geq: gen(opr,0,11);
                gtr: gen(opr,0,12);
                leq: gen(opr,0,13);
              end
           end
        end
     end {condition} ;
```

Program 5.6 (Continued)

```
begin {statement}
   if sym = ident then
   begin i := position(id);
      if i = 0 then error (11) else
      if table[i] .kind ≠ variable then
         begin {assignment to non-variable} error (12); i := 0
         end ;
      getsym; if sym = becomes then getsym else error (13);
      expression(fsys);
      if i ≠ 0 then
         with table[i] do gen(sto, lev-level, adr)
   end else
   if sym = callsym then
   begin getsym;
      if sym ≠ ident then error (14) else
         begin i := position(id);
            if i = 0 then error (11) else
            with table[i] do
               if kind = procedure then gen (cal, lev-level, adr)
               else error (15);
            getsym
         end
   end else
   if sym = ifsym then
   begin getsym; condition([thensym, dosym]+fsys);
      if sym = thensym then getsym else error (16);
      cx1 := cx; gen(jpc,0,0);
      statement(fsys); code[cx1].a := cx
   end else
   if sym = beginsym then
   begin getsym; statement([semicolon, endsym]+fsys);
      while sym in [semicolon]+statbegsys do
      begin
         if sym = semicolon then getsym else error (10);
         statement([semicolon, endsym]+fsys)
      end ;
      if sym = endsym then getsym else error (17)
   end else
   if sym = whilesym then
   begin cx1 := cx; getsym; condition([dosym]+fsys);
      cx2 := cx; gen(jpc,0,0);
      if sym = dosym then getsym else error (18);
      statement(fsys); gen(jmp,0,cx1); code[cx2].a := cx
   end ;
   test(fsys, [ ], 19)
end {statement} ;
```

```
begin {block} dx := 3; tx0 := tx; table[tx] .adr := cx; gen(jmp,0,0);
   if lev > levmax then error (32);
   repeat
      if sym = constsym then
      begin getsym;
         repeat constdeclaration;
            while sym = comma do
               begin getsym; constdeclaration
               end ;
            if sym = semicolon then getsym else error (5)
         until sym ≠ ident
      end ;
      if sym = varsym then
      begin getsym;
         repeat vardeclaration;
            while sym = comma do
               begin getsym; vardeclaration
               end ;
            if sym = semicolon then getsym else error (5)
         until sym ≠ ident;
      end ;
      while sym = procsym do
      begin getsym;
         if sym = ident then
            begin enter(procedure); getsym
            end
         else error (4);
         if sym = semicolon then getsym else error (5);
         block(lev+1,tx,[semicolon]+fsys);
         if sym = semicolon then
            begin getsym; test(statbegsys+[ident, procsym], fsys, 6)
            end
         else error (5)
      end ;
      test(statbegsys+[ident], declbegsys, 7)
   until ¬(sym in declbegsys);
   code[table[tx0].adr].a := cx;
   with table[tx0] do
      begin adr := cx; {start adr of code}
      end ;
   cx0 := cx; gen(int,0,dx);
   statement([semicolon, endsym]+fsys);
   gen(opr,0,0); {return}
   test(fsys, [ ], 8);
   listcode;
end {block} ;
```

```
procedure interpret;
   const stacksize = 500;
   var p,b,t: integer {program-, base-, topstack-registers}
      i: instruction; {instruction register}
      s: array [1 .. stacksize] of integer;   {datastore}
   function base(l: integer): integer;
      var b1: integer;
   begin b1 := b; {find base l levels down}
      while l > 0 do
         begin b1 := s[b1]; l := l-1
         end ;
      base := b1
   end {base} ;

begin writeln(' START PL/0');
   t := 0; b := 1; p := 0;
   s[1] := 0; s[2] := 0; s[3] := 0;
   repeat i := code[p]; p := p+1;
      with i do
      case f of
     lit: begin t := t+1; s[t] := a
          end ;
     opr: case a of       {operator}
          0: begin {return}
                   t := b-1; p := s[t+3]; b := s[t+2];
             end ;
          1: s[t] := -s[t];
          2: begin t := t-1; s[t] := s[t] + s[t+1]
             end ;
          3: begin t := t-1; s[t] := s[t] - s[t+1]
             end ;
          4: begin t := t-1; s[t] := s[t] * s[t+1]
             end ;
          5: begin t := t-1; s[t] := s[t] div s[t+1]
             end ;
          6: s[t] := ord(odd(s[t]));
          8: begin t := t-1; s[t] := ord(s[t]=s[t+1])
             end ;
          9: begin t := t-1; s[t] := ord(s[t]≠s[t+1])
             end ;
         10: begin t := t-1; s[t] := ord(s[t]<s[t+1])
             end ;
         11: begin t := t-1; s[t] := ord(s[t]≥s[t+1])
             end ;
```

Program 5.6 (Continued)

```
          12: begin t := t-1; s[t] := ord(s[t]>s[t+1])
              end ;
          13: begin t := t-1; s[t] := ord(s[t]≤s[t+1])
              end ;
          end ;
    lod: begin t := t+1; s[t] := s[base(l)+a]
         end ;
    sto: begin s[base(l)+a] := s[t]; writeln(s[t]); t := t-1
         end ;
    cal: begin {generate new block mark}
             s[t+1] := base(l); s[t+2] := b; s[t+3] := p;
             b := t+1; p := a
         end ;
    int: t := t+a;
    jmp: p := a;
    jpc: begin if s[t] = 0 then p := a; t := t-1
         end
      end {with, case}
   until p = 0;
   write(' END PL/0');
end {interpret} ;

begin {main program}
   for ch := 'A' to ';' do ssym[ch] := nul;
   word[ 1] := 'BEGIN ';   word[ 2] := 'CALL      ';
   word[ 3] := 'CONST';    word[ 4] := 'DO        ';
   word[ 5] := 'END   ';   word[ 6] := 'IF        ';
   word[ 7] := 'ODD   ';   word[ 8] := 'PROCEDURE';
   word[ 9] := 'THEN  ';   word[10] := 'VAR       ';
   word[11] := 'WHILE ';
   wsym[ 1] := beginsym; wsym[ 2] := callsym;
   wsym[ 3] := constsym; wsym[ 4] := dosym;
   wsym[ 5] := endsym;   wsym[ 6] := ifsym;
   wsym[ 7] := oddsym;   wsym[ 8] := procsym;
   wsym[ 9] := thensym;  wsym[10] := varsym;
   wsym[11] := whilesym;
   ssym['+'] := plus;    ssym['-'] := minus;
   ssym['*'] := times;   ssym['/'] := slash;
   ssym['('] := lparen;  ssym[')'] := rparen;
   ssym['='] := eql;     ssym[','] := comma;
   ssym['.'] := period;  ssym['≠'] := neq;
   ssym['<'] := lss;     ssym['>'] := gtr;
   ssym['≤'] := leq;     ssym['≥'] := geq;
   ssym[';'] := semicolon;
```

Program 5.6 (Continued)

```
    mnemonic[lit]  := 'LIT '; mnemonic[opr] := 'OPR';
    mnemonic[lod]  := 'LOD'; mnemonic[sto]  := 'STO';
    mnemonic[cal]  := 'CAL'; mnemonic[int]  := 'INT ';
    mnemonic[jmp] := 'JMP'; mnemonic[jpc]  := 'JPC';
    declbegsys := [constsym, varsym, procsym];
    statbegsys := [beginsym, callsym, ifsym, whilesym];
    facbegsys := [ident, number, lparen];
    page(output); err := 0
    cc := 0; cx := 0; ll := 0; ch := ' '; kk := al; getsym;
    block(0,0,[period]+declbegsys+statbegsys);
    if sym ≠ period then error (9);
    if err = 0 then interpret else write(' ERRORS IN PL/0 PROGRAM');
99: writeln
end .
```

Program 5.6 (Continued)

EXERCISES

5.1. Consider the following syntax.

$$\begin{aligned} S &::= A \\ A &::= B \mid \textbf{if } A \textbf{ then } A \textbf{ else } A \\ B &::= C \mid B{+}C \mid {+}C \\ C &::= D \mid C{*}D \mid {*}D \\ D &::= x \mid (A) \mid -D \end{aligned}$$

Which are the terminal and the non-terminal symbols? Determine the sets of leftmost and follow-symbols $L(X)$ and $F(X)$ for each non-terminal symbol X. Construct a sequence of parsing steps for the following sentences:

$x+x$
$(x+x)*(+-x)$
$(x*-+x)$
if $x+x$ **then** $x**x$ **else** $-x$
if x **then if** $-x$ **then** x **else** $x+x$ **else** $x**x$
if $-x$ **then** x **else if** x **then** $x+x$ **else** x

5.2. Does the grammar of Exercise 5.1 satisfy the Restrictive Rules 1 and 2 for one-symbol lookahead top-down parsing? If not, find an equivalent syntax which does satisfy these rules. Represent this syntax by a syntax graph and a data structure to be used by Program 5.3.

5.3. Repeat Exercise 5.2 for the following syntax:

$$
\begin{aligned}
S &::= A \\
A &::= B \mid \textbf{if } C \textbf{ then } A \mid \textbf{if } C \textbf{ then } A \textbf{ else } A \\
B &::= D = C \\
C &::= \textbf{if } C \textbf{ then } C \textbf{ else } C \mid D
\end{aligned}
$$

Hint: You may find it necessary to delete or replace some construct in order to allow for one-symbol top-down parsing to be applicable.

5.4. Given the following syntax, consider the problem of top-down parsing:

$$
\begin{aligned}
S &::= A \\
A &::= B{+}A \mid DC \\
B &::= D \mid D{*}B \\
D &::= x \mid (C) \\
C &::= +x \mid -x
\end{aligned}
$$

How many symbols do you have to look ahead at most in order to parse sentences according to this syntax?

5.5. Transform the description of PL/0 (Fig. 5.4) into an equivalent set of BNF-productions.

5.6. Write a program that determines the sets of initial and follow-symbols $L(S)$ and $F(S)$ for each non-terminal symbol S in a given set of productions.

Hint: Use part of Program 5.3 to construct an internal representation of the syntax in the form of a data structure. Then operate on this linked data structure.

5.7. Extend the PL/0 language and its compiler by the following features:

(a) A conditional statement of the form

⟨*statement*⟩ ::= **if** ⟨*condition*⟩ **then** ⟨*statement*⟩ **else** ⟨*statement*⟩

(b) A repetitive statement of the form

⟨*statement*⟩ ::= **repeat** ⟨*statement*⟩ {; ⟨*statement*⟩} **until** ⟨*condition*⟩

Are there any particular difficulties that might cause a change of form or interpretation of the given PL/0 features? You should not introduce any additional instructions in the repertoire of the PL/0 machine.

5.8. Extend the PL/0 language and compiler by introducing procedure parameters. Consider two possible solutions and chose one of them for your realization.

(a) *Value parameters.* The actual parameters in the call are expressions whose values are assigned to local variables represented by the formal parameters specified in the procedure heading.

(b) *Variable parameters.* The actual parameters are variables. Upon a call, they are substituted in place of the formal parameters. Variable parameters are implemented by passing the address of the actual parameter, storing it in the location denoted by the formal parameters. The actual parameters are then accessed indirectly via the transmitted address.

Hence, variable parameters provide access to variables defined outside the procedures, and the rules of scope may therefore be changed as follows: In every procedure only local variables may be accessed directly; non-local variables are accessible exclusively via parameters.

5.9. Extend the PL/0 language and compiler by introducing arrays of variables. Assume that the range of indices of an array variable *a* is indicated in its declaration as

var *a*(*low* : *high*)

5.10. Modify the PL/0 compiler to generate code for your available computer.
Hint: Generate symbolic assembly code in order to avoid problems with loader conventions. In a first step avoid trying to optimize code, for example, with respect to register usage. Possible optimizations should be incorporated in a fourth refinement step of the compiler.

5.11. Extend Program 5.5 into a program called "prettyprint." The purpose of this program is to read PL/0-texts and print them in a layout which naturally reflects the textual structure by appropriate line separation and indentation. First define accurate line separation and indentation rules based on the syntactic structure of PL/0; then implement them by superimposing write statements onto Program 5.5. (Write statements must be removed from the scanner, of course.)

REFERENCES

5-1. AMMANN, U., "The Method of Structured Programming Applied to the Development of a Compiler," *International Computing Symposium 1973*, A. Günther et al. eds., (Amsterdam: North-Holland Publishing Co., 1974), pp. 93–99.

5-2. COHEN, D. J. and GOTLIEB, C. C., "A List Structure Form of Grammars for Syntactic Analysis," *Comp. Surveys*, **2**, No. 1 (1970), 65–82.

5-3. FLOYD, R. W., "The Syntax of Programming Languages—A Survey," *IEEE Trans.*, EC-13 (1964), 346–53.

5-4. GRIES, D., *Compiler Construction for Digital Computers* (New York: Wiley, 1971).

5-5. KNUTH, D. E., "Top-down Syntax Analysis," *Acta Informatica*, **1**, No. 2 (1971), 79–110.

5-6. LEWIS, P. M. and STEARNS, R. E., "Syntax-directed Transduction," *J. ACM*, **15**, No. 3 (1968), 465–88.

5-7. NAUR, P., ed., "Report on the Algorithmic Language ALGOL 60," *ACM*, **6**, No. 1 (1963), 1–17.

5-8. SCHORRE, D. V., "META II, A Syntax-oriented Compiler Writing Language," *Proc. ACM Natl. Conf.*, **19**, (1964), D 1.3.1-11.

5-9. WIRTH, N., "The Design of a PASCAL Compiler," *Software-Practice and Experience*, **1**, No. 4 (1971), 309–33.

THE ASCII CHARACTER SET

y \ x	0	1	2	3	4	5	6	7
0	nul	dle		0	@	P	`	p
1	soh	dc1	!	1	A	Q	a	q
2	stx	dc2	"	2	B	R	b	r
3	etx	dc3	#	3	C	S	c	s
4	eot	dc4	$	4	D	T	d	t
5	enq	nak	%	5	E	U	e	u
6	ack	syn	&	6	F	V	f	v
7	bel	etb	'	7	G	W	g	w
8	bs	can	(	8	H	X	h	x
9	ht	em	)	9	I	Y	i	y
10	lf	sub	*	:	J	Z	j	z
11	vt	esc	+	;	K	[	k	{
12	ff	fs	,	<	L	\	l	|
13	cr	gs	—	=	M	]	m	}
14	so	rs	.	>	N	↑	n	~
15	si	us	/	?	O	_	o	del

The ordinal number of a character *ch* is computed from its coordinates in the table as

$$ord(ch) = 16*x + y$$

The characters with ordinal numbers 0 through 31 and 127 are so-called *control characters* used for data transmission and device control. The character with ordinal number 32 is the blank.

B PASCAL SYNTAX DIAGRAMS

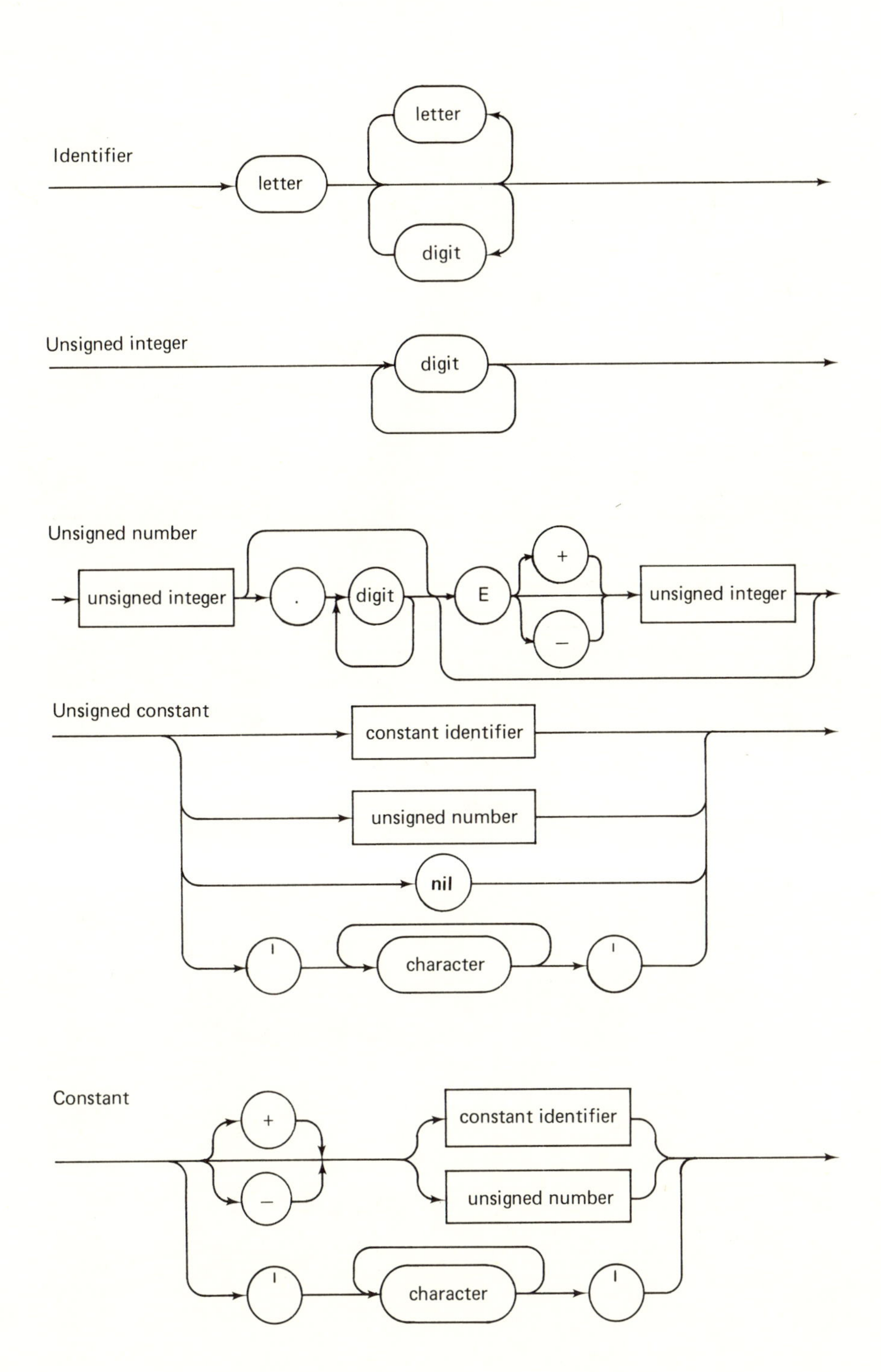
Identifier
letter
letter
digit
Unsigned integer
digit
Unsigned number
unsigned integer
.
digit
E
+
–
unsigned integer
Unsigned constant
constant identifier
unsigned number
nil
'
character
'
Constant
+
–
constant identifier
unsigned number
'
character
'

Variable

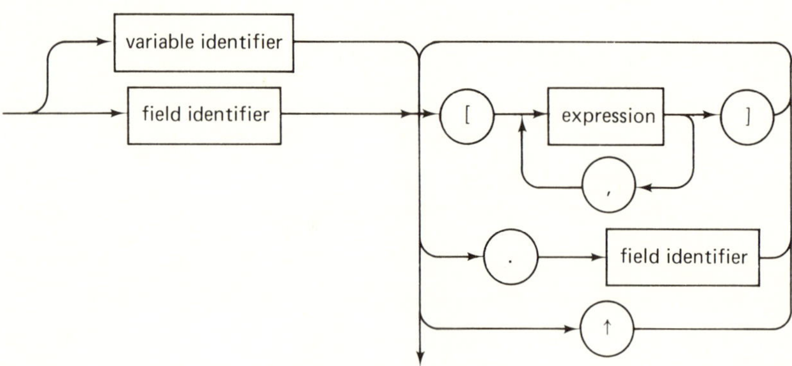
variable identifier
field identifier
[
expression
,
]
.
field identifier
↑

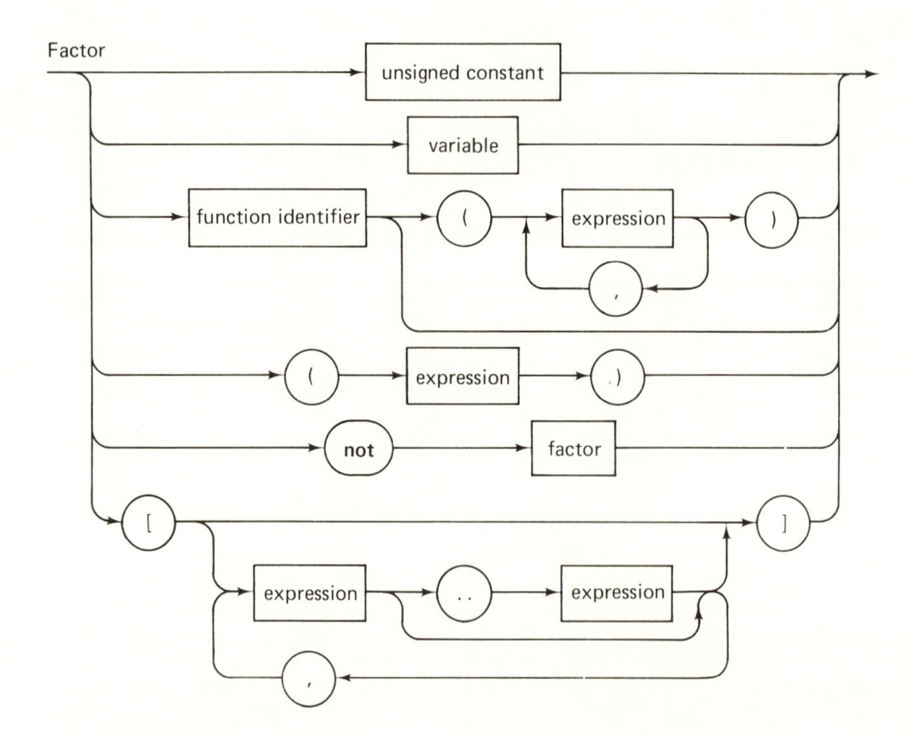
Factor
unsigned constant
variable
function identifier
(
expression
,
)
(
expression
.)
not
factor
[
expression
..
expression
,
]

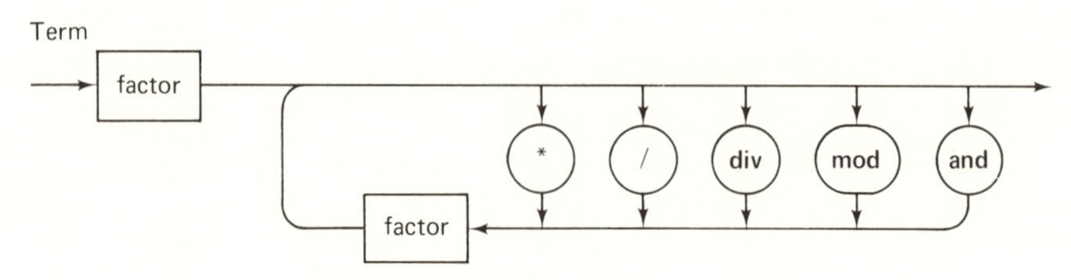
Term
factor
*
/
div
mod
and
factor

Simple expression

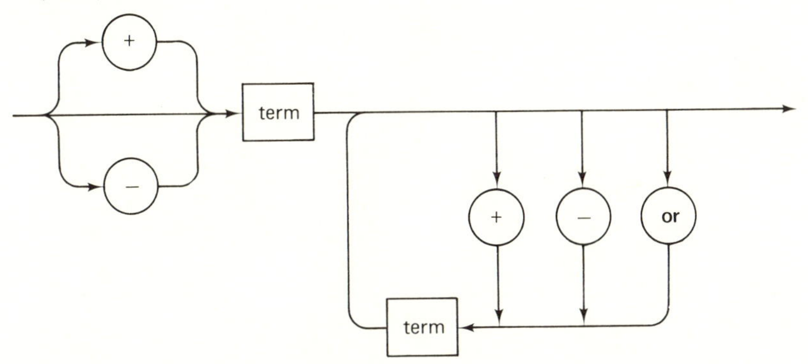

Expression

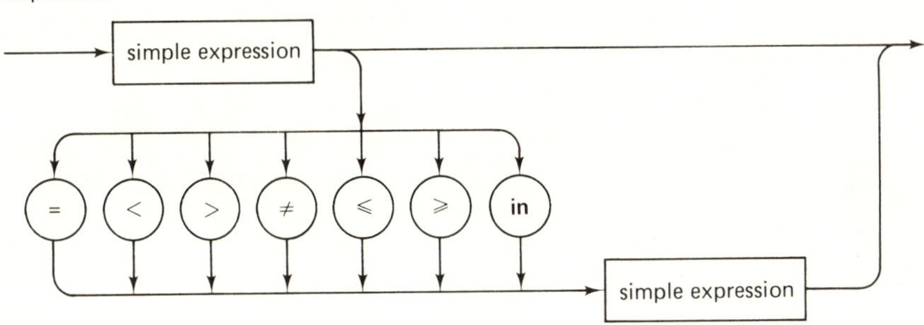

Parameter list

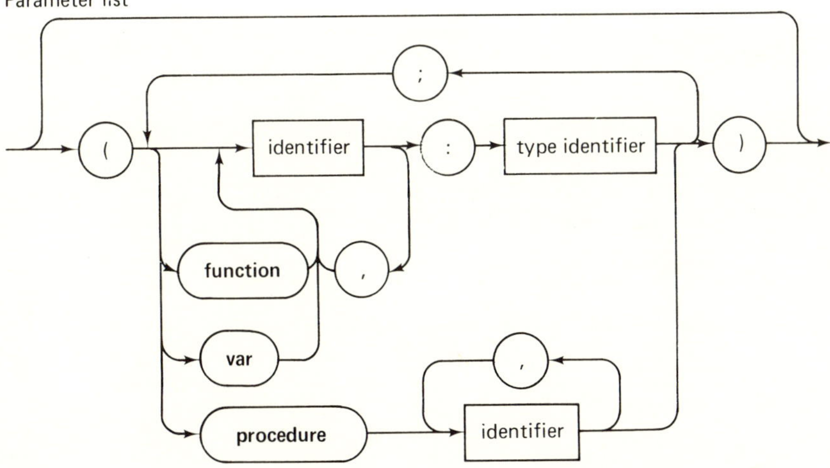

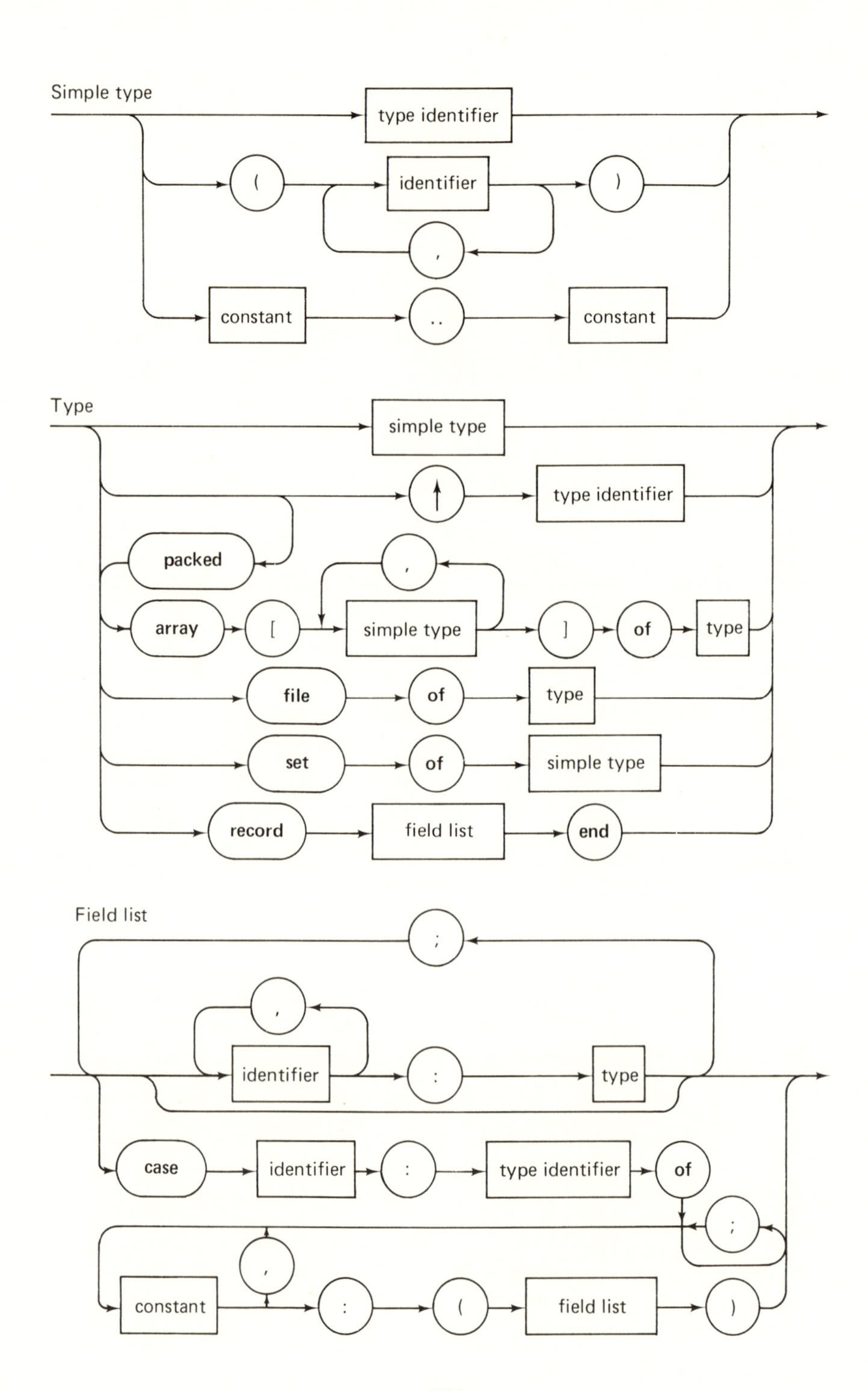
Simple type
type identifier
(
identifier
)
,
constant
..
constant
Type
simple type
↑
type identifier
packed
,
array
[
simple type
]
of
type
file
of
type
set
of
simple type
record
field list
end
Field list
;
,
identifier
:
type
case
identifier
:
type identifier
of
;
,
constant
:
(
field list
)

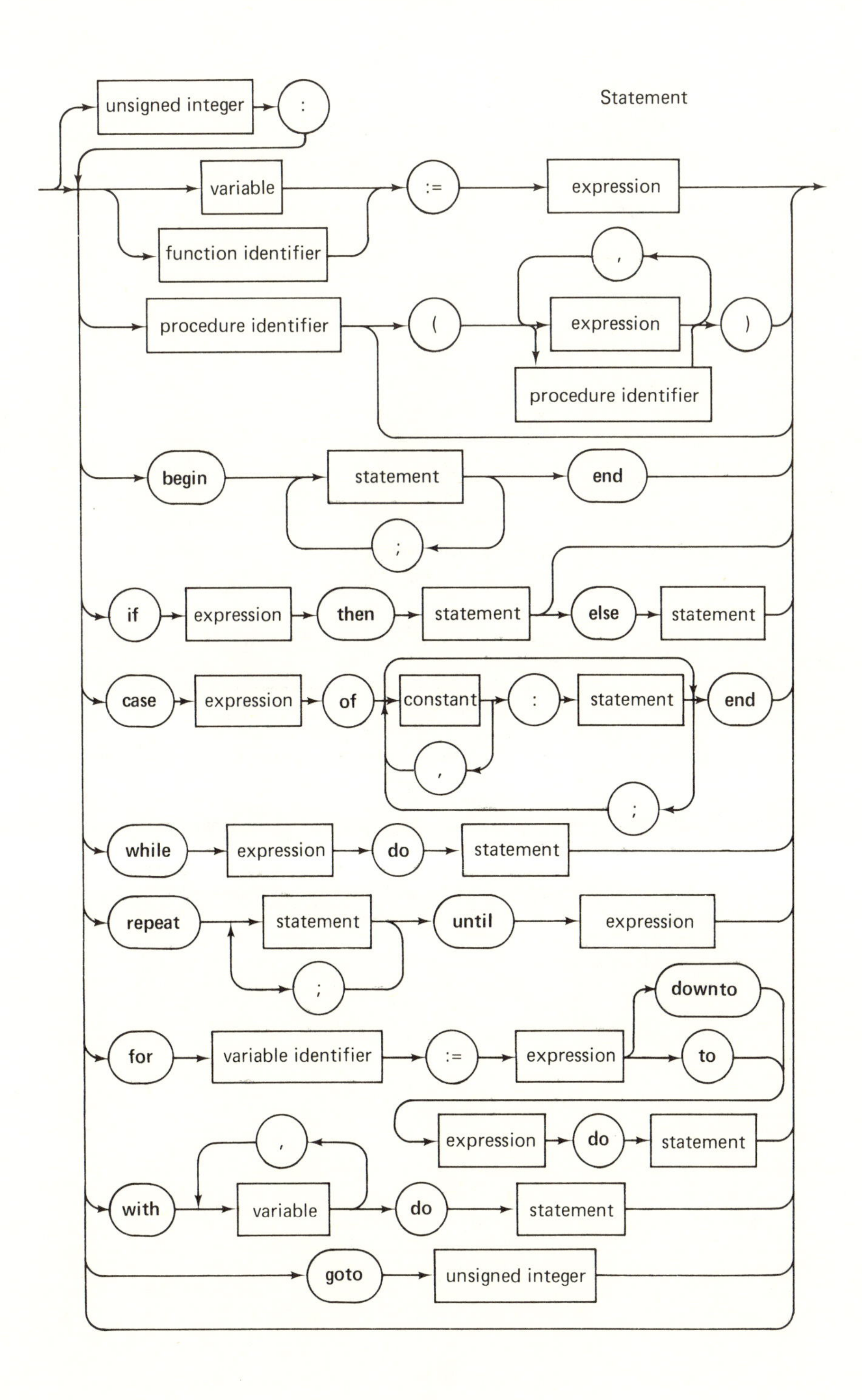
Statement
unsigned integer
:
variable
:=
expression
function identifier
procedure identifier
(
,
expression
)
procedure identifier
begin
statement
end
;
if
expression
then
statement
else
statement
case
expression
of
constant
:
statement
end
,
;
while
expression
do
statement
repeat
statement
until
expression
;
for
variable identifier
:=
expression
downto
to
expression
do
statement
with
,
variable
do
statement
goto
unsigned integer

Block

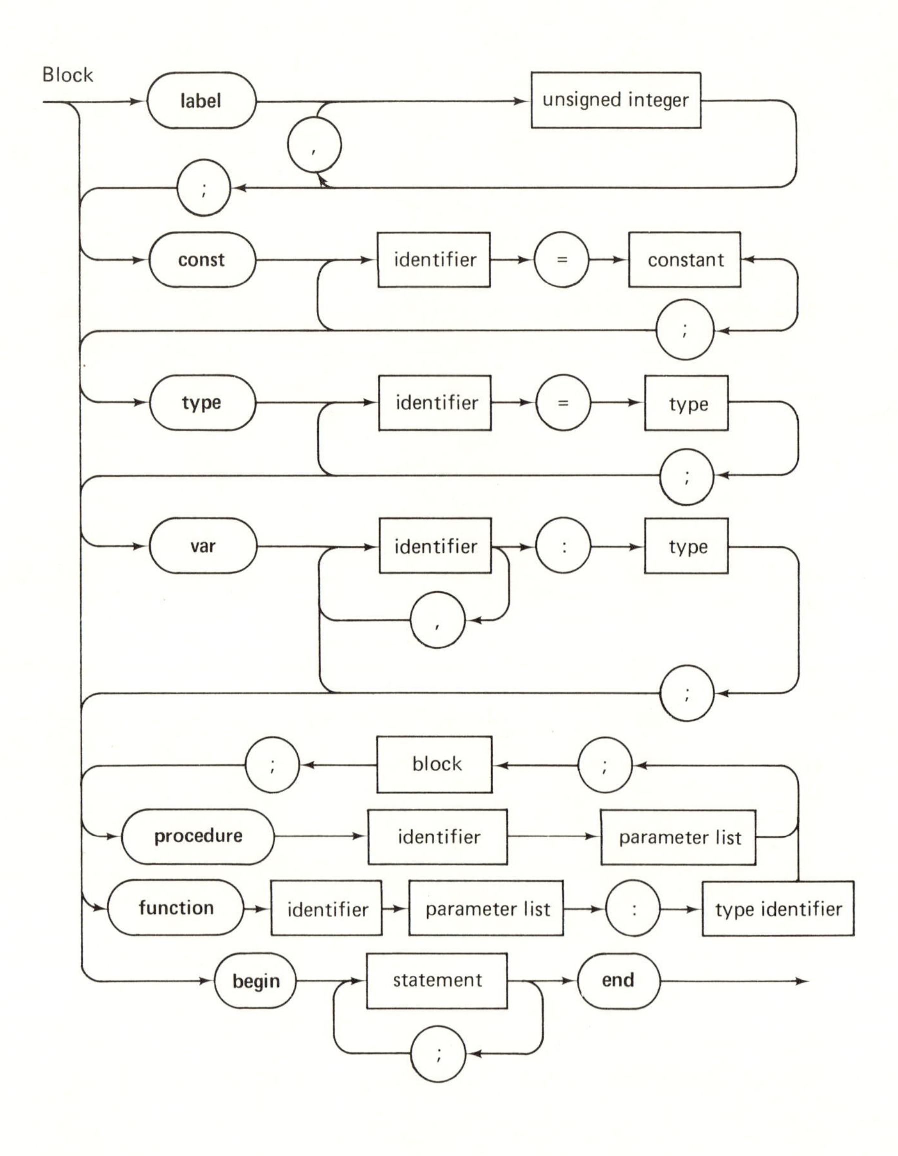

Program

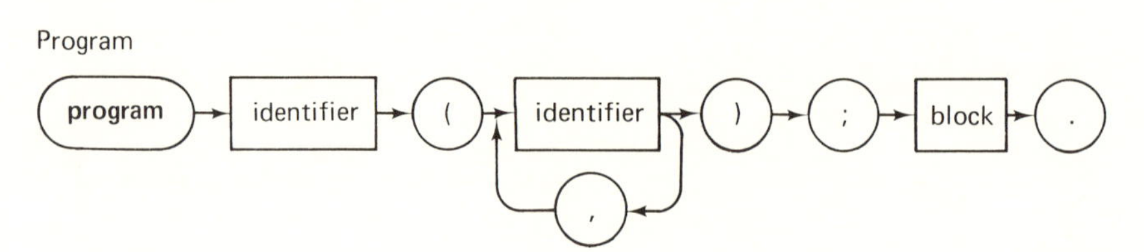

SUBJECT INDEX

Q

R

S

INDEX OF PROGRAMS